MATHEMATICS RESEARCH DEVELOPMENTS

FRACTIONAL CALCULUS

THEORY

MATHEMATICS RESEARCH DEVELOPMENTS

Additional books in this series can be found on Nova's website under the Series tab.

Additional e-books in this series can be found on Nova's website under the e-book tab.

MATHEMATICS RESEARCH DEVELOPMENTS

FRACTIONAL CALCULUS

THEORY

ROY ABI ZEID DAOU
AND
XAVIER MOREAU
EDITORS

nova publishers
New York

LIBRARY OF CONGRESS CATALOGING-IN-PUBLICATION DATA

Fractional calculus : theory / [edited by] Roy Abi Zeid Daou (Lebanese German University, Sahel Alma Campus, Keserwane--Lebanon) and Xavier Moreau (UniversitÃ©Bordeaux, Talence, France)
pages cm. -- (Mathematics research developments)
Includes bibliographical references and index.
ISBN 978-1-63463-002-3 (hardcover)
1. Fractional differential equations. 2. Fractional calculus. I. Abi Zeid Daou, Roy, editor. II. Moreau, Xavier, 1966- editor.
QC20.7.F73F72 2014
515'.83--dc23

2014035652

Published by Nova Science Publishers, Inc. † New York

CONTENTS

FOREWORD

Non-integer differentiation does not escape to the slogan “different operator, different properties and performances”. This is indeed the concise formula that is likely to explain the “why” of this operator, especially as most of its properties and performances favorably distinguish, not only the operator itself, but also the models that use it.

It is true that we have established non-integer models that overcome the mass-damping dilemma in mechanics and the stability-precision dilemma in automatic control, the technological achievements associated with these models have been made possible thanks to an adequate synthesis of the non-integer differentiation operator.

It is indeed the idea to synthesize non-integer differentiation (in a medium frequency range) through a recursive distribution of passive components, of transitional frequencies or of zeros and poles, which is at the origin of the non-integer differentiation operator real-time use and, therefore, of both analogical and numerical applications that arise from it. As for the corresponding dates, the synthesis as we have led it has been developed by stages and thus proposed and experimented in the 70s for half-integer orders, in the 80s for real non-integer orders and in the 90s for complex non-integer orders.

The first technological applications of this operator (henceforth usable in real-time) and notably the first application in 1975 of a “half-integer order controller” to the frequency control of a continuous dye laser, have widely contributed to take the non-integer differentiation out of the mathematician drawers and to arouse new developments likely to enrich the theoretical corpus of circuits and systems.

In this way France has been the first country to know a renewed interest on non-integer differentiation, this renewal having been well relayed thanks to the dynamism of the foreign scientific communities, at the European level as well as at the international level.

In this context, the French institutions have encouraged research in this field through the acknowledgement of major scientific advances and the support of initiatives or actions aiming to favor the synergies between the different themes and between the academic and industrial components, the University-Industry partnership having indeed been nationally rewarded by the AFCET’95 Trophy distinguishing the CRONE suspension as best technological innovation. Concerning the acknowledgements, let us cite the selection of the CRONE control as a “striking fact” of the Centre National de la Recherche Scientifique (CNRS) in 1997 and as “Flagship Innovation” of Alstom in 2000 (Hanover and Baden

Baden International Fairs, 2000), a Silver Medal of the CNRS in 1997 and the Grand Prix Lazare Carnot 2011 of the Science Academy (founded by the Ministry of Defense). Concerning the supports, let us cite the actions financially supported by the CNRS and the Ministry of Research: the edition of "La commande CRONE" (Hermès, 1991) with an exceptional help of the ministry; the International Summer School "Fractal and hyperbolic geometries, fractional and fractal derivatives in engineering, applied physics and economics" (Bordeaux, 1994); the national project of the CNRS, "Non-integer differentiation in vibratory insulation" (1997-1999); the colloquium "Fractional differential systems" (Paris, 1998); the launching in 1999 of the thematic action of the Ministry of Research "Systems with non-integer derivatives"; the launching in 2004 of the IFAC Workshop "Fractional Differentiation and its Applications" through the first Workshop FDA'04 (Bordeaux, 2004) with S. Samko as chairman of the International Program Committee; the magisterial lecture "From diversity to unexpected dynamic performances" initiated by the French Science Academy (Bordeaux, 5 January 2012).

But this academic support also found a guarantee in the industrial support brought by a strong partnership with major companies as PSA Peugeot-Citroën, Bosch (Stuttgart) and Alstom, such a partnership having indeed led to a high number of patents and technological transfers that have widely proved the industrial interest of non-integer approaches.

Alongside the shared efforts to inscribe these approaches in the realist frame of the University-Industry relations, our efforts have never stopped being shared with those of the international scientific community to develop the best relations and collaborations within this community. Without aiming for exhaustiveness, let us cite the involvement of European countries in the diffusion, promotion and animation within the community: the research group "Fracalmo", which originates from **Fra**ctional **cal**culus **mo**delling, started with a round table discussion in 1996 during the 2nd International Conference "Transform methods and special functions" held in Bulgaria; the journal "Fractional Calculus and Applied Analysis" (FCAA) started in 1998 with V. Kiryakova as managing editor; the survey on the "Recent history of fractional calculus" (*Communications in Nonlinear Science and Numerical Simulation*, 2010) at the initiative of J. T. Machado who desired to make an inventory of the major documents and events in the area of fractional calculus that had been produced or organized since 1974; the symposium "Fractional Signals and Systems" (FSS) launched by M. Ortigueira in 2009 at Lisbon, then held in 2011 at Coimbra (Portugal) and in 2013 at Ghent (Belgium). Both in and out of Europe, let us also recall the various events of the Workshop FDA after its launching at Bordeaux in 2004 (under the aegis of IFAC): Porto (Portugal) in 2006; Ankara (Turkey) in 2008; Badajoz (Spain) in 2010; Nanjing (China) in 2012; Grenoble (France) in 2013; Catania (Italy) in 2014 (under the aegis of IEEE).

Founder of the CRONE team that counts today about ten permanent researchers, I recognize this team to have always escorted me in the federative actions, launched within the national or international scientific community, to energize and harmonize the researches on both theoretical and applicative aspects of non-integer differentiation.

Xavier Moreau, member of the CRONE team, and Roy Abi Zeid Daou, associated member, who are implied in this book as coordinators well attest this will to be at the initiative of collective actions. It is then a pleasure for me, through this foreword, to warmly congratulate them for their efforts of scientific coordination at the international level, especially as the number of proposed contributions in this book well proves the dynamics of the fractional (or non-integer) community.

Alain Oustaloup
Institut National Polytechique de Bordeaux (INPB),
ENSEIRB-MATMECA, France

PREFACE

The first volume of this two-volume book, presents history, the mathematical modeling and the applications of fractional order systems, and contains lot of mathematical and theoretical studies and research related to this domain.

This volume is made up of 11 chapters. The first chapter presents an analysis of the Caputo derivative and the pseudo state representation with the infinite state approach. The second chapter studies the stability of a class of fractional Cauchy problems. The third chapter shows how to solve fractional order differential equations and fractional order partial differential equations using modern matrix algebraic approaches. Following this chapter, chapter four proposes another analytical method to solve differential equations with local fractional derivative operators.

Concerning chapter five, it presents the extended Borel transform and its related fractional analysis. After presenting the analytical resolution methods for fractional calculus, chapter six shows the essentials of fractional calculus on discrete settings. The initialization of such systems is shown in chapter seven. In fact, this chapter presents a generalized application of the Hankel operator for initialization of fractional order systems.

The last four chapters show some new studies and applications of non-integer calculus. In fact, chapter eight presents the fractional reaction-transport equations and evanescent continuous time random walks. Chapter nine shows a novel approach in the exponential integrators for fractional differential equations. Chapter ten presents the non-fragile tuning of fractional order PD controllers for integrating time delay systems. At the end, chapter eleven proposes a discrete finite-dimensional approximation of linear infinite dimensional systems.

To sum up, this volume presents a mathematical and theoretical study of fractional calculus along with a stability study and some applications. This volume ends up with some new techniques and methods applied in fractional calculus. This volume will be followed up by a second volume that focuses on the applications of fractional calculus in several engineering domains.

ABOUT THE EDITORS

Dr. Roy Abi Zeid Daou
Head of Biomedical Technologies Department
Lebanese German University
Sahel Alma Campus, keserwane – Lebanon
P.O. Box 206
Jounieh, Lebanon
Tel: 009619938938 (ext: 208); 009613396099
Email: r.abizeiddaou@lgu.edu.lb

Xavier Moreau, Professor
Laboratoire de l'Intégration du Matériau au Système (IMS) -
UMR 5218 CNRS Université Bordeaux 1 - IPB ENSEIRB-MATMECA
351 cours de la Libération - Bâtiment A31-B F33405
TALENCE Cedex - (33) FRANCE
Tel: (+33) (0)5 40 00 24 17
Fax: (+33) (0)5 40 00 66 44
Email: xavier.moreau@ims-bordeaux.fr

In: Fractional Calculus: Theory
Editors: R. Abi Zeid Daou and X. Moreau
ISBN: 978-1-63463-002-3

Chapter 1

ANALYSIS OF THE CAPUTO DERIVATIVE AND PSEUDO STATE REPRESENTATION WITH THE INFINITE STATE APPROACH

Jean-Claude Trigeassou[1,*], ***Nezha Maamri***[2,†] ***and Alain Oustaloup***[3,‡]
[1]Laboratory IMS, UMR 5218 CNRS, University of Bordeaux, INP Bordeaux, Talence Cedex, France
[2]LIAS, University of Poitiers, Poitiers Cedex, France
[3]Institut National Polytechique de Bordeaux (INPB), ENSEIRB-MATMECA, France

Abstract

This paper presents the interpretation of the Caputo derivative with the help of the infinite state approach, also known as the fractional integrator approach. After presentation of the modified Laplace transform equations, definition of state variables is discussed, as well as its application to the characterization of fractional system energy. Then, the infinite state approach is used to analyze system simulation based on usual Caputo initial conditions. Finally, a new formulation of the transition matrix is proposed, to take into account the true initial state of a fractional system.

Keywords: Caputo derivative, Riemann-Liouville integral, Fractional integrator, Infinite state approach, Initialization, Fractional transition matrix

* E-mail address: jean-claude.trigeassou@ims-bordeaux.fr; Address: IMS-LAPS, University of Bordeaux I, 351 Avenue de la Liberation, 33405 Talence cedex, France.
† E-mail address: nezha.maamri@univ-poitiers.fr; Address: LIAS, University of Poitiers, 40 Avenue du Recteur Pineau, 86022 Poitiers cedex, France.
‡ E-mail address: alain.oustaloup@ims-bordeaux.fr; Institut National Polytechique de Bordeaux (INPB), ENSEIRB-MATMECA, France

Introduction

A great number of research papers on fractional system theory rely on the Caputo derivative and the pseudo state space representation. Some recent papers have introduced a reasonable doubt on the physical initial conditions of this derivative and its applications to fractional system theory. So, the first objective of this paper is to analyze the Caputo derivative and to explain why the induced usual state space representation does not take into account the underlying infinite dimension of fractional differential systems. Consequently, the second objective is to provide solutions to this fundamental problem using the infinite state approach.

Several causes can explain this situation. Whereas only one definition of differentiation is used in integer order system theory, several definitions are available in fractional calculus. Secondly, fractional differentiation is always presented prior to fractional integration. Consequently, it is interpreted as a particular case of differentiation, corresponding to a negative value of the fractional order, whereas on the contrary integration is essential to perform Riemann-Liouville or Caputo differentiation. And above all, we have to take into account that fractional integration relies fundamentally on a convolution integral, i.e., it is completely different from conventional integration.

Three main definitions of fractional differentiation are available: Grünwald-Letnikov, Riemann-Liouville, Caputo (Caputo 1969), with different variants (Podlubny 1999, Ortigueira 2011)). In fact, Grünwald-Letnikov definition is the generalization of the numerical definition of the integer order derivative (Oustaloup 1995); so it is an approximate definition, requiring that the increment $h \to 0$. This derivative is not concerned by the present discussion. On the contrary, Riemann-Liouville and Caputo definitions are exact ones. Practically, users consider that they have to select the definition adapted to the treatment of their problem. Consequently, most of them prefer to use Caputo definition, because its initial conditions are said to have a nice physical interpretation. So, Caputo approach has become the reference in fractional calculus research (Diethelm 2010, Petras 2011 and others). Unfortunately, two initialization techniques - history function of Lorenzo and Hartley (Hartley 2009, Lorenzo 2011) and infinite state of Trigeassou and Maamri (Trigeassou 2010) - have shown that these Caputo initial conditions are wrong in the general case and have to be corrected, using either a time varying initialization or an appropriate initial infinite state. Moreover, it has been proved recently that these two different initialization techniques are equivalent (Hartley 2013).

The infinite state approach (Trigeassou 2013a) is based on the definition of a fractional integrator, generalizing the integer order case. Basically, it provides an efficient technique to perform fractional integration. Its frequency distributed infinite state $z_i(\omega,t)$ corresponds to the classical $x_i(t)$ state variable of the integer order integrator. So, a Fractional Differential System (FDS) or Fractional Differential Equation (FDE) can be interpreted as an integer order system, whose state is infinite dimensional.

Three recent papers present this approach and its application to the initialization of fractional derivatives and systems, as well as the analysis and the prediction of their transients (Trigeassou 2012a, 2012b, 2013a).

Another interest of this technique is to provide a tool for the analysis of Riemann-Liouville and Caputo derivatives, and to propose corrections to their usual initial conditions.

In this paper, we focus on the Caputo derivative and we intend to demonstrate that the usual approach leads to questionable conclusions, concerning the definition of state variables, the simulation of FDEs and FDSs, and the formulation of the transition matrix. Though not discussed in this paper, the Riemann-Louville derivative is concerned by the same drawbacks (refer to Trigeassou 2012a, 2012b).

This paper is composed of six sections. First section is the introduction. The Riemann-Liouville integral and the infinite state approach are reminded in section 2. The Laplace transform equations of the Caputo derivative are discussed in section 3. Fractional state variables are defined in section 4 using a system energy approach. Caputo initial conditions for FDE/FDS simulation are analyzed in section 5 and a formulation of the transition matrix for commensurate order systems is proposed in section 6.

Riemann-Liouville Integral and Infinite State Approach

Riemann-Liouville Integral

The nth fractional order Riemann-Liouville integral (n real positive) of a function $f(t)$ is defined by the relation (Podlubny 1999):

$$I^n\left(f(t)\right)=\frac{1}{\Gamma(n)}\int_0^t (t-\tau)^{n-1} f(\tau)d\tau \tag{1}$$

where $\Gamma(n)$ is the gamma function

$$\Gamma(n)=\int_0^\infty x^{n-1}e^{-x}dx \tag{2}$$

$I^n\left(f(t)\right)$ is the convolution of the function $f(t)$ with the impulse response:

$$h_n(t)=\frac{t^{n-1}}{\Gamma(n)}H(t) \tag{3}$$

of the fractional integration operator whose Laplace transform is:

$$I^n(s)=L\{h_n(t)\}=\frac{1}{s^n} \tag{4}$$

(where $H(t)$ is the Heaviside function).

Fractional differentiation is the dual operation of fractional integration. Consider the fractional integration operator $I^n(s)$ whose input and output are respectively $x(t)$ and $y(t)$. Then, with zero initial conditions:

$$y(t) = I^n\big(x(t)\big) \text{ or } Y(s) = \frac{1}{s^n} X(s) \tag{5}$$

Reciprocally, $x(t)$ is the nth order fractional derivative of $y(t)$ defined as:

$$x(t) = D^n\big(y(t)\big) \text{ or } X(s) = s^n Y(s) \tag{6}$$

This fractional derivative definition is based on the operator $I^n(s)$, without analytical formulation of $D^n\big(y(t)\big)$: it is the implicit definition of the fractional derivative. However, this implicit derivative exists only inside a closed loop, like in the simulation of fractional differential equations. Its main interest is to avoid the use of the Riemann-Liouville or Caputo derivatives.

Riemann-Liouville and Caputo Derivatives

Assume that the fractional order n is included between the two integer numbers $N-1$ and N, so $N-1 < n < N$. Then we can write:

$$D^n(s) = s^n = \frac{1}{s^N} s^n s^N = \frac{1}{s^{N-n}} s^N \tag{7}$$

where $\frac{1}{s^{N-n}}$ represents the Laplace transform of the fractional integration $I^{N-n}()$ and s^N corresponds to the integer order differentiation $\frac{d^N()}{dt^N}$. Then, with zero initial conditions:

$$L\{D^n(f)\} = D^n(s)F(s) = \frac{1}{s^{N-n}} s^N F(s) \tag{8}$$

Using inverse Laplace transform, we get two expressions of $D^n(f)$:

$$D_C^n(f) = L^{-1}\left\{\frac{1}{s^{N-n}}(s^N F(s))\right\} \tag{9}$$

$$D_{RL}^n(f) = L^{-1}\left\{ s^N \left(\frac{1}{s^{N-n}} F(s)\right)\right\} \tag{10}$$

The first expression (9) corresponds to:

$$D_C^n(f) = h_{N-n}(t) * \frac{d^N f(t)}{dt^N} \tag{11}$$

and the second one (10) to:

$$D_n^{RL}(f) = \frac{d^N}{dt^N}(h_{N-n}(t) * f(t)) \tag{12}$$

This first expression is known as the Caputo derivative (Caputo 1969):

$$D_C^n(f(t)) = \int_0^t \frac{(t-\tau)^{N-n-1}}{\Gamma(N-n)} \frac{d^N f(\tau)}{dt^N} d\tau \tag{13}$$

while the second one is the Riemann-Liouville derivative (Podlubny 1999):

$$D_{RL}^n(f(t)) = \frac{d^N}{dt^N}\left\{\int_0^t \frac{(t-\tau)^{N-n-1}}{\Gamma(N-n)} f(\tau)d\tau\right\} \tag{14}$$

Remark. It is important to notice that these two explicit derivatives rely on fractional integration $I^{N-n}(s) = \frac{1}{s^{N-n}}$, i.e., on convolution with the $h_{N-n}(t)$ impulse response.

Fractional Integrator

Practically, $I^n(f(t))$ is performed by a fractional integrator. This integrator is a linear frequency distributed system or diffusive system (Oustaloup 1983, Montseny 1998, Trigeassou 1999, 2012a, Heleschewitz 2000) with input $v(t)$ and output $x(t)$. Its frequency distributed state $z(\omega,t)$ verifies the differential equation (for the elementary frequency ω):

$$\frac{\partial z(\omega,t)}{\partial t} = -\omega\, z(\omega,t) + v(t) \tag{15}$$

and the output $x(t)$ of the fractional integrator is the weighted integral (with weight $\mu_n(\omega)$) of all the contributions $z(\omega,t)$ ranging from 0 to ∞:

$$x(t) = \int_0^\infty \mu_n(\omega)\, z(\omega,t)\, d\omega$$

$$\mu_n(\omega) = \frac{\sin n\pi}{\pi}\omega^{-n} \quad 0 < n < 1 \tag{16}$$

Notice that $z(\omega,t)$ is the true infinite dimension distributed state variable whereas $x(t)$ is only a pseudo state variable. The infinite state approach is so called by reference (Trigeassou 2013a) to the infinite dimension of $z(\omega,t)$. It is also called the fractional integrator approach.

If $v(t) = \delta(t)$, then $x(t)$ corresponds to the impulse response of the fractional integrator:

$$x(t) = h_n(t) = \int_0^{\infty} \mu_n(\omega)\, e^{-\omega t}\, d\omega \tag{17}$$

This distributed frequency model is not directly usable. It has to be replaced by a frequency discretized distributed modal model; refer to (Trigeassou 2011b, 2012a) for more details.

$$\underline{Z}^T(t) = [z_0 \; .... z_j \; ... z_J] \tag{18}$$

$$\begin{aligned} \frac{d\,\underline{Z}(t)}{dt} &= A\,\underline{Z}(t) + \underline{B}\,v(t) \\ x(t) &= \underline{C}^T\, \underline{X}(t) \end{aligned} \tag{19}$$

with:

$$A = \begin{bmatrix} 0 & & & 0 \\ & -\omega_1 & & \\ & & \ddots & \\ 0 & & & -\omega_J \end{bmatrix} \quad \underline{B} = \begin{bmatrix} 1 \\ 1 \\ \vdots \\ 1 \end{bmatrix} \tag{20}$$

$$\underline{C}^T = \begin{bmatrix} c_0 & c_1 & \cdots & c_J \end{bmatrix} \tag{21}$$

The frequency spectrum ranges from ω_1 to ω_J with the conditions: $\omega_1 \to 0$, $\omega_J \to \infty$, $J \gg 1$.

This model can be schematized by the graph of figure 1.

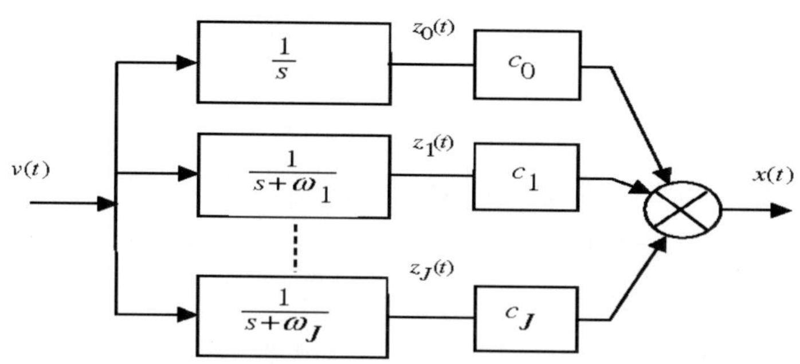

Figure 1. Finite dimension approximation of a fractional integrator.

Integer Order and Fractional Order Integration

The integer order integrator is a particular case ($n=1$) of the fractional integrator ($0 < n < 1$) with $\mu_1(\omega) = \delta(\omega)$, so:

$$h_1(t)=\int_0^{\infty}\delta(\omega)e^{-\omega t}\,d\omega=H(t) \tag{22}$$

where $H(t)$ is the Heaviside function. Then:

$$x_1(t)=h_1(t)*v(t)=\int_{-\infty}^{t}H(t-\tau)v(\tau)\,d\tau=\int_{-\infty}^{t}v(\tau)\,d\tau \tag{23}$$

because $H(t-\tau)=1$ for $\tau\le t$. So $x_1(t)$ is obtained by conventional integration. Notice that:

$$x_1(t)=\int_{-\infty}^{0}v(\tau)\,d\tau+\int_0^{t}v(\tau)\,d\tau=x_1(0)+\int_0^{t}v(\tau)\,d\tau \tag{24}$$

where $x_1(0)$ is the initial condition of:

$$I_1(v(t))=\int_0^{t}v(\tau)\,d\tau \tag{25}$$

In the fractional order case:

$$x_n(t)=h_n(t)*v(t)=\int_{-\infty}^{t}h_n(t-\tau)v(\tau)\,d\tau \tag{26}$$

$x_n(t)$ is the result of a convolution, thus it is not equal to:

$$\int_{-\infty}^{0}h_n(t-\tau)v(\tau)\,d\tau+\int_0^{t}h_n(t-\tau)v(\tau)\,d\tau \tag{27}$$

Because $h_n(t)$ is the weighted sum (17) of an infinity of exponential modes $e^{-\omega t}$, it is necessary to summarize the past behavior of the fractional integrator on $]-\infty;0]$ by $z(\omega,0)$. So we can express $x_n(t)$ with the usual linear system technique (Kailath 1980):

$$x_n(t)=\int_0^{\infty}\mu_n(\omega)\,z(\omega,0)\,e^{-\omega t}\,d\omega+\int_0^{t}h_n(t-\tau)v(\tau)\,d\tau \tag{28}$$

where the first term represents the free response initialized by $z(\omega,0)$ and the second term (Riemann-Liouville integral) is the forced response of the integrator (see also Sabatier 2010).

So, if the fractional integrator is not at rest at $t=0$ ($z(\omega,0)=0\ \forall\omega$), it is necessary to take into account its infinite distributed initial state $z(\omega,0)$. Consequently, the lower terminal of integral (1) can be equal to 0 in any case.

The response of the fractional integrator can also be expressed using the Laplace transform technique. Consider the Laplace transform of the distributed equation (15):

$$L\left\{\frac{\partial z(\omega,t)}{\partial t}\right\}=s\,Z(\omega,s)-z(\omega,0)=-\omega\,Z(\omega,s)+V(s) \tag{29}$$

where $z(\omega,0)$ is the initial state of the fractional integrator.

So $Z(\omega,s)=\dfrac{V(s)+z(\omega,0)}{s+\omega}$ and:

$$X(s)=\int_0^\infty \frac{\mu_n(\omega)\,z(\omega,0)}{s+\omega}d\omega+\int_0^\infty \frac{\mu_n(\omega)}{s+\omega}d\omega\,V(s) \tag{30}$$

The first term represents the free response of the integrator whereas the second one is the forced response with:

$$\int_0^\infty \frac{\mu_n(\omega)}{s+\omega}d\omega=\frac{1}{s^n} \tag{31}$$

Simulation of FDEs and FDSs

Consider the elementary one derivative system:

$$D^n(x(t))+a\,x(t)=u(t) \tag{32}$$

We can write:

$$D^n(x(t))=v(t)=u(t)-a\,x(t) \tag{33}$$

where $v(t)$ is the input of the fractional integration operator $I^n(s)=\dfrac{1}{s^n}$, and $x(t)$ is its output.

The simulation of this FDE can be interpreted as a closed loop system, corresponding to figure 2. Consequently, $v(t)$ is the implicit derivative of $x(t)$.

Simulation is performed using the equations:

$$\begin{aligned}&\frac{\partial z(\omega,t)}{\partial t}=-\omega\,z(\omega,t)+v(t)\\&x(t)=\int_0^\infty \mu_n(\omega)\,z(\omega,t)\,d\omega\end{aligned} \tag{34}$$

where the input $v(t)$ is defined by the feedback equation $v(t)=u(t)-a\,x(t)$.

Infinite State of FDEs and FDSs

The pseudo state space model of a FDS is (Trigeassou 2012b):

$$D^{\underline{n}}(\underline{X}(t)) = A\,\underline{X}(t) + \underline{B}\,u(t) \tag{35}$$

where:

$$D^{\underline{n}}(\underline{X}(t))^T = \left[D^{n_1}(x_1(t)) \ldots D^{n_i}(x_i(t)) \ldots D^{n_N}(x_N(t))\right] \tag{36}$$

A fractional integrator $I^{n_i}(s) = \frac{1}{s^{n_i}}$ is associated to each pseudo state variable $x_i(t)$:

$$\begin{aligned} &\frac{\partial z_i(\omega,t)}{\partial t} = -\omega\, z_i(\omega,t) + v_i(t) \\ &x_i(t) = \int_0^{\infty} \mu_{n_i}(\omega)\, z_i(\omega,t)\, d\omega \\ &v_i(t) = D^{n_i}(x_i(t)) \end{aligned} \tag{37}$$

The true state vector is $\underline{Z}(\omega,t)$ with:

$$\underline{Z}^T(\omega,t) = \left[z_1(\omega,t) \ldots z_i(\omega,t) \ldots z_N(\omega,t)\right] \tag{38}$$

The Laplace transform applied to the fractional integrator $I_{ni}(s)$ gives:

$$Z_i(\omega,s) = \frac{z_i(\omega,0) + V_i(s)}{s+\omega} \tag{39}$$

and for the state vector $\underline{Z}(\omega,s)$:

$$\underline{Z}(\omega,s) = \frac{\underline{z}(\omega,0)}{s+\omega} + \frac{\underline{V}(s)}{s+\omega} \tag{40}$$

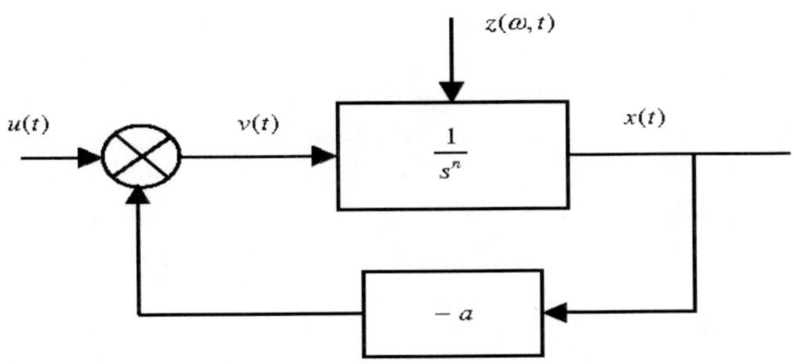

Figure 2. Simulation of a one derivative FDE.

Let us define:

$$\underline{X}(s)=\begin{bmatrix} \vdots \\ \int_0^{\infty} \mu_{n_i}(\omega)\, Z_i(\omega,s)d\omega \\ \vdots \end{bmatrix} \qquad \frac{1}{s^{\underline{n}}}=\begin{bmatrix} \frac{1}{s^{n_1}} & & 0 \\ & \frac{1}{s^{n_i}} & \\ 0 & & \frac{1}{s^{n_N}} \end{bmatrix} \tag{41}$$

and

$$s^{\underline{n}}=\left[\frac{1}{s^{\underline{n}}}\right]^{-1}=\begin{bmatrix} s^{n_1} & & 0 \\ & s^{n_i} & \\ 0 & & s^{n_N} \end{bmatrix} \tag{42}$$

Using these definitions, we get finally:

$$\underline{X}(s)=\left[s^{\underline{n}}-A\right]^{-1} s^{\underline{n}} \begin{bmatrix} \vdots \\ \int_0^{\infty} \frac{\mu_{n_i}(\omega)\, z_i(\omega,0)}{s+\omega} d\omega \\ \vdots \end{bmatrix} + \left[s^{\underline{n}}-A\right]^{-1} \underline{B}U(s) \tag{43}$$

Notice that the global free response of the FDS is composed of the free responses (30) of the elementary integrators.

Laplace Transform of the Caputo Derivative

Definition

According to the previous definition (13), the nth Caputo derivative of a function $f(t)$ is defined as the convolution of the Nth integer order derivative of $f(t)$ with the impulse response of the fractional integrator which order is $N-n$.

Because we use only $0<n<1$ with our previous presentation of FDEs and FDSs, we consider:

$$D_C^n(f)=\int_0^t \frac{(t-\tau)^{-n}}{\Gamma(1-n)} \frac{df(\tau)}{d\tau} d\tau \tag{44}$$

which can be interpreted as:

$$D_C^n(f)=I^{1-n}\left(\frac{df(t)}{dt}\right) \tag{45}$$

corresponding to the diagram of figure 3.

Two operations are required to calculate the Caputo derivative: integer order differentiation and fractional integration. Integer order differentiation is a classical operation. Fractional integration with order equal to $1-n$ is performed according to previous definition (15) (16):

$$\begin{aligned}\frac{\partial z_C(\omega,t)}{\partial t} &= -\omega\, z_C(\omega,t) + \frac{df(t)}{dt} \\ D_C^n(f) &= \int_0^\infty \mu_{1-n}(\omega)\, z_C(\omega,t)\, d\omega\end{aligned} \tag{46}$$

where $z_C(\omega,t)$ is the frequency distributed state of $I^{1-n}(s)$

Laplace Transform of the Caputo Derivative

The Laplace transform of equation (46) is:

$$\begin{aligned}sZ_C(\omega,s) - z_C(\omega,0) &= -w\,Z_C(\omega,s) + L\left\{\frac{df(t)}{dt}\right\} \\ L\left\{D_C^n(f)\right\} &= \int_0^\infty \mu_{1-n}(\omega)\, Z_C(\omega,s)\, d\omega\end{aligned} \tag{47}$$

where $z_C(\omega,0)$ is the initial state of $I^{1-n}(s)$ integrator.

The first equation gives:

$$Z_C(\omega,s) = \frac{z_C(\omega,0) + L\left\{\frac{df(t)}{dt}\right\}}{s+\omega} \tag{48}$$

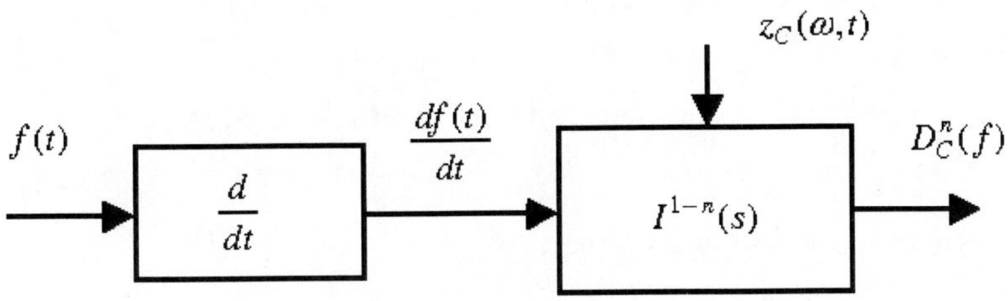

Figure 3. Calculation of the Caputo derivative.

and finally, because

$$L\left\{\frac{df(t)}{dt}\right\} = s\,F(s) - f(0) \tag{49}$$

and

$$\int_0^\infty \frac{\mu_{1-n}(\omega)}{s+\omega}\, d\omega = \frac{1}{s^{1-n}} \tag{50}$$

we obtain (refer to Trigeassou 2010, for more details):

$$L\left\{D_C^n(f)\right\} = s^n F(s) - s^{n-1} f(0) + \int_0^\infty \frac{\mu_{1-n}(\omega) z_C(\omega,0)}{s+\omega} d\omega \tag{51}$$

Discussion

The usual approach provides (Podlubny 1999):

$$L\left\{D_C^n(f)\right\} = s^n F(s) - s^{n-1} f(0) \tag{52}$$

Obviously, this result is wrong in the general case because we have to take into account the distributed initial state $z_C(\omega,0)$ of $I^{1-n}(s)$. Notice that $z_C(\omega,0)=0 \ \forall \omega$ implies $\frac{df}{dt}=0$ or $f(t)=cte$ in equation (46): this is the requirement for correctness of equation (52).

So, in the general case, the complete initial conditions of the Caputo derivative are $f(0)$ and $z_C(\omega,0)$: consequently, the Caputo derivative is no longer attractive because we have to include an additional infinite initial state, with no direct physical meaning.

The conclusion is the same with the Riemann-Liouville derivative. Because $D_{RL}^n(f) = \frac{d}{dt}\left\{h_{1-n}(t) * f(t)\right\}$ for $0 < n < 1$, we have also to take into account the initial state $z_{RL}(\omega,0)$ of $I_{1-n}(s)$. Refer to (Trigeassou 2010) for a global presentation and interpretation of fractional derivatives initial conditions.

Initial conditions are fundamental in system theory. The simple and physically attractive initial conditions of the Caputo derivative (equation 52) have induced the pseudo state space representation with apparently initial conditions in traditional form (Podlubny 1999). So, because the initial state $z_C(\omega,0)$ of the $I^{1-n}(s)$ integrator summarizes the past history of the pseudo state variable, we have to use another definition of fractional systems state variables.

State Variables and System Energy

Introduction

Consider the one derivative FDE (32). Because the usual approach gives

$$L\left\{D_C^n(x)\right\} = s^n X(s) - s^{n-1} x(0) \tag{53}$$

$x(t)$ has been regarded for a long time as the state variable of FDE (32) with initial value $x(0)$, according to the previous discussion.

Obviously, the fractional integrator technique tells us that $x(t) = \int_0^\infty \mu_n(\omega) z(\omega,t) d\omega$ is the weighted integral of the distributed variable $z(\omega,t)$. So it is straightforward that $x(t)$ is only a pseudo state variable (see also Sabatier 2010, 2013, Trigeassou 2013b) and $z(\omega,t)$ is the true infinite dimension state variable. This important result can also be derived from energy considerations.

State Variables and System Energy

A physical interpretation of state variables is related to system energy, as for system passivity (Ortega 1998). Consider the one derivative ODE:

$$\frac{dx(t)}{dt} + a\,x(t) = u(t) \tag{54}$$

which is simulated with the diagram of figure 2, replacing $\frac{1}{s^n}$ by $\frac{1}{s}$. An analogue realization of the integer order integrator $\frac{1}{s}$ corresponds to figure 4, using an operational amplifier, a resistor R and a capacitor C.

Electrostatic energy stored in capacitor C is expressed as

$$E_C(t) = \frac{1}{2} C\, v_C^2(t) = \frac{1}{2} C\, x^2(t) \tag{55}$$

because $v_C(t)$, voltage at the C terminals, is equal to $x(t)$. In the integer order case, $x(t)$ is directly used to express system energy $E_C(t)$: so $x(t)$ is said to be a true state variable. Notice also that $x(t)$ has memorization capability, thanks to energy storage by capacitor C.

Consider now

$$D^n(x(t)) + a\,x(t) = u(t) \tag{56}$$

which is simulated according to figure 2. Assume that we use the finite dimension approximation of the fractional integrator (figure 1). Consider the $z_j(t)$ variable, corresponding to the jth first order system:

$$\frac{dz_j(t)}{dt} + \omega_j\, z_j(t) = v(t) \tag{57}$$

This first order ODE is simulated with an integer order integrator (see figure 4), whose energy is:

$$E_{C_j}(t) = \frac{1}{2} C_j\, z_j^2(t) \tag{58}$$

So, the global energy of the FDE is expressed as:

$$E_C(t)=\sum_{j=0}^{J} E_{C_j}(t)=\sum_{j=0}^{J} \frac{1}{2} C_j z_j^2(t) \tag{59}$$

If we consider the continuous distribution $z(\omega,t)$, then $\sum_{j=0}^{J} E_{C_j}(t)$ has to be replaced by the continuous weighted integral (Trigeassou 2011a, 2013c):

$$E(t)=\int_0^{\infty} \mu_n(\omega) z^2(\omega,t)\, d\omega \tag{60}$$

So the conclusion is that $z(\omega,t)$ (or $z_j(t)$) is the true distributed state variable of the FDE whereas

$$V(x)=x^2(t)=\left[\int_0^{\infty} \mu_n(\omega) z(\omega,t)\, d\omega\right]^2 \tag{61}$$

cannot be related to the true FDE energy (Li 2009). So $x(t)$ is only a pseudo state variable.

Notice that $V(z)=E(t)$ is the true Lyapunov function characterizing FDE energy. So, $V(x)=x^2(t)$ is only a pseudo Lyapunov function.

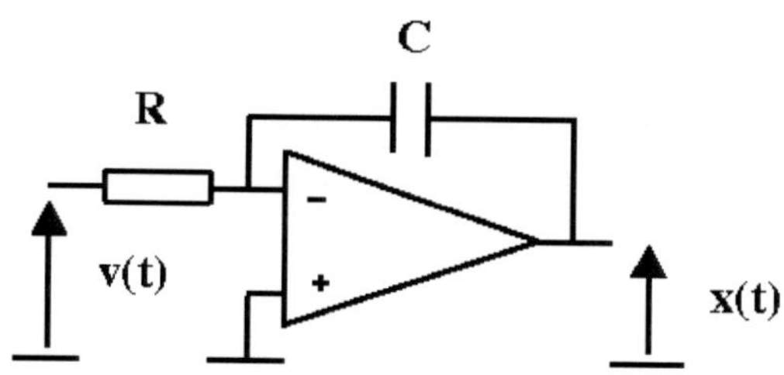

Figure 4. Analogue realization of the integer order integrator.

Illustrative Example

Consider again the one derivative system (56) with $a=1$ $n=0.5$ *and* $z(\omega,0)=0$ $\forall\omega$ where the input is:

$$\begin{cases} u(t)= U & for\ 0<t<5 \\ u(t)=-U & for\ 5<t<10 \\ u(t)= U & for\ t>10 \end{cases} \tag{62}$$

This system has been simulated with the following parameters:

$$\omega_b=0.001\, rd/s \quad \omega_h=1000\, rd/s \quad J=20 \quad Te=5\, ms$$

The response $x(t)$ is displayed on figure 5. We notice that there are two instants $t_1 = 5.25\,s$ and $t_2 = 10.2\,s$ corresponding to $x(t_1) = 0$ and $x(t_2) = 0$. Then, the system has been initialized at t_1 and t_2 using $z(\omega, t_1)$ and $z(\omega, t_2)$. The corresponding free responses have been displayed on the same figure 5 ($x_{init1}(t)$ $x_{init2}(t)$).

We notice that $x(t) = 0$ for $t_0 = 0$, $t_1 = 5.25\,s$, $t_2 = 10.2\,s$. Figure 6 displays the frequency distribution of $z(\omega, t_i)$ for these three instants (the index of ω is equal to that of the discretized frequency ω_j , j varying from 0 to J).

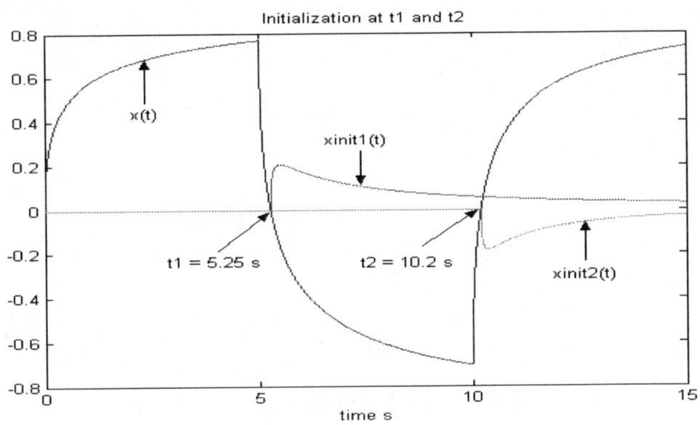

Figure 5. System forced response and initialized response.

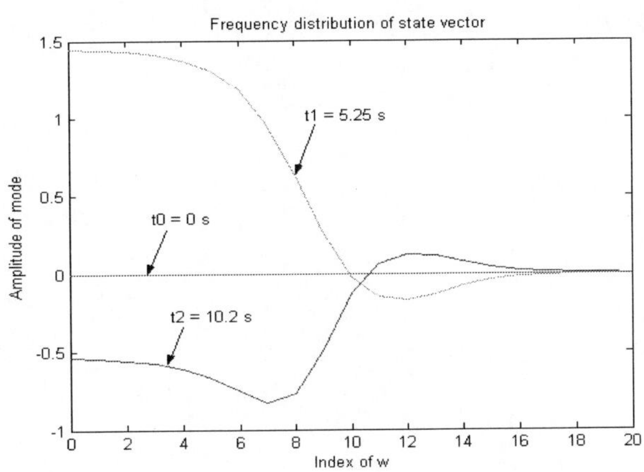

Figure 6. Comparison of distributed state vectors.

Though the output has the same value at these different instants, we can notice that the three distributions are completely different. These graphs illustrate the relevance of $z(\omega, t_i)$ for system free response analysis.

Using usual Caputo approach (Podlubny 1999), we can write:

$$X_{Cap}(s) = \frac{s^{n-1}\,x(0)}{s^n + a} \tag{63}$$

and

$$x_{Cap}(t) = L^{-1}\left\{X_{Cap}(s)\right\} = x(0)\,E_{n,1}(-at^n) \tag{64}$$

So, if $x(t_i)=0$, no transient can be predicted! Obviously, this numerical experiment has clearly demonstrated that the only knowledge of $x(t_i)$ is unable to predict transients. On the contrary, using $z(\omega,t_i)$, one can achieve perfect prediction of these transients (refer to Trigeassou 2012b for more details).

We have also represented on figure 7 the graphs of $V(x)$ and $V(z)$ with the free response starting at $t_1=5.25\,s$. It is obvious with this example that $V(x)$ is a semi definite positive function, whereas $V(z)$ is really positive definite: so, $V(z)$ is a decreasing Lyapunov function, able to characterize the energy of the system. Refer to (Trigeassou 2013c, 2013d) for a discussion on the influence of initial conditions on the decrease of the Lyapunov function, and more globally on fractional energy and Lyapunov stability.

Initial Conditions for FDE/FDS Simulation

Simulation of a One Derivative System

Consider again the one derivative system (56). The previous examples have demonstrated that usual Caputo approach leads to wrong transients.

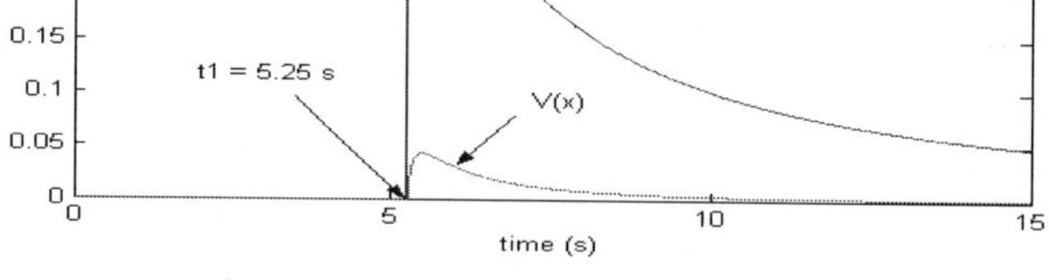

Figure 7. Comparison of Lyapunov functions.

So, numerical simulations based on Caputo initial conditions are certainly questionable. Nevertheless, we can ask the following question: are Caputo initial conditions always wrong?

The expression:

$$L\left\{D_C^n(x)\right\} = s^n X(s) - s^{n-1}x(0) \tag{65}$$

is true (refer to (51)) if

$$z_C(\omega,0)=0 \quad \forall\omega \tag{66}$$

Remind that $z_C(\omega,t)$ is the response of system (46) (see figure 3) to the excitation $\frac{dx(t)}{dt}$. The previous condition (66) is fulfilled if $\frac{dx(t)}{dt}=0$, or equivalently $x(t)=cte$.

However, because of long range memory phenomenon (remind that ω ranges from 0 to ∞),

$$\lim_{t\to\infty} z_C(\omega,t)=0 \ \ \forall\omega \quad if \ \frac{d(x(t)}{dt}=0 \ \ or \ \ x(t)=cte \ \ for \ t\in[0,\infty[\tag{67}$$

Reciprocally:

$$z_C(\omega,0)=0 \ \ \forall\omega \ if \ x(t)=cte \ \ for \ t\in]-\infty,0] \tag{68}$$

So, we can use equation (65) only if the system (46) is completely at rest (or at steady state) at the initial instant t_0: the condition system “at rest or at steady state” implies that

$$z_C(\omega,t_0)=0 \ \ \forall\omega \tag{69}$$

Remark. $z(\omega,t)$ is the distributed state of the fractional integrator $I^n(s)$ (do not confuse it with $z_C(\omega,t)$) and its output at $t=t_0$ is:

$$x(t_0)=\int_0^\infty \mu_n(\omega)\,z(\omega,t_0)\,d\omega \tag{70}$$

The system is at rest at

$$t=t_0 \ \ if \ z(\omega,t_0)=0 \ \forall \ \omega\neq 0 \tag{71}$$

and

$$x(t_0)=\int_0^\infty \mu_n(\omega)\,z(0,t_0)\,d\omega \tag{72}$$

Refer to (Trigeassou 2012b) for a general discussion on FDE/FDS physical initial conditions.

Simulation of a Two Derivatives FDE

Consider:

$$H(s)=\frac{1}{a_0+a_1 s^{n_1}+s^{n_1+n_2}} \tag{73}$$

corresponding to the FDS:

$$\begin{bmatrix} D_{n_1}(x_1(t)) \\ D_{n_2}(x_2(t)) \end{bmatrix} = \begin{bmatrix} 0 & 1 \\ -a_0 & -a_1 \end{bmatrix} \begin{bmatrix} x_1(t) \\ x_2(t) \end{bmatrix} + \begin{bmatrix} 0 \\ 1 \end{bmatrix} u(t) \tag{74}$$

If this system is simulated with Caputo initial conditions $x_1(t_0)$ and $x_2(t_0)$, it implies, according to the previous discussion, that:

$$x_1(t)=cte \neq 0 \quad for\ t \le t_0 \tag{75}$$

and

$$x_2(t)=cte \neq 0 \quad for\ t \le t_0 \tag{76}$$

So, the question is: are these initial conditions compatible with the previous condition (68)?

The Laplace transform technique relies on the hypothesis $t \in [0\,;+\infty[$. So consider that $u(t)=U=cte$ for $t>0$, which is equivalent to consider $u(t)=U=cte$ for $t<0$, with $x_1(\infty)$) corresponding to $x_1(0)$.

Then

$$X_1(s)=\frac{1}{a_0+a_1 s^{n_1}+s^{n_1+n_2}}\frac{U}{s} \tag{77}$$

and

$$x_1(\infty) = \begin{cases} \lim \\ s\to 0 \end{cases} s\,\frac{1}{a_0+a_1 s^{n_1}+s^{n_1+n_2}}\frac{U}{s}=\frac{U}{a_0} \tag{78}$$

Let

$$\varepsilon(t)=u(t)-a_0\,x_1(t)-a_1\,x_2(t) \tag{79}$$

Then

$$\varepsilon_1(\infty) = \begin{cases} \lim \\ s\to 0 \end{cases} s\,\frac{s^{n_1+n_2}}{a_0+a_1 s^{n_1}+s^{n_1+n_2}}\frac{U}{s}=0 \tag{80}$$

So, according to the long range memory phenomenon, the system is at rest at $t \to \infty$ when $x_1(\infty) = \dfrac{U}{a_0}$ and $\varepsilon(\infty) = 0$.

What can we say of $x_2(\infty)$?

According to (79):

$$a_1\, x_2(\infty) = U - a_0\, x_1(\infty) - \varepsilon(\infty) = 0 \tag{81}$$

So, this result is in contradiction with the hypothesis $x_2(\infty) = cte \neq 0$.

The conclusion is that it is impossible to verify the Caputo initial conditions with an arbitrary choice like:

$$\begin{cases} x_1(t_0) = cte \neq 0 \\ x_2(t_0) = cte \neq 0 \end{cases} \tag{82}$$

The only possibility is:

$$\begin{cases} x_1(t_0) = \dfrac{U}{a_0} \\ x_2(t_0) = 0 \end{cases} \tag{83}$$

Nevertheless, the system can be initialized with $x_1(t_0) \neq 0$ and $x_2(t_0) \neq 0$, but these values are no longer initial conditions. The true initial conditions are $z_1(\omega, t_0)$ and $z_2(\omega, t_0)$ which are related to $x_1(t_0)$ and $x_2(t_0)$ by:

$$\begin{cases} x_1(t_0) = \displaystyle\int_0^{\infty} \mu_{n1}(\omega)\, z_1(\omega, t_0)\, d\omega \\ x_2(t_0) = \displaystyle\int_0^{\infty} \mu_{n2}(\omega)\, z_2(\omega, t_0)\, d\omega \end{cases} \tag{84}$$

Refer to (Trigeassou 2012b) which presents a numerical simulation of a two derivatives system according to this definition of initial conditions.

Simulation of a N Derivatives FDS

The previous result (83) is easily generalized to a N derivatives system. Consider:

$$D^{\underline{n}}(\underline{X}(t)) = A\, \underline{X}(t) + \underline{B}\, u(t) \qquad \dim \underline{X} = N \tag{85}$$

The condition

$$\underline{X}(t_0) = \underline{cte} \text{ for } t \in \left]-\infty; t_0\right] \tag{86}$$

with respect to the long range memory phenomenon, means that the fractional system is completely at rest at $t = t_0$.

Thus

$$D^{\underline{n}}(\underline{X}(t_0)) = \underline{0} \tag{87}$$

If $u(t) = U$ has been applied to this system on $\left]-\infty; t_0\right]$, then we can express $\underline{X}(t_0)$ with:

$$A\underline{X}(t_0) + \underline{B}u(t_0) = 0 \tag{88}$$

Thus the condition $\underline{X}(t_0) = \underline{cte}$ for $t \in \left]-\infty; t_0\right]$ has to verify:

$$\underline{X}(t_0) = -A^{-1}\underline{B}U \tag{89}$$

This means that the components of $\underline{X}(t_0)$ cannot be chosen arbitrary in order to be coherent with Caputo hypotheses.

Remark. With the two derivatives example, we obtain:

$$\begin{bmatrix} x_1(t_0) \\ x_2(t_0) \end{bmatrix} = -\begin{bmatrix} 0 & 1 \\ -a_0 & -a_1 \end{bmatrix}^{-1} \begin{bmatrix} 0 \\ 1 \end{bmatrix} U \tag{90}$$

So we can verify that:

$$x_1(t_0) = \frac{U}{a} \; ; \; x_2(t_0) = 0 \tag{91}$$

The Transition Matrix of a Commensurate Order System

Introduction

Obviously, there is a challenge to generalize integer order system theory to fractional systems. Many results of classical theory (as observability and controllability) are based on the existence and the properties of the fractional transition matrix.

Apparent simplicity of usual Caputo initial conditions has encouraged many researchers to propose a transition matrix equation for commensurate order systems.

Unfortunately, due to the wrong definitions of Caputo state variables and initial conditions, this transition matrix equation is correct only for very particular cases; so it has to be corrected.

Formulation of the Transition Matrix Equation

Consider again the N derivatives commensurate order system (85), with $u(t)=0$ for $t>0$.

So

$$D^n(\underline{X}(t))=A\underline{X}(t) \tag{92}$$

where $\underline{X}(t)$ corresponds to the system free response initialized by state vector $\underline{Z}(\omega,0)$.

The Laplace transform of the free response (43) is:

$$\underline{X}(s)=(s^n I-A)^{-1} s^n \underline{f}(s) \tag{93}$$

with

$$\underline{f}^T(s)=[f_1(s)\ldots f_i(s)\ldots f_N(s)] \tag{94}$$

where

$$f_i(s)=\int_0^{\infty}\frac{\mu_n(\omega)\,z_i(\omega,0)}{s+\omega}d\omega \tag{95}$$

is the free response Laplace transform of the ith integrator $I_{n,i}(s)$ initialized by $z_i(\omega,0)$.

Assume that the system is completely at rest at $t=0$ and that $\underline{Z}(\omega,0)$ has been obtained according to the constant excitation

$$u(t)=U \text{ for } t\in\left]-\infty;0\right] \tag{96}$$

Then (89):

$$\underline{X}(0)=-A^{-1}\underline{B}U \tag{97}$$

where we have for each component:

$$x_i(0)=\int_0^{\infty}\mu_n(\omega)\,z_i(0,0)\,d\omega \tag{98}$$

with (Trigeassou 2012b)

$$z_i(\omega,0)=0\ \forall\ \omega\neq 0 \tag{99}$$

Thus:

$$f_i(s)=\int_0^{\infty}\frac{\mu_n(\omega)\,z_i(\omega,0)}{s+\omega}d\omega=\int_0^{\infty}\frac{\mu_n(\omega)\,z_i(0,0)}{s}d\omega=\frac{x_i(0)}{s} \tag{100}$$

So, with the help of this very restrictive case, we can write:

$$\underline{X}(s) = (s^n I - A)^{-1} s^n \frac{\underline{X}(0)}{s} = (s^n I - A)^{-1} s^{n-1} \underline{X}(0) \tag{101}$$

Then, let us define

$$L^{-1}\left\{(s^n I - A)^{-1} s^{n-1}\right\} = \Phi_{n,1}(t) \tag{102}$$

So, we can write

$$\underline{X}(t) = \Phi_{n,1}(t)\,\underline{X}(0) \tag{103}$$

This equation is the generalization to the N dimensional case of the Mittag-Leffler equation (Podlubny 1999):

$$x(t) = E_{n,1}(-at^n)\,x(0) \tag{104}$$

Consequently, we can define the transition matrix $\Phi_{n,1}(t)$ with the technique proposed in (Monje 2010).

Assume that $\Phi_{n,1}(t)$ can be expressed as an infinite series expansion:

$$\Phi_{n,1}(t) = A_0 + A_1 t^n + \ldots + A_k t^{kn} + \ldots. \tag{105}$$

where the matrices $A_0, A_1, \ldots, A_k$ are to be determined. Then equation (103) is expressed as:

$$\underline{X}(t) = (A_0 + A_1 t^n + \ldots + A_k t^{kn} + \ldots)\,\underline{X}(0) \tag{106}$$

$\Phi_{n,1}(t)$ has to verify equation (92):

So:

$$D^n(\underline{X}(t)) = A\,\underline{X}(t) = A(A_0 + A_1 t^n + \ldots + A_k t^{kn} + \ldots)\,\underline{X}(0) \tag{107}$$

and

$$D^n(\underline{X}(t)) = (A_0\, D^n(1) + A_1\, D^n(t^n) + \ldots + A_k\, D^n(t^{kn}) + \ldots)\,\underline{X}(0) \tag{108}$$

Let us remind (Oldham 1974) that:

$$D^n(t^\gamma) = \frac{\Gamma(\gamma+1)}{\Gamma(\gamma+1-n)} t^{\gamma-n} \tag{109}$$

Then

$$D^n(\underline{X}(t)) = (0 + A_1\,\Gamma(1) t^0 + \ldots + A_k \frac{\Gamma(2n+1)}{\Gamma(n+1)} t^{n(k-1)} + \ldots)\,\underline{X}(0) \tag{110}$$

Expressions (107) and (110) are identical if:

$$A_0 = I \quad A_1 = \frac{A}{\Gamma(1+n)} \quad \dots A_k = \frac{A^k}{\Gamma(1+kn)} \tag{111}$$

So

$$\Phi_{n,1}(t) = \sum_{k=0}^{\infty} \frac{A^k t^{kn}}{\Gamma(1+kn)} \tag{112}$$

which is the generalization of the exponential matrix:

$$e^{At} = \sum_{k=0}^{\infty} \frac{A^k t^k}{k!} \tag{113}$$

corresponding to $n=1$ and $\Gamma(k+1)=k!$

So

$$e^{At} = \Phi_{1,1}(t) \tag{114}$$

General Formulation of the Free Response

Obviously, equation (103) is too restrictive. Hopefully, it can be generalized to any free response initialized by $\underline{Z}(\omega,0)$.

We can also write (93) as:

$$\underline{X}(s) = (s^n I - A)^{-1} s^{n-1} s \, \underline{f}(s) \tag{115}$$

Let us define

$$\underline{g}(s) = s \, \underline{f}(s) \text{ so } \underline{g}(t) = \frac{d}{dt}\left(\underline{f}(t)\right) \tag{116}$$

Using the previous definition of $\Phi_{n,1}(t)$ (112), we can write:

$$\underline{X}(t) = L^{-1}\left\{(s^n I - A)^{-1} s^{n-1}\right\} * L^{-1}\left\{s \, \underline{f}(s)\right\} \tag{117}$$

So:

$$\underline{X}(t) = \Phi_{n,1}(t) * \underline{g}(t) \tag{118}$$

Remark 1. Let us assume steady state (or rest) for each component of $\underline{X}(t)$, so (Trigeassou 2012b):

$$z_i(\omega,0)=0 \ \forall \ \omega \neq 0 \tag{119}$$

and

$$x_i(0)=\int_0^{\infty} \mu_n(\omega)\, z_i(0,0)\, d\omega \tag{120}$$

Then

$$s\, f_i(s)=x_i(0) \tag{121}$$

So

$$L^{-1}\left\{s\, \underline{f}(s)\right\}=\underline{X}(0)\ \delta(t) \tag{122}$$

Then:

$$\underline{X}(t)=\Phi_{n,1}(t)*\underline{g}(t)=\Phi_{n,1}(t)*\underline{X}(0)\ \delta(t)=\ \Phi_{n,1}(t)\,\underline{X}(0) \tag{123}$$

which corresponds to equation (103).

Remark 2. In the one derivative case ($N=1$), we obtain:

$$x(t)=E_{n,1}(-at^n)*\frac{d}{dt}\left(\int_0^{\infty} \mu_n(\omega)\, z(\omega,0)\, e^{-\omega t}\, d\omega\right) \tag{124}$$

If the system is completely at rest at $t=0$, then:

$$x(0)=\int_0^{\infty} \mu_n(\omega)\, z(0,0)\, d\omega \tag{125}$$

and:

$$x(t)=E_{n,1}(-at^n)*x(0)\,\delta(t)=E_{n,1}(-at^n)\,x(0) \tag{126}$$

This means that equation (104) is correct only if the system is completely at rest at $t=0$ or at $t=t_0$.

So, we can conclude that in the general case, the correct free response is given by equation (118) which expresses that the free response is the result of a convolution between the transition matrix and the time derivative of each fractional order integrator free response. The main interest of equation (118) is to provide an analytic expression for the free response of fractional differential systems, using the infinite state approach. However, it is not an efficient way to compute numerically this free response; the relevant approach is to use direct simulation ((35) to (38)) with the fractional integrator discretized model ((18) to (21)).

Conclusion

For a long time, Caputo derivative has been regarded as the reference in fractional calculus, because of its claimed physical initial conditions. It is now well established that the correct use of this derivative in the general case requires a supplementary infinite dimension state variable. Infinite state approach has been used to analyze Caputo derivative and to provide solutions for its correct use in various domains such as definition of state variables and fractional energy characterization, true physical initial conditions for FDE and FDS simulation, and the derivation of a fractional transition matrix. In fact, Caputo derivative is not an appropriate tool for the analysis and the simulation of fractional systems because it does not take into account explicitly the infinite dimension of these systems.

A solution to this problem is provided by the implicit derivative and the associated infinite state variable which generalize integer order system theory to fractional systems.

References

Caputo, M., (1969). *Elasticita e Dissipazione*, Zanichelli, Bologna.

Diethelm, K., (2010). *The analysis of fractional differential equations*, Springer Verlag.

Hartley, T. T., C. F. Lorenzo, (2009). The error incurred in using the Caputo derivative Laplace transform. *Proceedings of the ASME IDET-CIE Conference,* San Diego, California, US.

Hartley, T. T., C. F. Lorenzo, J. C. Trigeassou, N. Maamri, (2013). Equivalence of history function based and infinite dimensional state initializations for fractional order operators, *ASME Journal of Computational and Nonlinear Dynamics*, Vol. 8 n°4.

Heleschewitz, D., (2000). Analyse et simulation de systèmes differentiels fractionnaires et pseudo differentiels sous représentation diffusive, *PhD thesis*, ENST Paris.

Kailath, T., (1980). *Linear systems*, Prentice-Hall, Englewood Cliffs, N.J.

Li, Y., Y. Q. Chen, I. Podlubny, (2009). Mittag Leffler stability of fractional order non linear dynamic systems, *Automatica* Vol. 45 pp. 1965-69.

Lorenzo, C. and T. Hartley, (2011). Time-varying initialization and Laplace transform of the Caputo derivative: with order between zero and one, *Proceedings of IDETC/CIE FDTA'2011 Conference*, Washington DC, US.

Monje, C. A., Y. Q. Chen, B. M. Vinagre, D. Xue, V. Feliu, (2010). *Fractional order systems and controls,* Springer-Verlag, London.

Montseny, G., (1998). Diffusive representation of pseudo differential time operators, *Proceedings ESSAIM*, Vol. 5, pp. 159-175.

Oldham, K. B., J. Spanier, (1974). *The fractional calculus*. Academic Press, New York.

Ortega, R., A. Loria., P. J. Nicklasson., H. Sira-Ramirez, (1998). *Passivity based control of Euler Lagrange systems*, Springer-Verlag, Berlin.

Ortigueira, M. D., (2011). *Fractional calculus for scientists and engineers*, Springer Science.

Oustaloup, A., (1983). *Systèmes asservis linéaires d'ordre fractionnaire: théorie et pratique*. Masson, Paris.

Oustaloup, A., (1995). *La dérivation non entière: théorie, synthèse et applications*. Hermès, Paris.

Petras, I., (2011). *Fractional order nonlinear systems: modeling, analysis and simulation*, Springer-Verlag.

Podlubny, I., (1999). *Fractional differential equations*. Academic Press, San Diego.

Sabatier, J., M. Merveillaut, R. Malti, A. Oustaloup, (2010). How to impose physically coherent initial conditions to a fractional system?, *Communications in Non Linear Science and Numerical Simulation*, Vol. 15, issue 5.

Sabatier, J., C. Farges, J. C. Trigeassou, (2013). Fractional systems state space description: some wrong ideas and proposed solutions. *Journal of Vibration and Control*, July 2, DOI: 10.1177/1077546313481839.

Trigeassou, J. C., T. Poinot, J. Lin, A. Oustaloup, F. Levron, (1999). Modeling and identification of a non integer order system. *ECC'99 European Control Conference*, Karlsruhe, Germany.

Trigeassou, J. C., N. Maamri, (2010). The initial conditions of Riemman-Liouville and Caputo derivatives: an integrator interpretation. *FDA'2010 Conference*, Badajoz, Spain.

Trigeassou, J. C., N. Maamri, J. Sabatier, A. Oustaloup, (2011a). A Lyapunov approach to the stability of fractional differential equations, *Signal Processing*, Vol. 91 Issue 3 pp. 437-445.

Trigeassou, J. C., A. Oustaloup, (2011b). Fractional integration: a comparative analysis of fractional integrators, *IEEE SSD'11*, Sousse, Tunisia.

Trigeassou, J. C., N. Maamri, J. Sabatier, A. Oustaloup, (2012a). Transients of fractional order integrator and derivatives, Special Issue: "Fractional Systems and Signals", In: *Signal, Image and Video Processing*, Volume 6, Issue 3, pp. 359-372.

Trigeassou, J. C., N. Maamri, A. Oustaloup, (2012b). State variables and transients of fractional order differential systems, *Computers and Mathematics with Applications*, Volume 64, pp. 3177-3140.

Trigeassou, J. C., N. Maamri, A. Oustaloup, (2013a). The infinite state approach: origin and necessity. *Computers and Mathematics with Application*, Volume 66 pp. 892-907.

Trigeassou, J. C., N. Maamri, A. Oustaloup, (2013b). The Caputo derivative and the infinite state approach, *IFAC Joint Conference FDA'13*, Grenoble France.

Trigeassou, J. C., N. Maamri, A. Oustaloup, (2013c). Lyapunov stability of linear fractional systems. Part 1: definition of fractional energy, *ASME IDETC-CIE Conference*, Portland Oregon, US.

Trigeassou, J. C., N. Maamri, A. Oustaloup, (2013d). Lyapunov stability of linear fractional systems. Part 2: derivation of a stability condition, *ASME IDETC-CIE Conference*, Portland Oregon, US.

In: Fractional Calculus: Theory
Editors: R. Abi Zeid Daou and X. Moreau

ISBN: 978-1-63463-002-3

Chapter 2

STABILITY OF A CLASS OF FRACTIONAL CAUCHY PROBLEM

Rabha W. Ibrahim*
Institute of Mathematical Sciences,
University Malaya, Kuala Lumpur, Malaysia

Abstract

In this work, we utilize the method of non-expansive operators to more general iterative and non-iterative fractional differential equations (Cauchy type) in sense of the Krasnoselskij iterations. The non-integer case is taken in sense of Riemann-Liouville fractional operators. Applications are illustrated. Ulam-Hyers stability for iterative fractional differential equation is defined and studied.

Keywords: Fractional differential equation, fractional calculus, Ulam-Hyers stability

Introduction

Fractional calculus (integrals and derivatives) of any order can be considered as a branch of mathematical physics, associated with differential equations, integral equations and integro-differential equations, where integrals are of convolution form with weak singular kernels of power law type. It has gained more and more interest in applications in several fields of applied sciences. Fractional differential equations (real and complex) are viewed as models for nonlinear differential equations; varieties of them play important roles, not only in mathematics, but also in physics, dynamical systems, control systems and engineering, to create the mathematical modeling of many physical phenomena.

Furthermore, they are employed in social science, such as, food upplement, climate and economics.

* E-mail address: rabhaibrahim@yahoo.com; Address: Institute of Mathematical Sciences, University Malaya, 50603, Kuala Lumpur, Malaysia.

Fractional models have been studied by many researchers, to sufficiently describe the operation of variety of computational, physical and biological processes and systems. Accordingly, considerable attention has been paid to the outcomes of fractional differential equations, integral equations and fractional partial differential equations of physical phenomena. Most of these fractional differential equations have analytic solutions, approximation and numerical techniques [1-3].

Existence and uniqueness are studied widely in the field of fractional differential equations. The main idea of this work is to establish the existence and uniqueness of nonlinear Cauchy problems of fractional order in the sense of Riemann-Liouville operators. Indeed, some results in the analytic continuation of solutions are imposed. In the fractional Cauchy problems, we replace the first order time derivative by a fractional derivative. This type of problems is helpful in physics [4].

Recently, the author intended the fractional Cauchy problems [5-7].

The Riemann-Liouville fractional differential operators may scarcely show the physical interpretation of the initial conditions required for the initial value problems including fractional differential equations.

On the other hand, these operator have advantages of convergence, higher stability and higher accuracy to derive different types of numerical algorithms (see [8, 9]).

The idea of the fractional calculus (that is, calculus of integrals and derivatives of any arbitrary real or complex order) was planted over 300 years ago. In 1823, Abel had investigated the generalized tautochrone problem, and he was the pioneer to apply fractional calculus techniques in a physical problem. Later, Liouville had applied fractional calculus to solve problems in potential theory. Since then, the fractional calculus has triggered the attention of many researchers in all area of sciences.

Definition 1.1. The fractional (arbitrary) order integral of the function g of order $\alpha>0$ is defined by

$$I_a^\nu g(t) = \int_a^t \frac{(t-\tau)^{\nu-1}}{\Gamma(\nu)} g(\tau)d\tau.$$

when $a = 0$, we write $I_a^\nu g(t) = g(t) * \chi_\nu(t)$, where $(*)$ denoted the convolution product (see [7]), $\chi_\nu(t) = \frac{t^{\nu-1}}{\Gamma(\nu)}, t > 0$ and $\chi_\nu(t) = 0, t \le 0$ and $\chi_\nu \to \Delta(t)$ as $\nu \to 0$ where $\Delta(t)$ is the delta function.

Definition 1.2. The fractional (arbitrary) order derivative of the function g of order $0 \le \nu < 1$ is defined by

$$D_a^\nu g(t) = \frac{d}{dt}\int_a^t \frac{(t-\tau)^{-\nu}}{\Gamma(1-\nu)} g(\tau)d\tau = \frac{d}{dt} I_a^{1-\nu} g(t).$$

Remark 1.1. From Definition 1.1. and Definition 1.2., we have

$$D^{\nu}t^{\mu} = \frac{\Gamma(\mu+1)}{\Gamma(\mu-\nu+1)}t^{\mu-\nu}, \mu > -1;\ 0 < \nu < 1$$

and

$$I^{\nu}t^{\mu} = \frac{\Gamma(\mu+1)}{\Gamma(\mu+\nu+1)}t^{\mu+\nu}, \mu > -1;\ \nu > 0.$$

Definition 1.3. The Caputo fractional derivative of order $\mu > 0$ is defined, for a smooth function $f(t)$ by

$$CD^{\mu}f(t) := \frac{1}{\Gamma(n-\mu)}\int_0^t \frac{f^{(n)}(\zeta)}{(t-\zeta)^{\mu-n+1}}d\zeta,$$

where $n = [\mu] + 1,$ (the notation $[\mu]$ stands for the largest integer, not greater than μ).

Note that there is a relationship between Riemann-Liouville differential operator and the Caputo operator

$$D_a^{\mu}f(t) = \frac{1}{\Gamma(1-\mu)}\frac{f(a)}{(t-a)^{\mu}} + cD_a^{\mu}f(t);$$

and they are equivalent in a physical problem (i.e., a problem which specifies the initial conditions i.e., if $f(0) = 0$, then the Riemann-Liouville derivative and the Caputo derivative of order $\alpha \in (0,1)$ coincide).

Methods

In this section, we impose the basic theory of non-expansive mappings in order to offer the notions and results that will be needed in the next sections of the paper. Let (Y, d) be a metric space. A mapping $Q: Y \to Y$ is said to be an ν-contraction if there exists $\nu \in [0,1)$ such that

$$d(Qx, Qy) \le \nu d(x, y), \forall\, x, y \in Y.$$

In the case where $\nu = 1$, the mapping P is said to be non-expansive. Let S be a non-empty subset of a real normed linear space E and $Q: S \to S$ be a map. In this putting, Q is non-expansive if

$$\| Qx - Qy \| \le \| x - y \| \quad \forall x, y \in S.$$

The following result is a fixed point theorem for non-expansive mappings, due to Browder, Ghode and Kirk, see e.g., [10]:

Theorem 2.1. Let S be a non-empty closed convex and bounded subset of a uniformly Banach space B. Then any non-expansive mapping $Q: S \to S$ has at least a fixed point.

Definition 2.1. Let S be a convex subset of a normed linear space N and let $Q: S \to S$ be a self-mapping. Given $u_0 \in S$ and a real number $\lambda \in [0,1]$, the sequence u_m defined by the formula

$$u_{m+1} = (1-\lambda)u_m + \lambda Q u_m, m = 0,1,2,\ldots$$

is usually called Krasnoselskij iteration or Krasnoselskij-Mann iteration.

Definition 2.2. Let S be a convex subset of a normed linear space N and let $Q: S \to S$ be a self-mapping. Given $u_0 \in S$ and a real number $\lambda_m \in [0,1]$, the sequence u_m defined by the formula

$$u_{m+1} = (1-\lambda_n)u_m + \lambda_n Q x_n, m = 0,1,2,\ldots$$

is usually called Mann iteration.

Edelstein [11] studies that strict convexity of B suffices for the Krasnoselskij iteration converge to a fixed point of Q. While, Egri and Rus [12] showed that for any subset of B, the Mann iteration converge to a fixed point of Q when Q is a non-expansive mapping.

Lemma 2.1. [13] Let S be a convex and compact subset of a Banach space B and let $Q: S \to S$ be a non-expansive mapping.

If the Mann iteration process u_m satisfies the assumptions

a $u_m \in S$ for all positive integers n,
b $0 \le \lambda_m \le b < 1$ for all positive integers n,
c $\sum_{m=0}^{\infty} \lambda_m = \infty$.

Then u_m converges strongly to a fixed point of Q.

Lemma 2.2. Let S be a closed bounded convex subset of a real normed space B and $Q: S \to S$ be a non-expansive mapping. If $I - Q$ maps closed bounded subset of B into closed subset of B and u_m is the Mann iteration, with λ_m satisfying assumptions (a)-(c) in Lemma 2.1, then u_m converges strongly to a fixed point of Q in S.

Existence of Solutions

In this section, we consider the existence and uniqueness results for the fractional differential equation

$$D^{\alpha}(u(t) - \mu\gamma_{\alpha}(t)) = f(t, u(g(u(t)))), \tag{1}$$

where $\gamma(t)$ is a positive real valued function satisfying $\gamma_{\alpha}(0) = 0$, and $\mu\gamma_{\alpha}(t) := u_0 \in J := [0,T]$ and $f \in C(J \times J, J)$. For example, one can utilize the function $\gamma_{\alpha}(t) = t^{\alpha}$. For $t \in J$ denote

$$M_t = \max\{t, T - t\}$$

and

$$C_{L,\alpha} = \{u: |u(t_1) - u(t_2)| \leq \frac{L}{\Gamma(\alpha+1)}|t_1 - t_2|^{\alpha}, \forall t_1, t_2 \in J\}, L > 0.$$

It is clear that $C_{L,\alpha}$ is a non-empty convex and a compact subset of the Banach space $(C[J], \|.\|)$, where $\| x \| = \sup_{t\in J}|x(t)|$.

Theorem 3.1. Let the following assumptions hold for (1):

(A0) $f \in C[J \times J, J]$;
(A1) $g \in C[J, J]$ S.T. $\exists \tilde{\ell} > 0: |g(u) - g(v)| \leq \tilde{\ell}|u - v|, \forall\, t, u, v \in J$;
(A2) $\exists \ell > 0: |f(t,u) - f(t,v)| \leq \ell|u - v|, \forall\, t, u, v \in J$;
(A3) If L is the Lipschitz constant such that $|u(t_1) - u(t_2)| \leq \frac{L}{\Gamma(\alpha+1)}|t_1 - t_2|^{\alpha}$, then

$$M = \max\{|f(t,u)|: (t,u) \in J \times J\} \leq \frac{L}{2};$$

(A4) One of the following conditions holds:

a $M\frac{T^{\alpha}}{\Gamma(\alpha+1)} \leq M_{u_0}$, where $M_{u_0} = \max\{u_0, T - u_0\}$;
b $u_0 = 0, M\frac{T^{\alpha}}{\Gamma(\alpha+1)} \leq T - u_0, f(t,u) \geq 0, \forall t, u \in J$;
c $u_0 = T, M\frac{T^{\alpha}}{\Gamma(\alpha+1)} \leq u_0, f(t,u) \geq 0, \forall t, u \in J$.

If

$$\frac{(\tilde{L}+1)T^{\alpha}\ell}{\Gamma(\alpha+1)} \leq 1, \quad (2)$$

then a unique solution exists of problem (1) in $C_{L,\alpha}$, which can be approximated by the Krasnoselskij iteration

$$u_{n+1} = (1-\lambda)u_n + \lambda u_0 + \lambda \int_0^t \frac{(t-\tau)^{\alpha-1}}{\Gamma(\alpha)} f(\tau, u(g(u(\tau))))d\tau,$$

where $\lambda \in (0,1)$ and $u_1 \in C_{L,\alpha}$ is arbitrary.

Proof. Consider the integral operator $P: C_{L,\alpha} \to C(J)$

$$Pu(t) = u_0 + \int_0^t \frac{(t-\tau)^{\alpha-1}}{\Gamma(\alpha)} f(\tau, u(g(u(\lambda\tau))))d\tau, t \in J, u \in C_{L,\alpha}.$$

Our aim is show that P has a fixed point in $C_{L,\alpha}$. We proceed to apply Schauder fixed point theorem or Banach fixed point theorem.

First, we show that $C_{L,\alpha}$ is invariant set with respect to $P, i.e. T(C_{L,\alpha}) \subset C_{L,\alpha}$. In virtue of condition (A4a) and for all $t \in J, u \in C_{L,\alpha}$, we have

$$\begin{aligned}|Pu(t)| &\leq |u_0| + |\int_0^t \frac{(t-\tau)^{\alpha-1}}{\Gamma(\alpha)} f(\tau, u(g(u(\lambda\tau))))d\tau| \\ &\leq |u_0| + M\frac{T^\alpha}{\Gamma(\alpha+1)} \leq T\end{aligned}$$

and

$$\begin{aligned}|Pu(t)| &\geq |u_0| - |\int_0^t \frac{(t-\tau)^{\alpha-1}}{\Gamma(\alpha)} f(\tau, u(g(u(\lambda\tau))))d\tau| \\ &\geq |u_0| - M\frac{T^\alpha}{\Gamma(\alpha+1)} \geq u_0 - M_{u_0} \geq 0.\end{aligned}$$

Thus $(Pu)(t) \in J, t \in J$. In a similar manner of (A4a), we treat the cases (A4b) and (A4c). Now, for every $t_1, t_2 \in J$, by (A3), we have

$$\begin{aligned}|(Pu)(t_1) - (Pu)(t_2)| &= |\int_0^{t_1} \frac{(t-\tau)^{\alpha-1}}{\Gamma(\alpha)} f(\tau, u(g(u(\lambda\tau))))d\tau \\ &- \int_0^{t_2} \frac{(t-\tau)^{\alpha-1}}{\Gamma(\alpha)} f(\tau, u(g(u(\lambda\tau))))d\tau| \\ &\leq M\frac{|t_1^\alpha - t_2^\alpha + 2(t_1-t_2)^\alpha|}{\Gamma(\alpha+1)} \\ &\leq 2M\frac{|t_1-t_2|^\alpha}{\Gamma(\alpha+1)} \leq L\frac{|t_1-t_2|^\alpha}{\Gamma(\alpha+1)}.\end{aligned}$$

Therefore, $(Pu) \in C_{L,\alpha}$ with $u \in C_{L,\alpha}$. Hence $P: C_{L,\alpha} \to C_{L,\alpha}$ (i. e., P is a self-mapping of $C_{L,\alpha}$). Let $u, v \in C_{L,\alpha}$ and $t \in J$, by applying (A1) and (A2), we receive

$$\begin{aligned}|(Pu)(t) - (Pv)(t)| &= |\int_0^{t} \frac{(t-\tau)^{\alpha-1}}{\Gamma(\alpha)} f(\tau, u(g(u(\lambda\tau))))d\tau \\ &- \int_0^{t} \frac{(t-\tau)^{\alpha-1}}{\Gamma(\alpha)} f(\tau, v(g(v(\lambda\tau))))d\tau| \\ &\leq \int_0^{t} \frac{(t-\tau)^{\alpha-1}}{\Gamma(\alpha)} |f(\tau, u(g(u(\lambda\tau)))) - f(\tau, v(g(v(\lambda\tau))))|d\tau \\ &\leq \ell\int_0^{t} \frac{(t-\tau)^{\alpha-1}}{\Gamma(\alpha)} |u(g(u(\lambda\tau))) - v(g(v(\lambda\tau)))|d\tau \\ &\leq \ell\int_0^{t} \frac{(t-\tau)^{\alpha-1}}{\Gamma(\alpha)} [\tilde{L}|g(u(\lambda\tau))) - g(v(\lambda\tau)))| \\ &+|u(g(v(\tau))) - v(g(v(\tau)))|]d\tau \\ &\leq \frac{T^\alpha \ell}{\Gamma(\alpha+1)}[\tilde{L}\tilde{\ell} + 1] \parallel u - v \parallel,\end{aligned}$$

where $1 \leq |g(u) - g(v)| \leq T$ and

$$\tilde{L} = \max \frac{L}{\Gamma(\alpha+1)}.$$

Now, by taking the supremum in the last assertion yields

$$\| Pu) - (Pv) \| \leq \frac{(\tilde{L}\tilde{\ell}+1)T^{\alpha}\ell}{\Gamma(\alpha+1)} \| u - v \|.$$

If $\frac{(\tilde{L}+1)T^{\alpha}\ell}{\Gamma(\alpha+1)} < 1$, then P is a contraction mapping. Hence, in view of the Banach fixed point theorem, Equation (1) has a unique solution. Now if

$$\frac{(\tilde{L}\tilde{\ell}+1)T^{\alpha}\ell}{\Gamma(\alpha+1)} = 1$$

then P is non-expansive and continuous; thus the Schauder fixed point theorem implies that Equation (1) has a solution in $C_{L,\alpha}$. Finally, in view of Lemmas 2.1. and 2.2., we obtain the second part of the theorem.

Next we establish the solution of Equation (1) in a subset of $C_{L,\alpha}$ defined by

$$C_{L,\alpha,\delta} = \{u \in C_{L,\alpha}: u(t) \leq \frac{\delta t^{\alpha}}{\Gamma(\alpha+1)}, \forall t \in J\}, \delta \in (0,1).$$

It is clear that $C_{L,\alpha,\delta}$ is non-empty, convex and compact in $C[J]$.

Theorem 3.2. Assume that the following conditions are satisfied:

(A5) $u_0 \leq \frac{\delta t_0^{\alpha}}{2\Gamma(\alpha+1)} t_0 (\neq 0) \in J$;

(A6) If L is the Lipschitz constant such that $|u(t_1) - u(t_2)| \leq \frac{L}{\Gamma(\alpha+1)} |t_1 - t_2|^{\alpha}$, then $M \leq \min\{\frac{\delta}{2}, \frac{L}{2}\}$;

(A7) There exists a $\bar{\tau} > 0$ such that $\bar{\tau} > -\frac{\ln(1-\delta)}{\delta(T-t_0)}, T \neq t_0$ and

$$\frac{T^{\alpha-1}\ell}{\Gamma(\alpha)\bar{\tau}}(\frac{1}{\delta} + \tilde{L}\tilde{\ell}) \max\{e^{\bar{\tau}t_0} - 1, 1 - e^{\bar{\tau}(t_0-T)}\} \leq 1 \tag{3}$$

If (A2), (A4) hold, then there exists at least one solution of problem (1) in $C_{L,\alpha,\delta}$ which can be approximated by the Krasnoselskij iteration

$$u_{n+1} = (1-\lambda)u_n + \lambda u_0 + \lambda \int_0^t \frac{(t-\tau)^{\alpha-1}}{\Gamma(\alpha)} f(\tau, u(u(\tau)))d\tau,$$

where $\lambda \in (0,1)$ and $u_1 \in C_{L,\alpha,\delta}$ is arbitrary.

Proof. We let the Banach space $C[J]$ • endowed with Bielecki's norm given by the formula

$$\| u \|_B = \max_{t\in J} |u(t)| e^{-s(t-t_0)}, s > 0, t > t_0$$

$$(t, s, t_0 \in J = [0,T], T < \infty).$$

Let P be defined as in the proof of Theorem 3.1. By assumptions (A2), (A4), and (A6), it follows that

$$P(C_{L,\alpha,\delta}) \subset C_{L,\alpha,\delta}.$$

Now we prove that $C_{L,\alpha,\delta}$ is an invariant set with respect to the operator P. Indeed, if $u \in C_{L,\alpha,\delta}$ and $t \in J$, then in view of (A5) and (A6), we have

$$\begin{aligned} Pu(t) &\le u_0 + M\frac{t^\alpha}{\Gamma(\alpha+1)} \\ &= u_0 + M\frac{(t^\alpha - t_0^\alpha) + t_0^\alpha}{\Gamma(\alpha+1)} \\ &\le \frac{\delta t_0^\alpha}{2\Gamma(\alpha+1)} + \frac{\delta t^\alpha}{2\Gamma(\alpha+1)} - \frac{\delta t_0^\alpha}{2\Gamma(\alpha+1)} + \frac{\delta t_0^\alpha}{2\Gamma(\alpha+1)} \\ &\le \frac{\delta t^\alpha}{\Gamma(\alpha+1)}, t > t_0, \end{aligned}$$

that is $Pu \in C_{L,\alpha,\delta}$.

Let $u, v \in C_{L,\alpha,\delta}$ and $t \in J$, we have

$$|(Pu)(t) - (Pv)(t)| \le \frac{T^{\alpha-1}\ell}{\Gamma(\alpha)} \Big| \int_0^t (\tilde{L}|u(\tau) - v(\tau)| + |u(v(\lambda\tau)) - v(v(\lambda\tau))|) d\tau \Big|$$
$$\le \frac{T^{\alpha-1}\ell}{\Gamma(\alpha)} \Big(\Big| \int_0^t \tilde{L}\tilde{\ell} e^{s(\tau-t_0)} d\tau \Big| + \Big| \int_0^t e^{s(\delta\tau - t_0)} d\tau \Big| \Big) \| u - v \|_B$$
$$\le \frac{T^{\alpha-1}\ell}{\Gamma(\alpha)} \Big(\Big| \frac{\tilde{L}\tilde{\ell}}{s} (e^{s(t-t_0)} - 1) \Big|$$
$$+ \frac{1}{\delta s} |e^{s(\delta t - t_0)} - e^{s(\delta t_0 - t_0)}| \Big) \| u - v \|_B.$$

This yields

$$\begin{aligned} |(Pu)(t) - (Pv)(t)| e^{-s(\tau - t_0)} &\le \frac{T^{\alpha-1}\ell}{s\Gamma(\alpha)} \Big(\tilde{L}\tilde{\ell} |1 - e^{-s(t-t_0)}| \\ &+ \frac{1}{\delta} |e^{s(\delta-1)t} - e^{s(\delta t_0 - t)}| \Big) \| u - v \|_B \end{aligned}$$

$$:= L(t) \parallel u - v \parallel_B$$

where $L(t)$ is a continuous function. Then there exists a constant $\hat{L} > 0$ such that

$$\max_{t \in J} |L(t)| \leq \hat{L}.$$

Thus we have

$$\parallel Pu - Pv \parallel_B \leq \hat{L} \parallel u - v \parallel_B,$$

which shows that P is Lipschitzian, hence continuous. By Schauder's fixed point theorem, it follows that T has at least one fixed point which is actually a solution of the initial value problem (1).

We proceed to show that P is non-expansive function. The function

$$g(t) = 1 - e^{-s(t-t_0)}, s > 0, t > t_0$$

is strictly increasing on J and $g(t_0) = 0$; furthermore,

$$\max_{t \in J} g(t) = \max\{e^{\tilde{\tau} t_0} - 1, 1 - e^{\tilde{\tau}(t_0 - T)}\}.$$

Similarly for the function

$$h(t) = e^{s(\delta-1)t} - e^{s(\delta t_0 - t)}$$

then

$$h'(t) = s e^{s(\delta-1)t}[(\delta - 1) + e^{s\delta(t-t_0)}].$$

Now the function

$$k(t) = (\delta - 1) + e^{s\delta(t-t_0)}$$

is strictly decreasing on J; hence,

$$k(t) \geq k(T) = (\delta - 1) + e^{s\delta(T-t_0)}. \quad (4)$$

For $\delta \in (0,1)$ and $T \neq t_0$ then by the assumption (A7) there exists a $\overline{\tau} > 0$ such that

$$\overline{\tau} > -\frac{\ln(1-\delta)}{\delta(T - t_0)}, T \neq t_0$$

which implies that $k(T) > 0$ and hence, h is strictly increasing on J. If we put $s = \overline{\tau}$ we have

$$\max_{t\in J}|h(t)| = \max\{|1-e^{s\delta t_0}|, |e^{s(\delta-1)T}-e^{s(\delta t_0-T)}|\}.$$

Since $\delta \in (0,1)$ then, we obtain

$$\begin{aligned} |e^{s(\delta-1)T}-e^{s(\delta t_0-T)}| &= e^{s(\delta-1)T}|1-e^{s\delta(t_0-T)}| \\ &\le 1-e^{s\delta(t_0-T)} \end{aligned}$$

for sufficient s, δ, T and t_0. Moreover, we have

$$|1-e^{s\delta t_0}| \le e^{s\delta t_0}-1.$$

Consequently, we receive

$$L(t) \le \max\{e^{s\delta t_0}-1, 1-e^{s\delta(t_0-T)}\}\frac{T^{\alpha-1}\ell}{s\Gamma(\alpha)}(\frac{1}{\delta}+\tilde{L}\tilde{\ell}).$$

This shows that P is a non-expansive operator.

A similar argument holds when $T = t_0$ in Equation (4), we have $k(T) = \delta > 0$. Hence, h is strictly increasing on J. Finally, one can use Lemmas 2.1. and 2.2. to obtain the second part of the theorem. This completes the proof.

Note that iterative differential equations are studied widely in [14-16].

Example 3.1. Consider the following initial value problem associated to the fractional iterative differential equation

$$\begin{cases} D^{0.5}u(t) = -\frac{1}{3}+\frac{1}{4}u(u(t)), t \in [0,1] \\ u(0) = \frac{1}{3} \end{cases} \tag{5}$$

where $g(u(t)) = u(t) \in C^1([0,1],[0,1])$. We focus on the solutions $u \in C^1([0,1],[0,1])$ belonging to the set

$$\begin{aligned} C_{1,0.5} &= \{u: |u(t_1)-u(t_2)| \le \frac{1}{\Gamma(\frac{3}{2})}|t_1-t_2|^{0.5}, \forall t_1, t_2 \in [0,1]\} \\ &= \{u: |u(t_1)-u(t_2)| \le \frac{1}{0.886}\sqrt{|t_1-t_2|}, \forall t_1, t_2 \in [0,1]\} \\ &= \{u: |u(t_1)-u(t_2)| \le 1.1\sqrt{|t_1-t_2|}, \forall t_1, t_2 \in [0,1]\}. \end{aligned}$$

To satisfy (A4a), we have $M \le \frac{L}{2} \simeq \frac{1}{2}, M_{\frac{1}{3}} = \max\{\frac{1}{3}, \frac{2}{3}\} = \frac{2}{3} = 0.666$ and

$$M\frac{T^{\alpha}}{\Gamma(\alpha+1)}=\frac{1}{2}\times\frac{1}{0.886}=0.56<0.666.$$

Hence (A4a) is satisfied. The function $f(t)=-\frac{1}{3}+\frac{1}{4}u$ is Lipschitzian with the Lipschitz constant $\ell=\frac{1}{4}$. This shows that

$$\frac{(\tilde{L}+1)T^{\alpha}\ell}{\Gamma(\alpha+1)}=\frac{2.1\times 0.25}{0.886}=0.592<1.$$

Therefore, by Theorem 3.1., we obtain information on the existence and approximation of the solutions of the initial value problem (5).

If we consider the function $f(t)=-\frac{1}{3}+\frac{422}{1000}u$ in Example 3.1., then we have

$$\frac{(\tilde{L}+1)T^{\alpha}\ell}{\Gamma(\alpha+1)}=\frac{2.1\times 0.422}{0.886}\simeq 1.$$

Therefore, again by Theorem 3.1. we pose the existence and approximation of the solutions of the initial value problem (5).

Again, we consider the problem (5) on the interval $[\frac{3}{4},1]$ for $\ell=0.025$, where $u\in C^{1}([\frac{3}{4},1],[\frac{3}{4},1])$. We are interested in the solutions $u\in C^{1}([\frac{3}{4},1],[\frac{3}{4},1])$ belonging to the set

$$\begin{aligned}C_{1,\frac{1}{2},\frac{3}{4}} &=\{u\in C_{1,\frac{1}{2}}: u(t)\leq\frac{\delta t^{\alpha}}{\Gamma(\alpha+1)},\forall t\in J\},\delta\in(0,1)\}\\ &=\{u: u(t)\leq\frac{\frac{3}{4}t^{\frac{1}{2}}}{\Gamma(\frac{3}{2})},\forall t\in[\frac{3}{4},1]\}\\ &=\{u: u(t)\leq 0.846\sqrt{t}, t\in[\frac{3}{4},1]\}.\end{aligned}$$

Our aim is to satisfy the assumptions of Theorem 3.2. (A2) and (A4) are valid. Since $u_0=\frac{1}{3}$ and $t_0=\frac{3}{4}$, we have

$$u_0\leq\frac{\delta t_0^{\alpha}}{2\Gamma(\alpha+1)}\Rightarrow\frac{1}{3}<\frac{3}{8};$$

hence (A5) is satisfied. Moreover, a computation gives

$$M\leq\min\{\frac{\delta}{2},\frac{L}{2}\}=\min\{\frac{\delta}{2},\frac{L}{2}\}=\{\frac{3}{8},\frac{1}{2}\}=\frac{3}{8}$$

thus (A6) is satisfied. Now, we proceed to satisfy (A7); since

$$-\frac{\ln(1-\delta)}{\delta(T-t_0)} = \frac{-\ln\frac{1}{4}\times 16}{3} = 6.933$$

and

$$\max\{e^{\overline{\tau}t_0}-1, 1-e^{\overline{\tau}(t_0-T)}\} = \max\{189.5, .826\}$$

Then, for $\overline{\tau} = 7$, we impose

$$\begin{aligned}\frac{T^{\alpha-1}\ell}{\Gamma(\alpha)\overline{\tau}}\left(\frac{1}{\delta}+\tilde{L}\right)\max\{e^{\overline{\tau}t_0}-1, 1-e^{\overline{\tau}(t_0-T)}\} &= \frac{0.175}{37.17}\max\{189.5, .826\} \\ &= 0.758 < 1.\end{aligned}$$

Hence, in view of Theorem 3.2., problem (5) has a solution in the set $C_{1,\frac{1}{2},\frac{3}{4}}$.

We can observe that problem (5) doesn't satisfy Theorem 3.2. on the set $C_{1,\frac{1}{2},\frac{1}{2}}$ over the interval $[\frac{1}{2}, 1]$:

$$\begin{aligned}C_{1,\frac{1}{2},\frac{1}{2}} &= \{u \in C_{1,\frac{1}{2}}: u(t) \le \frac{\delta t^{\alpha}}{\Gamma(\alpha+1)}, \forall t \in J\}, \delta \in (0,1)\} \\ &= \{u: u(t) \le \frac{\frac{1}{2}t^{\frac{1}{2}}}{\Gamma(\frac{3}{2})}, \forall t \in [\frac{1}{2}, 1]\} \\ &= \{u: u(t) \le 0.564\sqrt{t}, t \in [\frac{1}{2}, 1]\}.\end{aligned}$$

For $u_0 = \frac{1}{3}, t_0 = \frac{1}{2}, \alpha = \frac{1}{2}, \delta = \frac{1}{2}$, a calculation poses

$$u_0 \le \frac{\delta t_0^{\alpha}}{2\Gamma(\alpha+1)} \Rightarrow \frac{1}{3} > \frac{0.35}{1.772};$$

therefore, condition (A5) can not be achieved.

Finally, problem (5) doesn't satisfy Theorem 3.2. on the set $C_{1,\frac{1}{2},\frac{1}{2}}$ over the interval $[\frac{3}{4}, 1]$:

$$\begin{aligned}C_{1,\frac{1}{2},\frac{1}{2}} &= \{u \in C_{1,\frac{1}{2}}: u(t) \le \frac{\delta t^{\alpha}}{\Gamma(\alpha+1)}, \forall t \in J\}, \delta \in (0,1)\} \\ &= \{u: u(t) \le \frac{\frac{1}{2}t^{\frac{1}{2}}}{\Gamma(\frac{3}{2})}, \forall t \in [\frac{3}{4}, 1]\} \\ &= \{u: u(t) \le 0.5\sqrt{t}, t \in [\frac{3}{4}, 1]\}.\end{aligned}$$

For $u_0 = \frac{1}{3}, t_0 = \frac{3}{4}, \alpha = \frac{1}{2}, \delta = \frac{1}{2}$, a calculation yields

$$u_0 \leq \frac{\delta t_0^\alpha}{2\Gamma(\alpha+1)} \Rightarrow \frac{1}{3} > \frac{1}{4};$$

therefore, condition (A5) can not be achieved.

Ulam-Hyers Stability

The analysis on stability of fractional differential equations is more complicated than the classical differential equations, since fractional derivatives are non-local and have weakly singular kernels. Recently, Li and Zhang [17] provided an overview on the stability results of the fractional differential equations. Particular, Li et al. [18] devoted to study the Mittag–Leffler stability and the Lyapunov's methods; Deng [19] derived sufficient conditions for the local asymptotical stability of nonlinear fractional differential equations and Li et al. studied the stability of fractional-order nonlinear dynamic systems using the Lyapunov direct method and generalized Mittag-Leffler stability [20]. Furthermore, there are few works on the Ulam stability of fractional differential equations, which may provide a new way for the researchers to investigate the stability of fractional differential equations from different perspectives. First, the Ulam stability and data dependence for fractional differential equations with Caputo derivative has been posed by Wang et al. [21] while Ibrahim [22] with Riemann-Liouville derivative. Finally, the author generalized the Ulam-Hyers stability for fractional differential equation including infinite power series [23-26].

In this section, we continue our study by imposing the Ulam-Hyers stability for iterative Cauchy fractional differential equation.

Definition 4.1. Problem (1) has the Ulam-Hyers stability if there exists a positive constant K with the following property: For every $\varepsilon > 0, u \in C_{L,\alpha}$, if

$$|D^\alpha u(t) - f(t, u(g(u(t))))| < \varepsilon$$

Then, there exists some $v \in C_{L,\alpha}$ satisfying

$$D^\alpha v(t) = f(t, v(g(v(t))))$$

with $v(0) = v_0$ such that

$$|u(t) - v(t)| < K\varepsilon.$$

Theorem 4.1. Assume that $\max|f(t,u)| := M\max|u|$ where $t \in J = [0,T], M \in (0, \frac{\Gamma(\alpha+1)}{T^\alpha})$. If

$$\max|D^\alpha u(t)| \geq \frac{|u_0|}{1 - \frac{T^\alpha M}{\Gamma(\alpha+1)}}$$

then problem (1) has the Ulam-Hyers stability in $C_{L,\alpha}$.

Proof. For every $\varepsilon > 0, u \in C_{L,\alpha}$, we let

$$|D^{\alpha}u(t) - f(t, u(g(u(\lambda t))))| < \varepsilon$$

with $u(0) = u_0$. Under the assumptions of Theorem 3.1., Equation (1) has a unique solution in $C_{L,\alpha}$. Since

$$|u(t)| \quad \leq |u_0| + \frac{T^{\alpha}|f|}{\Gamma(\alpha+1)}$$

yields

$$\max|u| \leq \frac{|u_0|}{1 - \frac{T^{\alpha}M}{\Gamma(\alpha+1)}} \leq \max|D^{\alpha}u(t)|.$$

Thus by (A2), we observe that

$$\begin{aligned} \max|u(t) - v(t)| \quad &\leq \max|D^{\alpha}(u(t) - v(t))| \\ &\leq \max|D^{\alpha}u(t) - D^{\alpha}v(t) - f(t,u) + f(t,v)| \\ &+ \max|f(t,u) - f(t,v)| \\ &\leq \varepsilon + \ell \max|u(t) - v(t)|; \end{aligned}$$

hence, we receive

$$\max|u(t) - v(t)| \leq \frac{\varepsilon}{1-\ell} := K\varepsilon.$$

It is clear that $v(0) = u_0 = v_0$ yields

$$|u(t) - v(t)| \leq K\varepsilon.$$

Thus (1) has the Ulam-Hyers stability.

Conclusion

We have employed the the concept of the Krasnoselskij iterations (see Definition 2.1.) to approximate the solution of fractional differential equations of Cauchy types. In the same manner of this work one can apply the Mann iteration (Definition 2.2.).

References

[1] R. Lewandowski, B. Chorazyczewski, Identification of the parameters of the Kelvin–Voigt and the Maxwell fractional models, used to modeling of viscoelastic dampers, *Computers and Structures* 88 (2010) 1-17.

[2] F. Yu, Integrable coupling system of fractional soliton equation hierarchy, *Physics Letters A* 373 (2009) 3730-3733.

[3] H. A. Jalab, R. W. Ibrahim, Denoising algorithm based on generalized fractional integral operator with two parameters, *Discrete Dynamics in Nature and Society*, (2012) 1-14.

[4] R. Hilfer, Fractional diffusion based on Riemann-Liouville fractional derivatives, *J. Phys. Chem. Bio.* 104(2000) 3914-3917.

[5] R. W. Ibrahim, Existence and uniqueness of holomorphic solutions for fractional Cauchy problem, *J. Math. Anal. Appl.* 380 (2011) 232-240.

[6] R. W. Ibrahim, Existence of fractional parametric Cauchy problem, *Applied Sciences*, 14(2012) 74-81.

[7] R. W. Ibrahim, Modified fractional Cauchy problem in a complex domain, *Advances in Difference Equations,* 2013, 2013:149.

[8] I. Podlubny, *Fractional Differential Equations*, Acad. Press, London, 1999.

[9] A. A. Kilbas, H. M. Srivastava and J. J. Trujillo, Theory and applications of fractional differential equations. *North-Holland, Mathematics Studies*, Elsevier 2006.

[10] V. Berinde, *Iterative Approximation of Fixed Points*, 2nd Ed., Springer Verlag, Berlin Heidelberg New York, 2007.

[11] M. Edelstein, A remark on a theorem of M. A. Krasnoselskij, *Amer. Math. Monthly*, 73 (1966) 509-510.

[12] E. Egri, I. Rus, First order iterative functional-diferential equation with parameter, *Stud. Univ. Babes-Bolyai Math.* 52 (2007) 67-80.

[13] C. Chidume, *Geometric Properties of Banach spaces and nonlinear Iterations*, Springer Verlag, Berlin, Heidelberg, New York, 2009.

[14] Yang, D. and Zhang, W., Solution of equivariance for iterative differential equations, *Appl. Math. Lett.* 17(2004) 759-765.

[15] A. Ronto, M. Ronto, Succsesive approximation method for some linear boundary value problems for differential equations with a special type of argument deviation, *Miskolc Math. Notes,* 10(2009) 69-95.

[16] V. Berinde, Existence and approximation of solutions of some first order iterative differential equations, *Miskolc Math. Notes,* Vol. 11 (2010) pp. 1326.

[17] C. P. Li, F. R. Zhang, A survey on the stability of fractional differential equations. *Eur. Phys. J. Special Topics* 193 (2011) 27-47.

[18] Y. Li, Y. Chen, I. Podlubny, Mittag-Leffler stability of fractional order nonlinear dynamic systems. *Automatica* 45(2009) 1965-1969.

[19] W. Deng, Smoothness and stability of the solutions for nonlinear fractional differential equations. *Nonlinear Anal.* 72 (2010) 1768-1777.

[20] Y. Li, Y. Chen, I. Podlubny, Stability of fractional-order nonlinear dynamic systems: Lyapunov direct method and generalized Mittag-Leffler stability. *Comput. Math. Appl.* 59 (2010) 1810-1821.

[21] J. Wang, L. Lv, Y. Zhou, Ulam stability and data dependence for fractional differential equations with Caputo derivative. *Electron. J. Qualit. Th. Diff. Equat*. 63(2011) 1-10.

[22] R. W. Ibrahim, Approximate solutions for fractional differential equation in the unit disk, *Electron. J. Qualit. Th. Diff. Equat.* 64 (2011) 1-11.

[23] R. W. Ibrahim, Ulam stability for fractional differential equation in complex domain, *Abstract and Applied Analysis* (2012) 1-10.

[24] R. W. Ibrahim, Generalized Ulam-Hyers stability for fractional differential equations, *Int. J. Math*., 23(5) (2012) 1-9.

[25] R. W. Ibrahim, Ulam-Hyers stability for Cauchy fractional differential equation in the unit disk, *Abstract and Applied Analysis*, (2012) 1-10.

[26] R. W. Ibrahim, On generalized Hyers-Ulam stability of admissible functions, *Abstract and Applied Analysis* (2012) 1-10.

In: Fractional Calculus: Theory
Editors: R. Abi Zeid Daou and X. Moreau

ISBN: 978-1-63463-002-3

Chapter 3

NUMERICAL SOLUTION OF FRACTIONAL ORDER DIFFERENTIAL EQUATIONS VIA MATRIX-BASED METHODS

Matthew Harker* and Paul O'Leary
University of Leoben, Institute for Automation, Leoben, Austria

Abstract

This chapter presents a new matrix-based method for the numerical solution of fractional order differential equations (FDE). A new discretization of fractional order derivatives is presented, based on integer order numerical differentiation, and fractional order numerical integration. The discretized FDE and its associated initial-, boundary-, or interior-values are formulated as a linear constrained least squares optimization problem; with the proposed approach, the discretization of FDE becomes a straightforward one-to-one mapping of symbols. In this vein, a large class of linear FDE can be solved under the proposed framework. The method is demonstrated on common examples such as the Bagley-Torvik Equation, and the Relaxation-Oscillation Equation.

Keywords: Fractional Differential Equations, Matrix Algebra, Numerical Methods

AMS Subject Classification: 65L05, 65L10, 65F05

1. Introduction

Given the number of fractional order differential equations which cannot be solved analytically, developing algorithms which can accurately solve them numerically is essential to the study of such equations. This chapter presents a matrix based framework for the numerical solution of fractional order differential equations (FDE). Due to the memory principles of fractional calculus, the discretization of FDE naturally falls into a convenient matrix form. Classical matrix methods are based on fractional differences (discretized Grünwald-Letnikov derivatives), or on Taylor matrix methods. These methods both suffer from numerical instability due respectively to the problem formulation, and instability of Vandermonde

*E-mail address: matthew.harker@unileoben.ac.at

matrices. The present work outlines modern methods for the numerical approximation of fractional order integrals and derivatives, providing approximations to both Riemann-Liouville as well as Caputo derivatives. Further it details the discretization of FDE along with initial-, boundary-, and interior-value conditions; these conditions impose linear constraints on the numerical solution. Finally, the solution of the system is obtained as a least squares optimization problem with linear constraints (LSE). The LSE approach provides a stable solution to the FDE, mainly due to the fact that the approximation is consistent, i.e., in the limit, the discretization approaches the continuous equation it approximates. Classical finite difference type approaches of the past suffer in terms of stability, due to the fact that they are sometimes inconsistent. This chapter outlines the matrix methods of the past, and gives a detailed description of the modern matrix methods for the numerical solution of FDE. The techniques are demonstrated and compared by computing numerical solutions to classical FDE such as IVP for the Relaxation-Oscillation Equation, and BVP for the Bagley-Torvik equation.

1.1. Summary

This chapter outlines a new approach to the discretization and numerical solution of fractional order differential equations based on matrix algebra. Matrix methods for the solution of ODE and PDE are by no means new [1, 2, 3]; however, there are limitations to the traditional methods which have left them relatively unused. The greatest deficit of these matrix methods is the treatment (or lack thereof) of the boundaries. The classical formulas assume zero boundary conditions, mainly for the convenience of using an identical derivative formula at all points. The discretized equation is then solved at the interior points of the solution interval, that is to say, that the differential equation is *not* solved at the boundaries. This approach stems from early work on finite difference methods [4, 5], and has persisted in the literature ever since [6]. Podlubny [7], has taken a similar tack at the numerical solution of fractional order differential equations [8]. We present a new matrix method for the solution of FDE, with three key factors leading to vastly better numerical results than in the published literature:

1. New discretization of fractional integration based on the discretization of Volterra Integral Equations in matrix form.

2. Numerical differentiation matrices that are of consistent accuracy at all points, and do not assume any boundary conditions whatsoever.

3. Formulation of the numerical solution of the FDE as a least squares problem with equality constraints (LSE) to enforce initial-, boundary-, and interior-values on the solution.

Firstly, the proposed numerical differentiation formulas enable the solution of the FDE over the entire interval of interest, including the boundary points. Secondly, the new discretization of fractional integrals provides an alternative to fractional difference approaches, which require a relatively small step size to be effective. Thirdly, the LSE formulation allows for a global approximation to the solution of the FDE in the least squares sense, improving the accuracy over methods which simply solve an unconstrained linear system of

equations.
When equipped with the newly proposed differentiation/integration matrices, the discretization of an FDE becomes a simple one-to-one mapping of symbols. Further, the implementation of non-homogenous initial/boundary values is done directly, without the need to modify the original problem to a homogenous one as in [7]. The new formulas also work identically for the case arbitrarily spaced nodes (non-constant node spacing, h), which could be used to advantage in adaptive-type solutions. The literature on the numerical solution of differential equations is necessarily vast, however, there is a smattering of matrix based methods [5, 7, 9, 10, 11], some pertaining to FDE. The most relevant appears to be the method of Podlubny [7]; the method proposed here differs fundamentally from that of Podlubny in the discretization of fractional integration and the numerical solution of the FDE as an LSE problem. The new method is demonstrated numerically with a selection of FDE which have known analytic solutions. It is shown to yield accurate results with a relatively small number of abscissae, which is a weak point of fractional difference methods such as [7]. Further, because the discretization is *consistent* (See [6]) the solutions become more accurate with a larger number of abscissae, which is not the case for Taylor matrix methods such as [11]. The new method can be trivially extended to FDE of variable order, as well as the solution of fractional order partial differential equations [9]. A further future application will be in the numerical solution of inverse problems.

1.2. Preliminaries

1.2.1. Matrix Fractional Differences

Perhaps the most classic and most often applied method for numerically solving FDE is by using the method of "fractional differences". These provide an approximation to the fractional derivative based on the Grünwald-Letnikov definition of the fractional derivative. These are defined for the set of evenly spaced points,

$$x_k = kh \quad \text{for} \quad k = 0, \ldots, n \quad \text{with} \quad h = \frac{b-a}{n} \tag{1}$$

as

$${}_aD^{\alpha}_{x_k} f(x) \approx \frac{\nabla^{\alpha} f(x_k)}{h^{\alpha}} = h^{-\alpha} \sum_{j=1}^{k} (-1)^j \binom{\alpha}{j} f_{i-j} \tag{2}$$

Podlubny noticed that these coefficients can be conveniently arranged in matrix form to yield the structured matrix,

$$\mathsf{F}^{(\alpha)} = \frac{1}{h^{\alpha}} \begin{bmatrix} \omega_0^{(\alpha)} & 0 & 0 & \cdots & 0 \\ \omega_1^{(\alpha)} & \omega_0^{(\alpha)} & 0 & \ddots & \vdots \\ \omega_2^{(\alpha)} & \omega_1^{(\alpha)} & \omega_0^{(\alpha)} & \ddots & 0 \\ \vdots & \ddots & \ddots & \ddots & \vdots \\ \omega_{n-1}^{(\alpha)} & \cdots & \omega_2^{(\alpha)} & \omega_1^{(\alpha)} & \omega_0^{(\alpha)} \end{bmatrix}, \tag{3}$$

where the coefficients are given as,

$$\omega_j^{(\alpha)} = (-1)^{(j)} \binom{\alpha}{j} \qquad \text{for} \qquad j = 0, \ldots, n, \tag{4}$$

and are generated by means of the binomial recurrence. Podlubny refers to the structure of these matrices as triangular strip matrices[1], and elucidates their many useful properties. Such important properties include the fact that they are commutative (analogous to differential operators) and that their inverses are numerical fractional integrals (also analogous to their analytic counterparts).
Given the aforementioned matrices for numerical fractional differentiation it is straightforward to use these to directly discretize an FDE (as will be seen in detail in the following) as,

$$\mathsf{L}\boldsymbol{y} = \boldsymbol{g}. \tag{5}$$

The approach taken by Podlubny is to approach only FDE with zero initial conditions, generally,

$$f(0) = 0, \quad f'(0) = 0 \quad \ldots \quad f^{(p-1)}(0) = 0 \tag{6}$$

Discretizing these initial conditions according to the fractional difference formulæ, and solving for the first p points yields,

$$f_0 = f_1 = \ldots = f_{p-1} = 0. \tag{7}$$

The remainder of the points, i.e., $f_p, \ldots, f_n$ are then solved for using the linear differential operator, reduced in size for the particular problem, i.e.,

$$\mathsf{L}_{n-p}\boldsymbol{f}_{n-p} = \boldsymbol{g}_{n-p}. \tag{8}$$

Upon solving this equation for the remaining unknowns, one obtains a plausible solution for the FDE; however, there remain some unresolved issues:

1. The numerical accuracy of fractional differences is not well established, as it appears there is an implicit constraint that the (pseudo-)points $f_{-n}, \ldots, f_{-1}$ are all constrained to be zero.

2. There is an inconsistency in the discretization introduced by treating the initial conditions and the linear operator separately.

3. There is no means of treating problems other that initial value problems (i.e., boundary value problems are not applicable).

1.2.2. Taylor Collocation

The collocation method is a classic algorithm for solving ODE [12]. This typically involves approximating the solution by a piecewise polynomial function, and ensuring that the function satisfies the differential equation at a finite number of nodes.
In terms of the numerical solution of FDE a solution has been given in terms of simple polynomial approximation [11] (as opposed to piecewise polynomial). The derivation of

[1]Such matrices are now typically called Toeplitz matrices; in this case they are lower triangular Toeplitz matrices.

the method is quite involved, however, one of the mechanisms on which it depends is a "fractional" variation of the the Vandermonde matrix.

$$\mathsf{V}_\alpha(\boldsymbol{x}) = \begin{bmatrix} 1 & x_1^\alpha & x_1^{2\alpha} & \cdots & x_1^{(n-1)\alpha} \\ 1 & x_2^\alpha & x_2^{2\alpha} & \cdots & x_2^{(n-1)\alpha} \\ \vdots & \vdots & \vdots & \ddots & \vdots \\ 1 & x_n^\alpha & x_n^{2\alpha} & \cdots & x_n^{(n-1)\alpha} \end{bmatrix} \tag{9}$$

This modification is a variation of the standard Vandermonde matrix, which is known to become increasingly poorly numerically conditioned with the increasing number of abscissae. While the introduction of the factor α may either increase or decrease the condition number of the matrix, it in no way improves the situation. The only means of improving this situation (which by no means solves the problem) is to use higher precision in the arithmetic. While this is possible with some symbolic computation packages such as Maple, it is more difficult with other computational packages, or, for example on a PLC. While the method in itself seems promising, it has the following deficiencies:

1. Only low degree polynomial approximations can be made over a small number of abscissae, usually also over a very short interval, say, $x \in [0, 1]$. Very seldom is the solution to an FDE given by a polynomial.

2. Accuracy of the method is low due to the use of the Vandermonde matrix.

1.2.3. Volterra Equations

Some inspiration can be found for improving methodology in solving FDE numerically in the literature for Integral Equations. Fractional integration is clearly related to Abel's Integral Equation [13, V.II,p.511], i.e.,

$$\frac{1}{\sqrt{\pi}} \int_0^x \frac{f(\xi)}{\sqrt{x-\xi}} \, \mathrm{d}\xi = \phi(x) \tag{10}$$

This is clearly a special case of the Volterra Equation of the first kind,

$$\int_a^x K(x,\xi) f(\xi) \, \mathrm{d}\xi = \phi(x), \tag{11}$$

for which there exist numerical algorithms for approximating the solution, $f(x)$. A typical method is to divide the interval $[a, b]$ into evenly spaced nodes, as,

$$x_i = a + (i-1)h, \quad \text{for} \quad i = 1, \ldots, n, \quad \text{with} \quad h = \frac{b-a}{n-1}, \tag{12}$$

and then apply the trapezoidal rule for numerical integration to yield the approximation,

$$\int_a^{x_i} K(x_i,\xi) f(\xi) \, \mathrm{d}\xi \approx h \left(\frac{1}{2} K(x_i, x_0) f_1 + \sum_{j=2}^{i-1} K(x_i, x_j) f_j + \frac{1}{2} K(x_i, x_i) f_i \right) \tag{13}$$

Using the initial value,

$$f_1 = g_1, \tag{14}$$

we then obtain a recurrence relation for the subsequent values, as,

$$\frac{1}{2}hK(x_i, x_i)f_i = \phi(x_i) - h\left(\frac{1}{2}K(x_i, x_0)f_1 + \sum_{j=2}^{i-1} K(x_i, x_j)f_j\right) \quad \text{for} \quad i = 2, \ldots, n. \tag{15}$$

Thus, given the initial value, we obtain the remaining values in terms of all previous values, and the sampled kernel. While this provides us with blueprint for the process of numerical fractional integration, the approach has a number of difficulties which need to be addressed:

1. There is a cumulative propagation of errors along the solution. Namely, the approximation error over the first interval propagates through to all subsequent solution points, not to mention roundoff error. Whereas in some cases there may be some cancellation, in practice we should not be left hoping for such a case.

2. The kernel of interest for fractional integration, $K(x, \xi) = (x - \xi)^{\nu-1}$ has a singularity at $x = \xi$. Clearly the method requires the evaluation of the kernel at the points (x_i, x_i) which is undefined for the case when $\nu \in [0, 1]$. The method can therefore not be directly applied.

3. We cannot implement initial- or boundary-conditions with this method. Clearly for the given example, an initial value is sufficient, but if we wish to solve FDE, we must be able to implement multiple initial- or boundary-values, or even perhaps interior values on the solution.

1.3. Matrix Operations and Identities

Throughout the remainder of this chapter, some rather specialized matrix notation is used, which we summarize here (See [14] for further details.). The element-wise product or Hadamard product of matrices is given as the matrix C such that,

$$\mathsf{C} = \mathsf{A} \circ \mathsf{B} \qquad \rightarrow \qquad c_{i,j} = a_{i,j} b_{i,j} \tag{16}$$

The corresponding Hadamard power (element-wise power) of a matrix is denoted as,

$$\mathsf{C} = \mathsf{A}^{\circ\nu} \qquad \rightarrow \qquad c_{i,j} = a_{i,j}^{\nu} \tag{17}$$

Further, we define the "diag" operator, which converts a vector into a diagonal matrix, i.e., if

$$\boldsymbol{u} = \begin{bmatrix} u_1 \\ u_2 \\ \vdots \\ u_n \end{bmatrix} \tag{18}$$

then the diag operator produces,

$$\operatorname{diag}\{\boldsymbol{u}\} = \begin{bmatrix} u_1 & 0 & \dots & 0 \\ 0 & u_2 & \dots & 0 \\ \vdots & \vdots & \ddots & \vdots \\ 0 & 0 & \dots & u_n \end{bmatrix} \tag{19}$$

For manipulation purposes, we frequently use the vector of ones, i.e.,

$$\mathbf{1} = \begin{bmatrix} 1 \\ 1 \\ \vdots \\ 1 \end{bmatrix} \tag{20}$$

Together, these clearly satisfy the relation

$$\operatorname{diag}\{\boldsymbol{u}\}\,\mathbf{1} = \boldsymbol{u}, \tag{21}$$

an identity which is used in subsequent derivations. A further useful identity involving the Hadamard product and diagonal matrices is for general matrices A and B and diagonal matrices D and E we have,

$$\mathsf{D}\,(\mathsf{A} \circ \mathsf{B})\,\mathsf{E} = (\mathsf{DAE}) \circ \mathsf{B} = \mathsf{A} \circ (\mathsf{DBE})\,, \tag{22}$$

among other identities, see [14, p.304]. With regards to notation, a numerical fractional integration matrix of order ν, with $\nu > 0$, is denoted as

$$\mathsf{S}^{(-\nu)} \tag{23}$$

whereby the parentheses distinguish it from a matrix power. The derivative matrix of integer order, n, is denoted as

$$\mathsf{D}^{(n)}, \tag{24}$$

which computes the nth numerical derivative of a vector. Since we have arrived at numerical fractional integration and integer order differentiation by different mechanisms, we have opted for different notations for the two to avoid any confusion.

2. Fractional Integration and Differentiation Matrices

In the previous section, we described some previous approaches to the numerical solution of FDE, and pointed out a number of their difficulties. As an alternative approach, we derive numerical differentiation of integer order in terms of piecewise interpolating functions in matrix form. Further, we develop a discretization of fractional integration based directly on the multiple integration formula. Together these tools allow us to discretize FDE in a consistent manner.

2.1. Integer Order Differentiation Matrices

A function whose values $f_1, \ldots, f_N$, where $f_k = f(x_k)$, are given at the N abscissae $x_1, \ldots, x_N$ can be interpolated by a set of interpolating functions, $\varphi_k(x)$, in the form,

$$\phi(x) = \varphi_1(x) f_1 + \varphi_2(x) f_2 + \ldots + \varphi_N(x) f_N. \tag{25}$$

In this case, the interpolating functions have the property, $\varphi_i(x_j) = \delta_{ij}$, where δ_{ij} is the Kronecker delta, implying that the ith interpolating function has the value $\varphi_i(x_i) = 1$ at x_i, and the value $\varphi_i(x_j) = 0$ at the remaining nodes, $j \neq i$. Functions with this property include, e.g., Lagrange Interpolating Polynomials [15], Gauss's Interpolating Sines [16], the Cardinal Sine Function [17], etc. The appropriate choice of interpolating function will depend on the particular problem being solved. For example, if the solution is known to be sinusoidal, the appropriate choice would be Interpolating Sines. The following arguments apply to all aforementioned interpolating functions, but due to the familiarity of polynomials, we use them as the example of choice. To obtain numerical differentiation formulas, we differentiate the interpolating function[2] yielding,

$$\phi'(x) = \varphi_1'(x) f_1 + \varphi_2'(x) f_2 + \ldots + \varphi_N'(x) f_N \tag{26}$$

By evaluating this derivative at each node x_k, we obtain an expression for the numerical derivatives of the function, which naturally takes the matrix form,

$$\begin{bmatrix} \phi'(x_1) \\ \vdots \\ \phi'(x_N) \end{bmatrix} = \begin{bmatrix} \varphi_1'(x_1) & \ldots & \varphi_N'(x_1) \\ \vdots & \ddots & \vdots \\ \varphi_1'(x_N) & \ldots & \varphi_N'(x_N) \end{bmatrix} \begin{bmatrix} f_1 \\ \vdots \\ f_N \end{bmatrix}. \tag{27}$$

For demonstrative purposes, we may take Lagrange Interpolating Polynomials for the case $N = 3$. For the arbitrary abscissae, we obtain the 3-point derivative matrix,

$$\mathsf{D}_3^{(1)} = \begin{bmatrix} -\frac{2h_1+h_2}{h_1(h_1+h_2)} & \frac{h_1+h_2}{h_1 h_2} & -\frac{h_1}{h_2(h_1+h_2)} \\ -\frac{h_2}{h_1(h_1+h_2)} & -\frac{h_1-h_2}{h_1 h_2} & \frac{h_1}{h_2(h_1+h_2)} \\ \frac{h_2}{h_1(h_1+h_2)} & -\frac{h_1+h_2}{h_1 h_2} & \frac{h_1+2h_2}{h_2(h_1+h_2)} \end{bmatrix}. \tag{28}$$

Practically, extending this approach to longer support lengths will be limited by the numerical conditioning of the matrices that it generates. Higher order formulas require higher precision arithmetic; the matrix formulation of numerical differentiation can expose the devastating effects of roundoff error. Furthermore, for polynomials, high degree polynomials are generally not appropriate for approximating derivatives due to large amplitude oscillations between nodes, known as the Runge Phenomenon. Thankfully, by using consistent formulas, we can still obtain highly accurate results using a low order piecewise polynomial representation of the function. Specifically, the aforementioned matrices can be extended to arbitrary support lengths by noting that the derivative is a local property of a function, and using the derived formulas locally. Extending such formulae to a support of

[2] In this vein, we may also define derivatives of arbitrary integer order, whereby, for polynomials, the Nth derivative will vanish for an N-point sequence.

m-points can be done intuitively by using the end-point formulas (forward- and backward-type formulas) at the respective end-points and the central formula at all internal points; however, it may be done rigorously for the general case by turning to the truncation error of the derived formulae. For example, the case of $N = 3$ and $m = 5$ and $h_k = h$ for $k = 1, \ldots, 4$, we have the differentiation matrix,

$$\begin{bmatrix} f_1' \\ f_2' \\ f_3' \\ f_4' \\ f_5' \end{bmatrix} = \frac{1}{2h} \begin{bmatrix} -3 & 4 & -1 & 0 & 0 \\ -1 & 0 & 1 & 0 & 0 \\ 0 & -1 & 0 & 1 & 0 \\ 0 & 0 & -1 & 0 & 1 \\ 0 & 0 & 1 & -4 & 3 \end{bmatrix} \begin{bmatrix} f_1 \\ f_2 \\ f_3 \\ f_4 \\ f_5 \end{bmatrix} \tag{29}$$

This matrix is second order polynomial accurate at all points[3], with no assumed boundary conditions. Such differentiation matrices thus define the numerical derivative of a vector in matrix form as,

$$\boldsymbol{f}' = \mathsf{D}^{(1)} \boldsymbol{f}, \tag{30}$$

which is a fundamental relation on which we base the discretization of FDE. The corresponding matrix for a second order accurate second derivative is obtained by twice differentiating the interpolating polynomial in Equation (25) to obtain,

$$\begin{bmatrix} f_1'' \\ f_2'' \\ f_3'' \\ f_4'' \\ f_5'' \end{bmatrix} = \frac{1}{h^2} \begin{bmatrix} 1 & -2 & 1 & 0 & 0 \\ 1 & -2 & 1 & 0 & 0 \\ 0 & 1 & -2 & 1 & 0 \\ 0 & 0 & 1 & -2 & 1 \\ 0 & 0 & 1 & -2 & 1 \end{bmatrix} \begin{bmatrix} f_1 \\ f_2 \\ f_3 \\ f_4 \\ f_5 \end{bmatrix} \quad \text{or} \quad \boldsymbol{f}'' = \mathsf{D}^{(2)} \boldsymbol{f}, \tag{31}$$

Note that the first two formulas are identical, as are the last two; this is due to the fact that the second derivative of a parabola is constant; this also indicates that the second order derivative matrix is rank two deficient, as it should be. Equations (29) and (31) are consistent in the sense that they are derived from the same piecewise polynomial interpolating function. If the differential operators are derived in a consistent manner, then the discretization of the FDE will consequently be consistent. The use of formulas with higher order accuracy for differentiation is essential to obtaining high accuracy matrices for computing Caputo derivatives. For reference, a five point derivative matrix (fourth order accurate) is given as,

$$\begin{bmatrix} f'(x_0) \\ f'(x_1) \\ f'(x_2) \\ f'(x_3) \\ f'(x_4) \end{bmatrix} \approx \frac{1}{12h} \begin{bmatrix} -25 & 48 & -36 & 16 & -3 \\ -3 & -10 & 18 & -6 & 1 \\ 1 & -8 & 0 & 8 & -1 \\ -1 & 6 & -18 & 10 & 3 \\ 3 & -16 & 36 & -48 & 25 \end{bmatrix} \begin{bmatrix} f(x_0) \\ f(x_1) \\ f(x_2) \\ f(x_3) \\ f(x_4) \end{bmatrix} \tag{32}$$

[3]In contrast, in MATLAB many functions use the central formula produced here, however, uses only the first order accurate forward and backward differences at the endpoints. Such matrices are inconsistent in terms of accuracy, which clearly affects solutions as a whole to least squares type inverse problems. The reasoning behind this albeit inconsistent approach appears likely due to the fact that it defines a tridiagonal matrix which can be solved relatively quickly.

2.2. Fractional Order Integral Matrices

The fractional order integral, $f^{(-\nu)}(x)$, is defined by extending the multiple integration formula of a function $f(x)$ to non-integer values [8], as,

$$f^{(-\nu)}(x) \triangleq \frac{1}{\Gamma(\nu)} \int_a^x (x-\xi)^{\nu-1} f(\xi)\,\mathrm{d}\xi, \tag{33}$$

where $\nu > 0$. In order to develop a numerical matrix operator for fractional integration, we recall (cf. Section 1.2.3) that the fractional integral is essentially an integral transform over a Volterra-type domain (See Fig. 1), since it is identical in form to,

$$\phi(x) = \int_a^x K(x,\xi) f(\xi)\,\mathrm{d}\xi, \tag{34}$$

which is a Volterra Equation of the first kind [18, 19]. The fractional integral $f^{(-\nu)}(x)$ is dependent on $f(x)$ by means of the linear operation of integration; hence, this problem has a discretization that takes the form of a matrix multiplication. Clearly, we can treat the function space as an infinite dimensional vector space; in this space linear operators take the form of infinite dimensional matrices. In the discrete space, the functions and linear operators take the form of finite dimensional vectors and matrices respectively. An issue which must be considered is the fact that when the value of ν lies in the interval $[0,1]$ the kernel of the integral equation has a singularity along the line, $\xi = x$. The case $\nu \in (0,1)$ is critical to both the Caputo and Riemann-Liouville definitions of fractional derivatives, since both definitions involve integrals strictly within this range. A common approach [19] is to eliminate the singularity by adding and subtracting the term $\int_a^x K(x,\xi) f(x)\,\mathrm{d}\xi$, which for the fractional integral yields,

$$\begin{aligned} f^{(-\nu)}(x) = {} & \frac{1}{\Gamma(\nu)} \int_a^x (x-\xi)^{\nu-1} (f(\xi) - f(x))\,\mathrm{d}\xi \\ & + \frac{1}{\Gamma(\nu)} \int_a^x (x-\xi)^{\nu-1} f(x)\,\mathrm{d}\xi. \end{aligned} \tag{35}$$

The integrand of the first term is now zero along the line $\xi = x$; the integrand of the second term depends only on the kernel (i.e., it is independent of $f(x)$), and thus can be integrated analytically by straightforward invocation of the monomial rule. By integrating the second term, the fractional integral can then be written as,

$$f^{(-\nu)}(x) = \int_a^x G(x,\xi)\,\mathrm{d}\xi + \frac{(x-a)^\nu}{\Gamma(\nu+1)} f(x), \tag{36}$$

where

$$G(x,\xi) = \frac{1}{\Gamma(\nu)} (x-\xi)^{\nu-1} (f(\xi) - f(x)). \tag{37}$$

The linear operator is thereby separated into the sum of two operators; consequently, the discretization of the fractional integral can be written as the linear combination of two matrices,

$$\boldsymbol{f}^{(-\nu)} = \mathsf{L}_1 \boldsymbol{f} + \mathsf{L}_2 \boldsymbol{f}, \tag{38}$$

where the two terms correspond respectively to the two terms in Equation (36). Discretization of the second term is straightforward as,

$$\mathsf{L}_2 = \mathrm{diag}\left\{\frac{(\boldsymbol{x} - \mathbf{1}a)^{\circ\nu}}{\Gamma(\nu+1)}\right\} \tag{39}$$

where the "diag" operator converts a vector into a diagonal matrix, and $\circ$ is the Hadamard power (See Section 1.3 or [14]). As for the first term in Equation (36), we require two steps: the discretization of the function $G(x, \xi)$, and the discretization of an integral over a Volterra-type domain. Discretization of the function $G(x, \xi)$ defined in Equation (37) is given by the matrix,

$$\mathsf{G} = \mathsf{K}\left(\mathbf{1}\boldsymbol{f}^{\mathrm{T}} - \boldsymbol{f}\mathbf{1}^{\mathrm{T}}\right), \tag{40}$$

where,

$$\mathsf{K} = \frac{1}{\Gamma(\nu)}\left[\mathsf{M} \circ \left(\boldsymbol{x}\mathbf{1}^{\mathrm{T}} - \mathbf{1}\boldsymbol{x}^{\mathrm{T}}\right)\right]^{\circ(\nu-1)}. \tag{41}$$

In this case, the matrix M is a mask selecting the sub-diagonal portion of the matrix it is multiplied with in the Hadamard sense, i.e.,

$$\mathsf{M} = [m_{i,j}] \quad \text{with} \quad m_{i,j} = \begin{cases} 0 & i \leq j \\ 1 & j > i. \end{cases} \tag{42}$$

As for the integral of the function $G(x, \xi)$, the domain of integration is shown in Fig. 1. If we denote the resulting integral as,

$$g(x) = \int_a^x G(x, \xi)\, \mathrm{d}\xi, \tag{43}$$

then by applying the trapezoidal rule to obtain the approximate values, g_k, of $g(x)$, for $k = 1, \ldots, m$, we have,

$$\begin{aligned} g_1 &= 0 \\ g_2 &= \frac{h_1}{2}G_{2,1} + \frac{h_1}{2}G_{2,2} \\ g_3 &= \frac{h_1}{2}G_{2,1} + \frac{h_1 + h_2}{2}G_{2,2} + \frac{h_2}{2}G_{2,3} \\ &\vdots \\ g_m &= \frac{h_1}{2}G_{m,1} + \sum_{k=2}^{m-1}\frac{h_{k-1} + h_k}{2}G_{m,k} + \frac{h_{m-1}}{2}G_{m,m} \end{aligned} \tag{44}$$

where the $G_{i,j}$ are the entries of the matrix $\mathsf{G} = [G_{i,j}]$. In matrix form, Equations (44) can then be written as,

$$\boldsymbol{g} = (\mathsf{G} \circ \mathsf{S})\mathbf{1} \tag{45}$$

where the matrix S is the first order integration matrix,

$$\mathsf{S} = \frac{1}{2}\begin{bmatrix} 0 & 0 & 0 & \cdots & 0 \\ h_1 & h_1 & 0 & \cdots & 0 \\ h_1 & h_1 + h_2 & h_2 & \cdots & 0 \\ \vdots & \vdots & \ddots & \ddots & \vdots \\ h_1 & h_1 + h_2 & \cdots & h_{n-2} + h_{n-1} & h_{n-1} \end{bmatrix}. \tag{46}$$

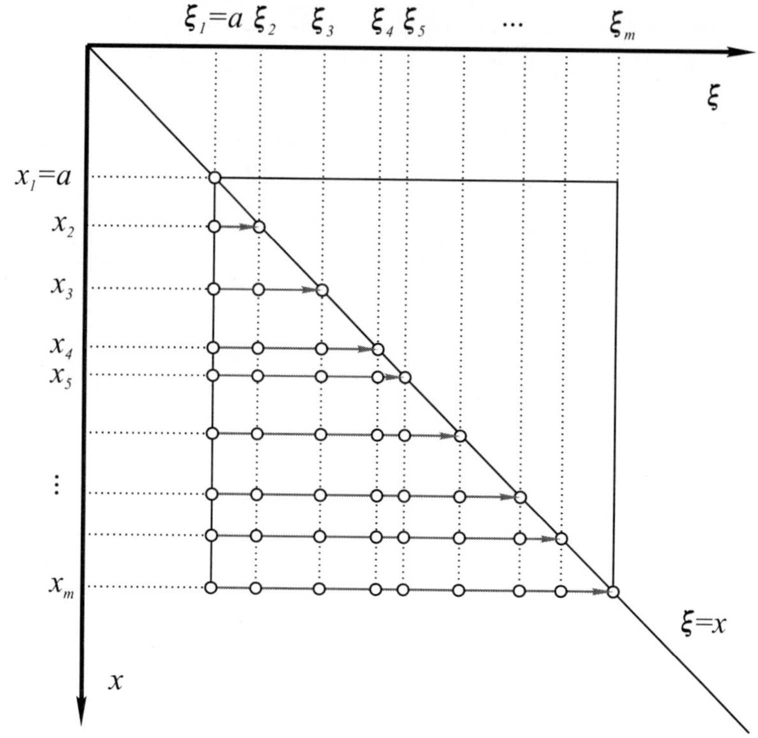

Figure 1. Domain for the integral equation and its integration paths (red arrows). The kernel is singular along the line $\xi = x$. The orientation of the axes is to facilitate the discretization in conventional matrix-vector form.

By combining Equations (40) and (45), the first term in Equation (38) is then,

$$\mathsf{L}_1 \boldsymbol{f} = \left(\left[\mathsf{K}(\mathbf{1}\boldsymbol{f}^{\mathrm{T}} - \boldsymbol{f}\mathbf{1}^{\mathrm{T}}) \right] \circ \mathsf{S} \right) \mathbf{1}. \tag{47}$$

In order to develop a fractional integration matrix, it is necessary to extricate the vector $\boldsymbol{f}$ on the r.h.s. of this relation, thereby obtaining the expression for its coefficient matrix, L_1. Defining the intermediate matrix, $\mathsf{F} = \operatorname{diag}\{\boldsymbol{f}\}$, the expression simplifies (cf. [14]) to

$$\begin{aligned} \mathsf{L}_1 \boldsymbol{f} &= ((\mathsf{KF} - \mathsf{FK}) \circ \mathsf{S})\, \mathbf{1} & (48) \\ &= ((\mathsf{KF}) \circ \mathsf{S} - (\mathsf{FK}) \circ \mathsf{S})\, \mathbf{1} & (49) \end{aligned}$$

Since the matrix F is diagonal, we may invoke the identity given in Equation (22) to yield,

$$\mathsf{L}_1 \boldsymbol{f} = ((\mathsf{K} \circ \mathsf{S})\, \mathsf{F} - \mathsf{F}\, (\mathsf{K} \circ \mathsf{S}))\, \mathbf{1}. \tag{50}$$

Finally the vector $\boldsymbol{f}$ can be factored out simplifying this relation to,

$$\mathsf{L}_1 \boldsymbol{f} = ((\mathsf{K} \circ \mathsf{S}) - \operatorname{diag}\{(\mathsf{K} \circ \mathsf{S})\, \mathbf{1}\})\, \boldsymbol{f}. \tag{51}$$

Discrete fractional integration can therefore be written in the matrix form,

$$\boldsymbol{f}^{(-\nu)} = \mathsf{S}^{(-\nu)} \boldsymbol{f} \quad \text{with} \quad \mathsf{S}^{(-\nu)} = \mathsf{L}_1 + \mathsf{L}_2. \tag{52}$$

Combining the terms in Equations (51) and (39) yields the formula for the fractional integration matrix as,

$$\mathsf{S}^{(-\nu)} = (\mathsf{K} \circ \mathsf{S}) - \mathrm{diag}\left\{ (\mathsf{K} \circ \mathsf{S})\,\mathbf{1} - \frac{(\boldsymbol{x} - \mathbf{1}a)^{\circ\nu}}{\Gamma(\nu+1)} \right\} \tag{53}$$

Note that the Gamma function could alternatively factored out of K, which is the formula used in the MATLAB code provided in Section 4.

2.3. Fractional Order Caputo Derivatives and Their Accuracy

Combining the integer order differentiation matrices with the fractional order integration matrix, we yield an approximation for the Caputo derivative in the form,

$$\boldsymbol{y}^{(\alpha)} = \mathsf{S}^{(-\nu)}\mathsf{D}^{(n)}\boldsymbol{y} \tag{54}$$

Firstly, we can demonstrate the accuracy of the fractional integration matrix; namely, that is integrates a constant function exactly. If we multiply the matrix, $\mathsf{S}^{(-\nu)}$, by the vector of ones, we obtain,

$$\mathsf{S}^{(-\nu)}\mathbf{1} = (\mathsf{K} \circ \mathsf{S})\,\mathbf{1} - \mathrm{diag}\left\{(\mathsf{K} \circ \mathsf{S})\,\mathbf{1}\right\}\mathbf{1} + \mathrm{diag}\left\{ \frac{(\boldsymbol{x} - \mathbf{1}a)^{\circ\nu}}{\Gamma(\nu+1)} \right\}\mathbf{1} \tag{55}$$

Due to the property in Equation (21), the first two terms cancel, yielding the relation,

$$\mathsf{S}^{(-\nu)}\mathbf{1} = \mathrm{diag}\left\{ \frac{(\boldsymbol{x} - \mathbf{1}a)^{\circ\nu}}{\Gamma(\nu+1)} \right\}\mathbf{1} \tag{56}$$

$$= \frac{(\boldsymbol{x} - \mathbf{1}a)^{\circ\nu}}{\Gamma(\nu+1)} \tag{57}$$

Given that the exact integral solution is,

$${}_aD_x^{-\nu}1 = \frac{1}{\Gamma(\nu+1)}(x-a)^{\nu}, \tag{58}$$

we see that the numerical result is exact for constant functions. Thus, if we use this matrix in conjunction with a second order accurate integer order differentiation matrix to compute a Caputo derivative of order $\alpha = 1.8$, we see that the solution will be exact for polynomials up to degree two. Specifically, applying the derivative matrix to a second order polynomial of the form $y(x) = A + Bx + Cx^2$, it results in the constant function $y''(x) = 2C$; next, applying the fractional integral matrix, we obtain the function

$$y^{(\alpha)}(x) = \frac{2C}{\Gamma(\nu+1)}(x-a)^{\nu}, \tag{59}$$

which is the exact solution.

3. Discretization and Solution of FDE

3.1. Discretization of Differential Equations

The derivative and fractional integration matrices developed thus far enable the numerical solution of p-term fractional differential equations with variable coefficients of the form,

$$a_1(x)y^{(\alpha_1)} + a_2(x)y^{(\alpha_2)} + \ldots + a_p(x)y^{(\alpha_p)} = g(x) \tag{60}$$

In terms of the Caputo definition of the fractional derivative, we have,

$$y^{(\alpha)}(x) = \frac{1}{\Gamma(\nu)} \int_a^x (x-\xi)^{(\nu-1)} y^{(n)}(\xi)\, \mathrm{d}\xi \tag{61}$$

where $\alpha = n - \nu$ such that n is the integer part of α rounded towards $+\infty$, and ν is in the range $(0, 1)$. In terms of the ceiling function, α can be decomposed as,

$$n = \lceil \alpha \rceil \qquad \text{and} \qquad \nu = \lceil \alpha \rceil - \alpha. \tag{62}$$

The order α Caputo derivative can therefore be seen as the νth-order integral of the integer order n^{th} derivative of the function $y(x)$. Correspondingly, the discretization of a Caputo derivative of a discrete function $\boldsymbol{y}$ is given as the matrix product,

$$\boldsymbol{y}^{(\alpha)} = \mathsf{S}^{(-\nu)}\mathsf{D}^{(n)}\boldsymbol{y} \tag{63}$$

whereby the matrix $\mathsf{D}^{(n)}$ was developed in Section 2.1, and the matrix $\mathsf{S}^{(-\nu)}$ was developed in Section 2.2. In this vein, the FDE (60) is discretized as,

$$\begin{aligned} \mathsf{A}_1\mathsf{S}^{(-\nu_1)}\mathsf{D}^{(n_1)}\boldsymbol{y} + \mathsf{A}_2\mathsf{S}^{(-\nu_2)}\mathsf{D}^{(n_2)}\boldsymbol{y} + \ldots \\ + \mathsf{A}_p\mathsf{S}^{(-\nu_p)}\mathsf{D}^{(n_p)}\boldsymbol{y} = \boldsymbol{g} \end{aligned} \tag{64}$$

where

$$\mathsf{A}_k = \operatorname{diag}\left\{a_k(x_1), a_k(x_2), \ldots, a_k(x_n)\right\}. \tag{65}$$

For convenience, we may therefore define the generalized differential operator as,

$$\begin{aligned} \mathsf{L} = \mathsf{A}_1\mathsf{S}^{(-\nu_1)}\mathsf{D}^{(n_1)} + \mathsf{A}_2\mathsf{S}^{(-\nu_2)}\mathsf{D}^{(n_2)} + \ldots \\ + \mathsf{A}_p\mathsf{S}^{(-\nu_p)}\mathsf{D}^{(n_p)} \end{aligned} \tag{66}$$

whereby the discretized differential equation then takes the form of the linear system of equations,

$$\mathsf{L}\boldsymbol{y} = \boldsymbol{g}. \tag{67}$$

3.2. Discretization of Boundary Conditions

As with integer order ODE, FDE defined in terms of Caputo derivatives are solved with specified integer order initial or boundary values, such as,

$$y(a) = d_0, \quad y^{(1)}(a) = d_1, \quad \ldots, \quad y^{(q-1)}(a) = d_{q-1} \tag{68}$$

In vector form, the discretization of an initial value of the function $y(x)$ takes the form,

$$\boldsymbol{e}_1^{\mathrm{T}} \boldsymbol{y} = d_0, \tag{69}$$

where $\boldsymbol{e}_1$ is the first coordinate vector. In a similar manner, initial conditions on the integer order derivatives can be discretized as,

$$\boldsymbol{e}_1^{\mathrm{T}} \mathsf{D}^{(k)} \boldsymbol{y} = d_k \tag{70}$$

In this vein, we may place q constraints on the solution vector, which can be written in the form,

$$\mathsf{C}^{\mathrm{T}} \boldsymbol{y} = \boldsymbol{d} \tag{71}$$

where the matrix C for an IVP is given as,

$$\mathsf{C} = \begin{bmatrix} \boldsymbol{e}_1 & \mathsf{D}^{\mathrm{T}} \boldsymbol{e}_1 & \cdots & \left(\mathsf{D}^{(q-1)}\right)^{\mathrm{T}} \boldsymbol{e}_1 \end{bmatrix} \tag{72}$$

In a similar manner, constraints for boundary values over the interval $[a, b]$, such as,

$$\begin{array}{llll} y(a) = d_0, & y^{(1)}(a) = d_1, & \ldots, & y^{(q-1)}(a) = d_{q-1} \\ y(b) = d_q, & y^{(1)}(b) = d_{q+1}, & \ldots, & y^{(r-1)}(b) = d_{q+r-1} \end{array} \tag{73}$$

can be implemented with

$$\mathsf{C} = \begin{bmatrix} \boldsymbol{e}_1 & \mathsf{D}^{\mathrm{T}} \boldsymbol{e}_1 & \cdots & \left(\mathsf{D}^{(q-1)}\right)^{\mathrm{T}} \boldsymbol{e}_1 & \boldsymbol{e}_m & \mathsf{D}^{\mathrm{T}} \boldsymbol{e}_m & \cdots & \left(\mathsf{D}^{(r-1)}\right)^{\mathrm{T}} \boldsymbol{e}_m \end{bmatrix} \tag{74}$$

And generally, any constraint at a point c in the interval $[a, b]$ of the form

$$a_0 y(c) + a_1 y^{(1)}(c) + \ldots + a_q y^{(q-1)}(c) = d, \tag{75}$$

i.e., a Robin type condition, can be implemented without difficulty. The proposed method thus provides a general framework for solving IVP, BVP, and general multi-point FDE.

3.3. Solution via LSE

With the differential equation itself discretized as per Equation (67), and the boundary conditions discretized per Equation (71), the numerical solution of the differential equation amounts to solving the linear least squares problem with equality constraints (LSE), that is,

$$\min_{\mathsf{C}^{\mathrm{T}} \boldsymbol{y} = \boldsymbol{d}} \|\mathsf{L} \boldsymbol{y} - \boldsymbol{g}\|_2^2 \tag{76}$$

An efficient solution based on QR decomposition is outlined in [3, §12.4.1]. The formulation of the solution of the FDE as a Linear Constrained Least Squares Problem is fundamental to the versatility of the new method. Specifically, any type of initial, boundary or interior values can be specified for the solution in any number ($< n$) and in any combination.

A brief synopsis of the algorithm demonstrates a link to the classic theory of the solution of ordinary differential equations. Let

$$\mathsf{C} = \mathsf{QR} = \begin{bmatrix} \hat{\mathsf{Q}} & \tilde{\mathsf{Q}} \end{bmatrix} \begin{bmatrix} \hat{\mathsf{R}} \\ 0 \end{bmatrix} \tag{77}$$

be the QR decomposition of C. The LS solution can therefore be written as a linear combination of the orthonormal basis Q, as,

$$\boldsymbol{y} = \hat{\mathsf{Q}}\boldsymbol{u} + \tilde{\mathsf{Q}}\boldsymbol{v}. \tag{78}$$

Substituting this relation into the constraint equations, the solution for $\boldsymbol{u}$ is given by,

$$\boldsymbol{u} = \hat{\mathsf{R}}^{-\mathrm{T}}\boldsymbol{d} \tag{79}$$

Subsequently, the solution for $\boldsymbol{v}$ is obtained from the differential equation as,

$$\boldsymbol{v} = \left(\mathsf{L}\tilde{\mathsf{Q}}\right)^{+} \left(\boldsymbol{g} - \mathsf{L}\hat{\mathsf{Q}}\hat{\mathsf{R}}^{-\mathrm{T}}\boldsymbol{d}\right) \tag{80}$$

With these relations, the explicit solution to the differential equation is given as,

$$\boldsymbol{y}_{\mathrm{LS}} = \tilde{\mathsf{Q}}(\mathsf{L}\tilde{\mathsf{Q}})^{+}\boldsymbol{g} + \left(\mathsf{I} - \tilde{\mathsf{Q}}(\mathsf{L}\tilde{\mathsf{Q}})^{+}\mathsf{L}\right)\hat{\mathsf{Q}}\hat{\mathsf{R}}^{-\mathrm{T}}\boldsymbol{d} \tag{81}$$

In relation to the classic theory of ODEs, the first term is a particular solution to the differential equation satisfying homogeneous boundary conditions. The coefficient matrix of the second term (i.e., of vector of initial values $\boldsymbol{d}$) is an $m \times p$ matrix of solutions satisfying unit boundary conditions,

$$\begin{array}{llll} f_1(a) = 1, & f_1'(a) = 0, & \ldots, & f_1^{(p-1)}(a) = 0 \\ f_2(a) = 0, & f_2'(a) = 1, & \ldots, & f_2^{(p-1)}(a) = 0 \\ \vdots & \vdots & \ddots & \vdots \\ f_p(a) = 0, & f_p'(a) = 0, & \ldots, & f_p^{(p-1)}(a) = 1 \end{array} \tag{82}$$

4. Numerical Examples

The following MATLAB code generates the fractional integration matrix proposed in this paper. The input is a vector of the abscissa x, and the order, nu,

```
m = length(x) ;
a = x(1) ;
z = ones(m,1) ;
S = cumtrapz( x, eye(m) ) ;
K = ((x-a)*z' - z*(x-a)').^(nu-1) ;
KS = tril(K,-1) .* S ;
S = (KS - diag(KS * z - ((x - a).^nu )/nu)) / gamma(nu) ;
```

Computationally, it requires only matrix multiplication as well as Hadamard products and powers. The dominating factor in the computational complexity of the algorithm as a whole is therefore the LSE solution, i.e., $\mathcal{O}(m^3)$ [3].

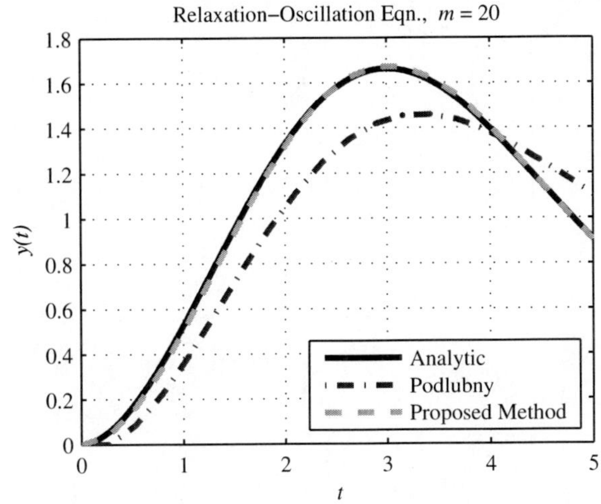

Figure 2. Numerical solution of the Relaxation-Oscillation Equation with homogeneous initial conditions $y(0) = 0$, $y'(0) = 0$ and the number of abscissa $m = 20$. For such a small number of points, the fractional difference method of Podlubny deviates greatly from the true solution.

4.1. Relaxation-Oscillation Equation

A particular instance of the Relaxation-Oscillation Equation [8, 7] is the FDE of the form,

$$y^{(\alpha)}(t) + y(t) = 1 \qquad \text{with} \qquad \begin{matrix} y(0) = d_0 \\ y'(0) = d_1 \end{matrix} \tag{83}$$

It has the analytic solution,

$$y(t) = d_0 E_{\alpha,1}(-t^\alpha) + d_1 t E_{\alpha,2}(-t^\alpha) + t^\alpha E_{\alpha,\alpha+1}(-t^\alpha) \tag{84}$$

where $E_{\alpha,\beta}(t)$ is the Mittag-Leffler function of two arguments. As in [7], we solve the FDE for the value $\alpha = 1.8$. According to the proposed framework, this is discretized as,

$$\left(\mathsf{S}^{(-0.2)}\mathsf{D}^{(2)} + \mathsf{I}\right) \boldsymbol{y} = \boldsymbol{1}, \tag{85}$$

subject to the linear constraints (initial values),

$$\mathsf{C}^{\mathrm{T}}\boldsymbol{y} = \boldsymbol{d}, \quad \text{with} \quad \mathsf{C} = \begin{bmatrix} e_1 & \mathsf{D}^{\mathrm{T}} e_1 \end{bmatrix}. \tag{86}$$

The numerical solution is shown in both Fig. 2 for $m = 20$ for the homogenous initial conditions $\boldsymbol{d} = \begin{bmatrix} 0 & 0 \end{bmatrix}^{\mathrm{T}}$, and Fig. 3 for $m = 200$. The case for $m = 20$ shows that even for a small number of points, the new solution yields quite accurate result in comparison to Podlubny's method. While the results of Podlubny's method improve dramatically for the case $m = 200$, the result of the new method is visually indistinguishable from the analytic solution at the plotting scale (maximum absolute error is 1.1×10^{-3}).

In Fig. 4, the numerical solutions with $m = 100$ for the non-homogenous initial conditions $\boldsymbol{d} = \begin{bmatrix} 1 & -1 \end{bmatrix}^{\mathrm{T}}$ are shown. Similar results are obtained as for the homogeneous case.

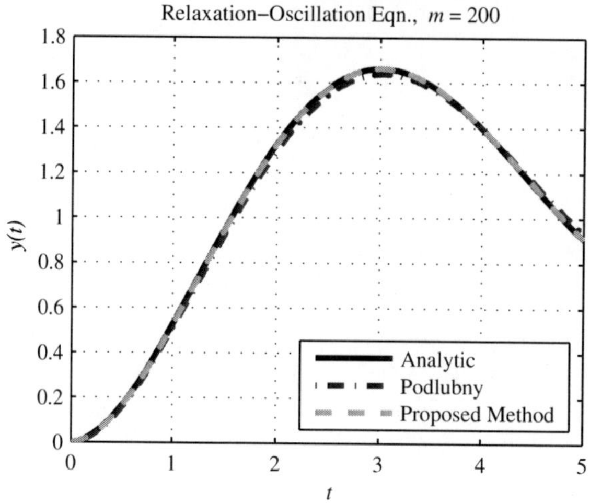

Figure 3. Numerical solution of the Relaxation-Oscillation Equation with homogeneous initial conditions $y(0) = 0$, $y'(0) = 0$ and the number of abscissa at $m = 200$. For a large number of points, the fractional difference method is much improved, yet still exhibits a systematic deviation from the true solution.

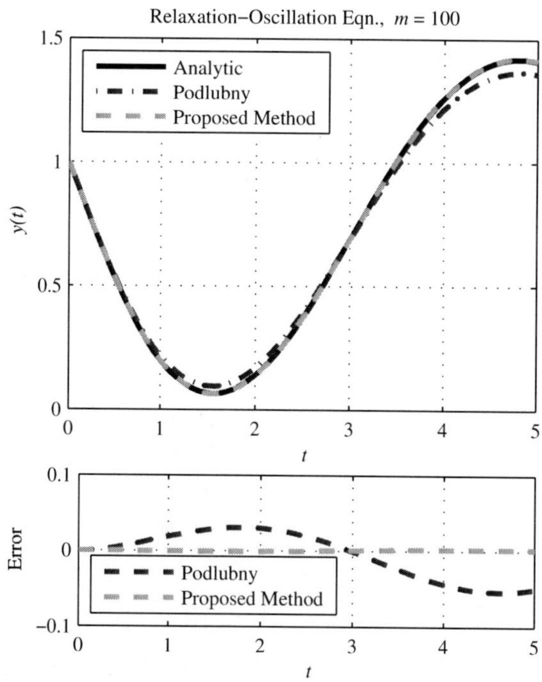

Figure 4. Numerical solution of the Relaxation-Oscillation Equation for with non-homogeneous initial conditions $y(0) = 1$, $y'(0) = -1$ with $m = 100$. The maximum absolute error for Podlubny's Method is 0.0539, whereas for the new method it's 0.0023, that is, an order of magnitude smaller.

However, in this case, it should be recalled that the method of Podlubny can only be applied to homogeneous initial conditions. Thus, the numerical solution involves the analytic transformation of the non-homogeneous problem to a homogeneous one. In contrast, for the newly proposed method based on constrained least squares, the solution methodology is identical in both cases. The numerical residuals w.r.t. the analytical solution are also shown in Fig. 4. For the newly proposed solution, the residuals are all smaller than 2.3×10^{-3}, and are roughly an order of magnitude smaller than the Podlubny solution.

4.2. Bagley-Torvik Equation

The Bagley-Torvik Equation [8] is an FDE involving a derivative of order $\frac{3}{2}$, of the form,

$$a_1 y''(t) + a_2 y^{(\frac{3}{2})}(t) + a_3 y(t) = g(t) \tag{87}$$

A particular case of the Bagley-Torvik Equation is the BVP,

$$\begin{aligned} & y''(t) + y^{(\frac{3}{2})}(t) + y(t) = 2 + 4\sqrt{\frac{t}{\pi}} + t^2 \\ & y(0) = 0, \quad y'(0) = 0, \quad y(1) = 1, \quad y'(1) = 2, \end{aligned} \tag{88}$$

which has the analytical solution [20], $y(t) = t^2$. The numerical solution of this BVP using the new method is shown in Fig. 5. Even for the modest case of $m = 50$, the results are difficult to visually distinguish from the analytical solution; the numerical residuals are

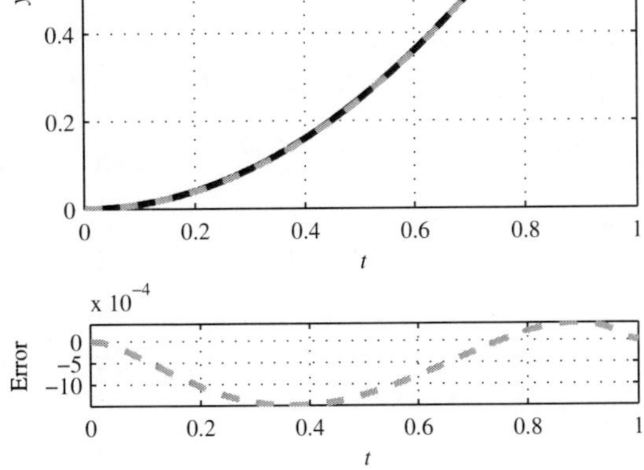

Figure 5. Numerical solution of the Bagley-Torvik Equation with boundary conditions $y(0) = 0$, $y'(0) = 0$, $y(1) = 1$, $y'(1) = 2$ with $m = 50$. Note that the error is zero at the end points, because the boundary conditions are enforced exactly via the LSE optimization. The maximum absolute error is 0.0015 or 1.5×10^{-3}.

shown with respect to the analytical solution, whereby the maximum deviation is about 1.5×10^{-3}. The results in this case are significant, since the method of Podlubny cannot be adapted to solve BVP, i.e., with constraints other than initial values. These results are far better than those presented in [20], in spite of the fact that the proposed method is also far simpler.

5. Conclusion

We presented a new approach to the discretization and numerical solution of fractional order differential equations based on matrix algebra. The method is computationally efficient, and yields results which are an order of magnitude more accurate than modern methods. In particular, the method works exceptionally well in extreme cases such as very small/large number of nodes, where other methods fail completely.

Acknowledgments

The authors would like to thank Igor Podlubny and Tomas Skovranek, for many long helpful discussions, and for directing us to suitable FDE for numerical testing.

References

[1] C. Lanczos. An iteration method for the solution of the eigenvalue problem of linear differential and integral operators. *J. of Research of the NBS*, 45(4):255–282, 1950.

[2] F. W. Dorr. The direct solution of the discrete Poisson equation on a rectangle. *SIAM Rev.*, 12(2):248–263, 1970.

[3] G. H. Golub and C. F. Van Loan. *Matrix Computations*. The Johns Hopkins University Press, Baltimore, 3rd edition, 1996.

[4] R. Courant, K. Friedrichs, and H. Lewy. Über die partiellen differenzengleichungen der mathematischen physik. *Mathematische Annalen*, 100:32–74, 1928.

[5] C. Lanczos. *Linear Differential Operators*. Dover, Mineola, NY, 1997.

[6] J. C. Strikwerda. Finite Difference Schemes and Partial Differential Equations. *SIAM*, Philadelphia, 2nd edition, 2004.

[7] I. Podlubny. Matrix approach to discrete fractional calculus. *Fractional Calculus and Applied Analysis*, 3(4):359–386, 2000.

[8] I. Podlubny. *Fractional Differential Equations*, volume 198 of *Mathematics in Science and Engineering*. Academic Press, San Diego, 1999.

[9] I. Podlubny, A. Chechkin, T. Skovranek, Y.-Q. Chen, and B. M. Vinagre Jara. Matrix approach to discrete fractional calculus II: partial fractional differential equations. *Journal of Computational Physics*, 228(8):3137–3153, 2009.

[10] M. Harker and P. O'Leary. Least squares surface reconstruction from measured gradient fields. In *CVPR 2008*, pages 1–7, Anchorage, AK, 2008. IEEE.

[11] Y. Keskin, O. Karaoglu, S. Servi, and G. Oturanç. The approximate solution of high-order linear fractional differential equation with variable coefficients in terms of generalized Taylor polynomials. *Mathematical and Computational Applications*, 16(3):617–629, 2011.

[12] E. Hairer, S. P. Nørsett, and G. Wanner. *Solving Ordinary Differential Equations I: Nonstiff Problems*. Springer, second edition, 2008.

[13] R. Courant and F. John. *Introduction to Calculus and Analysis*, volume I & II of *Classics in Mathematics*. Springer, Berlin, 1989.

[14] R. A. Horn and C. R. Johnson. *Topics in Matrix Analysis*. Cambridge University Press, New York, 2006.

[15] R. L. Burden and J. D. Faires. *Numerical Analysis*. Thomson Learning, Inc., eighth edition, 2005.

[16] M. Abramowitz and I. A. Stegun, editors. Handbook of Mathematical Functions. *Applied Mathematics Series*. National Bureau of Standards, Washington, D.C., 1972.

[17] R. N. Bracewell. *The Fourier Transform and its Applications*. McGraw-Hill, second edition, 1986.

[18] F. G. Tricomi. *Integral Equations*. Dover Publications, New York, 1985.

[19] L. M. Delves and J. L. Mohamed. *Computational Methods for Integral Equations*. Cambridge University Press, Cambridge, 1985.

[20] M. A. Z. Raja, J. A. Khan, and I. M. Qureshi. Solution of fractional order system of Bagley-Torvik equation using evolutionary computational intelligence. *Mathematical Problems in Engineering*, 2011:18, 2011.

In: Fractional Calculus: Theory
Editors: R. Abi Zeid Daou and X. Moreau

ISBN: 978-1-63463-002-3

Chapter 4

ON ANALYTICAL METHODS FOR DIFFERENTIAL EQUATIONS WITH LOCAL FRACTIONAL DERIVATIVE OPERATORS

Xiao-Jun Yang[1,*], Dumitru Baleanu[2,3,4] and J. A. Tenreiro Machado[5]

[1]Department of Mathematics and Mechanics,
China University of Mining and Technology, Xuzhou Campus,
Xuzhou, Jiangsu, China
[2]Department of Chemical and Materials Engineering, Faculty of Engineering,
King Abdulaziz University, Jeddah, Saudi Arabia
[3]Institute of Space Sciences, Magurele, RO, Bucharest, Romania
[4]Department of Mathematics and Computer Sciences,
Faculty of Arts and Sciences, Cankaya University, Ankara, Turkey
[5]Department of Electrical Engineering, Institute of Engineering,
Polytechnic of Porto, Rua Dr. Antonio Bernardino de Almeida,
Porto, Portugal

Abstract

This chapter reviews new analytical methods for a family of differential equations with local fractional derivatives. We concentrate mainly on Fourier series, and Fourier and Laplace transforms with local fractional operators. The potential applications of the reported results are discussed.

Keywords: Differential equations, analytical methods, local fractional derivatives, local fractional Fourier series method, local fractional Fourier transforms, local fractional Laplace transforms

---

* E-mail address: dyangxiaojun@163.com; Address: Department of Mathematics and Mechanics, China University of Mining and Technology, Xuzhou Campus, Xuzhou, Jiangsu 221008, China (Corresponding author)

Introduction

The classical integral transforms, such as Fourier series, and Fourier and Laplace transforms were applied to analyze the differential equations (Davies, 2002; Brown and Churchill, 1993; Bracewell, 1986; Schiff, 1999). These transforms play important roles in the areas of mathematical physics (Tranter, 1966), control theory (Komornik and Loreti, 2005), time series (Brockwell and Davis, 2009), finance (Cherubini et al., 2010), signal processing (Lathi, 1998), digital holography (Schnars and Jueptner, 2005), and economics (Grubbström, 1967).

The fractional Fourier and Laplace transforms (Podlubny, 1999; Baleanu et al., 2012) are the generalizations of the fractional order based on the fractional calculus (FC) theory (Mainardi, 2010; Kilbas et al., 2006; Magin, 2006; Baleanu, 2011; Klafter et al., 2012; West et al., 2003; Tarasov, 2011). The fractional differential equations (FDEs) are important in many scientific areas (Podlubny, 1999; Baleanu et al., 2012; Kilbas et al., 2006). For example, the fractional Schrödinger (Laskin, 2002), Burgers (Momani, 2006), evolution (Zhou and Jiao, 2010), Klein–Gordon (Golmankhaneh, 2011) and telegraph (Zhao and Li, 2012) equations were investigated during the last years. The fractional integral transforms were used to obtain the analytical solutions for the FDEs (Podlubny, 1999; Baleanu et al., 2012; Zayed, 2002). Other techniques were proposed, such as the Adomian decomposition (Momani and Odibat, 2006), finite difference (Meerschaert and Tadjeran, 2004), homotopy (Jafari and Seifi, 2009; Dehghan et al., 2010) and heat-balance integral (Hristov, 2010) methods.

For the non-differentiable functions defined on Cantor sets (Yang, 2011; Yang, 2012), the classical and fractional calculus do not work properly. Local fractional calculus (Yang, 2011; Yang, 2012; Kolwankar, 1998; Carpinteri and Cornetti, 2002; Carpinteri, 2004; Ma et al., 2013; Hao et al., 2013; Liao et al., 2013; Adda and Cresson, 2001; Babakhani and Gejji, 2002; Chen et al., 2010; Yang, 2013a; Yang, 2013b; Zhao et al., 2013a; Yang, 2013c) was suggested for modeling the fractal problems. Also, the local fractional differential equations (LFDEs) were developed. Solutions for LFDEs (Baleanu, 2014; Yang and Baleanu, 2013; Liu et al., 2013; Yang, 2014a,b; Zhao, 2013b; Zhang, 2013; Su, 2013; Wang, 2014; Hu, 2012; Yang, 2013d, e, f, g) are the non-differential function, which are defined on Cantor sets. We recall that the local fractional integral transforms (LFITs) concluding the local fractional Fourier series and the local fractional Fourier and Laplace transforms (Yang, 2011; Liao et al., 2013; Liu et al., 2013; Yang, 2014b, Zhao, 2013b, Zhang, 2013; Wang, 2014; Hu, 2012; Yang, 2013 e, f) are generalizations of integral transforms based upon local fractional operators.

The main aim of this chapter is to investigate the solutions for LFDEs by using the local fractional integral transforms. The present study is structured as follows. In Section 2, the results for local fractional integrals and derivatives are formulated. The basic notions and properties for (LFITs) are presented in Section 3. Applications of the local fractional integral transforms to LFDEs are studied in Section 4. Finally, the conclusions are outlined in Section 5.

Local Fractional Derivatives and Integrals

In order to investigate the LFITs, below we introduce the local fractional derivatives and integrals.

Lemma 1 (Yang, 2011; Yang, 2012).

If (Ω,d) and (Ω',d') are metric spaces, $E\subset R$ and $f:E\to\Omega'$ satisfies

$$\rho d(x,y)\le d'(f(x),f(y))\le \tau d(x,y) \tag{1}$$

where ρ and τ are positives and finite constants, then

$$\rho^s H^s(E)\le H^s(f(E))\le \tau^s H^s(E) \tag{2}$$

for $s\ge 0$ and the s-dimensional Hausdorff measures H^s.

Suppose (Ω,d) and (Ω',d') are metric spaces. A bijection $f:(\Omega,d)\to(\Omega',d')$ is said to be a bi-Lipschitz mapping, if there are constants $\rho,\tau>0$ such that for all $x_1,x_2\in\Omega$, [see (Yang, 2012)]

$$\rho d(x_1,x_2)\le d'(f(x_1),f(x_2))\le \tau d(x_1,x_2). \tag{3}$$

Lemma 2 (Yang, 2012).

If $f:(\Omega,d)\to(\Omega',d')$ is a bi-Lipschitz mapping, then for all $A\in\Omega$ we have [20, 34]

$$\dim_H(A)=\dim_H(f(A)). \tag{4}$$

Lemma 3 (Yang, 2012).

Let F be a subset of the real line and be a fractal. If $f:(F,d)\to(\Omega',d')$ is a bi-Lipschitz mapping, then there is for constants $\rho,\tau>0$ and $F\subset\mathbb{R}$,

$$\rho^s H^s(F)\le H^s(f(F))\le \tau^s H^s(F) \tag{5}$$

such that for all $x_1,x_2\in F$,

$$\rho^\alpha |x_1-x_2|^\alpha \le |f(x_1)-f(x_2)|\le \tau^\alpha |x_1-x_2|^\alpha. \tag{6}$$

Making use of ***Lemma 2*** and ***Lemma 3***, we get

$$\dim_H(F)=\dim_H(f(F))=\alpha=s. \tag{7}$$

Lemma 4 (Yang, 2012).

Let F be a subset of the real line and be a fractal. If $f:((\zeta,\xi),d)\rightarrow((\eta,\upsilon),d^{'})$ is a bi-Lipschitz mapping, then there is for constants $\rho,\tau>0$, [20, 22]

$$\rho^{\alpha}|x_1-x_2|^{\alpha}\le|f(x_1)-f(x_2)|\le\tau^{\alpha}|\zeta-\xi|^{s}, \tag{8}$$

where $E=(\eta,\upsilon)$.

Lemma 5 (Yang, 2012).

Let F be a subset of the real line and be a fractal. If $f(\Omega)$ is a bi-Lipschitz mapping, then there are any $x_1,x_2\in\Omega\subset R$ and positive constant υ such that [20, 22]

$$|f(x_1)-f(x_2)|\le\upsilon|x_1-x_2|^{s}. \tag{9}$$

Definition 1 (Yang, 2012).

If for $|x-x_0|<\delta$, $\varepsilon,\delta>0$ with $\varepsilon,\delta\in R$ there is a local fractional continuity, namely,

$$\lim_{x\rightarrow x_0}f(x)=f(x_0), \tag{10}$$

then we have

$$|f(x)-f(x_0)|<\varepsilon^{\alpha}. \tag{11}$$

From Eq.(11) we have

$$\dim_H(F\cap(a,b))=\dim_H(C_\alpha(a,b))=\alpha. \tag{12}$$

Here, we note $C_\alpha(a,b)=\{f:f(x)$ is local fractional continuous, $x\in F\cap(a,b)\}$.

Definition 2 (Yang, 2011; Yang, 2012; Baleanu, 2014; Yang and Baleanu, 2013; Liu et al., 2013; Yang, 2014a,b; Zhao, 2013b; Zhang, 2013; Su, 2013; Wang, 2014; Hu, 2012; Yang, 2013d, e, f, g).

Let $f(x)\in C_\alpha(a,b)$. The local fractional derivative of $f(x)$ of order α at $x=x_0$ is defined as follows:

$$f^{(\alpha)}(x_0)=\frac{d^{\alpha}f(x)}{dx^{\alpha}}\Big|_{x=x_0}=\lim_{x\rightarrow x_0}\frac{\Delta^{\alpha}(f(x)-f(x_0))}{(x-x_0)^{\alpha}}, \tag{13}$$

where $\Delta^{\alpha}\left(f(x)-f(x_0)\right)\cong\Gamma(1+\alpha)\Delta\left(f(x)-f(x_0)\right)$.

For $0<\alpha\le 1$, the expression (See (Yang, 2012))

$$\Delta^{\alpha} f(x)=f^{(\alpha)}(x)(\Delta x)^{\alpha}+\lambda(\Delta x)^{\alpha} \tag{14}$$

is called the increment of $f(x)$, where Δx is increment of x and $\lambda\to 0$ as $\Delta x\to 0$.

For $0<\alpha\le 1$, the expression (Yang, 2012)

$$d^{\alpha} f=f^{(\alpha)}(x)(dx)^{\alpha} \tag{15}$$

is called the α -local fractional differential of $f(x)$.

The following formulas are valid (Yang, 2012):

$$D_x^{(\alpha)} c=0 \ (c \text{ is a constant}), \tag{16}$$

$$D_x^{(\alpha)} x^{\alpha}/\Gamma(1+\alpha)=1, \tag{17}$$

$$\frac{d^{\alpha}x^{k\alpha}}{dx^{\alpha}}=\frac{\Gamma(1+k\alpha)}{\Gamma(1+(k-1)\alpha)}x^{(k-1)\alpha}, \tag{18}$$

$$\frac{d^{\alpha}E_{\alpha}(x^{\alpha})}{dx^{\alpha}}=E_{\alpha}(x^{\alpha}), \tag{19}$$

$$\frac{d^{\alpha}E_{\alpha}(cx^{\alpha})}{dx^{\alpha}}=cE_{\alpha}(cx^{\alpha}), \tag{20}$$

$$\frac{d^{\alpha}\cos_{\alpha}(x^{\alpha})}{dx^{\alpha}}=-\sin_{\alpha}(x^{\alpha}), \tag{21}$$

$$\frac{d^{\alpha}\sin_{\alpha}(x^{\alpha})}{dx^{\alpha}}=\cos_{\alpha}(x^{\alpha}), \tag{22}$$

where

$$E_{\alpha}(x^{\alpha})=\sum_{k=0}^{\infty}\frac{x^{\alpha k}}{\Gamma(1+k\alpha)}, \tag{23}$$

$$\sin_\alpha x^\alpha = \sum_{k=0}^{\infty} (-1)^k \frac{x^{\alpha(2k+1)}}{\Gamma[1+\alpha(2k+1)]}, \tag{24}$$

$$\cos_\alpha x^\alpha = \sum_{k=0}^{\infty} (-1)^k \frac{x^{2\alpha k}}{\Gamma(1+2\alpha k)}. \tag{25}$$

Definition 3 (Yang, 2011; Yang, 2012; Liao et al., 2013; Liu et al., 2013; Yang, 2014b, Zhao, 2013b, Zhang, 2013; Wang, 2014; Hu, 2012; Yang, 2013 e, f).

Let $f(x) \in C_\alpha(a,b)$. The local fractional integral of $f(x)$ of order α in the interval $[a,b]$ is defined by

$$_aI_b^{(\alpha)} f(x) = \frac{1}{\Gamma(1+\alpha)} \int_a^b f(t)(dt)^\alpha = \frac{1}{\Gamma(1+\alpha)} \lim_{\Delta t \to 0} \sum_{j=0}^{j=N-1} f(t_j)(\Delta t_j)^\alpha, \tag{26}$$

where $\Delta t_j = t_{j+1} - t_j$, $\Delta t = \max\{\Delta t_0, \Delta t_1, \Delta t_3, ...\}$ and $[t_j, t_{j+1}]$, $j = 0,...,N-1$, $t_0 = a$, $t_N = b$, is a partition of the interval $[a,b]$.

Here, the following formulas are valid (Yang, 2012):

$$_0I_x^{(\alpha)} \frac{x^{n\alpha}}{\Gamma(1+n\alpha)} = \frac{x^{(n+1)\alpha}}{\Gamma(1+(n+1)\alpha)}, \tag{27}$$

$$_0I_x^{(\alpha)} \cos_\alpha(x^\alpha) = \sin_\alpha(x^\alpha), \tag{28}$$

$$_0I_x^{(\alpha)} \sin_\alpha(x^\alpha) = 1 - \cos_\alpha(x^\alpha), \tag{29}$$

$$_0I_x^{(\alpha)} E_\alpha(x^\alpha) = E_\alpha(x^\alpha) - 1, \tag{30}$$

$$_{-\pi}I_\pi^{(\alpha)} \sin_\alpha(nx)^\alpha = 0, \tag{31}$$

$$_{-\pi}I_\pi^{(\alpha)} \cos_\alpha(nx)^\alpha = 0. \tag{32}$$

Local Fractional Integral Transforms

In this section we introduce the local fractional Fourier series, Fourier and Laplace transforms.

Local Fractional Fourier Series

Definition 4 (Yang, 2011; Liao et al., 2013; Liu et al., 2013; Yang, 2014b, Zhao, 2013b, Zhang, 2013; Wang, 2014; Hu, 2012; Yang, 2013 f).

Local fractional trigonometric Fourier series of $f(t)$ of order is given by

$$f(t) = a_0 + \sum_{i=1}^{\infty} a_k \sin_\alpha\left(k^\alpha t^\alpha\right) + \sum_{i=1}^{\infty} b_k \cos_\alpha\left(k^\alpha t^\alpha\right) \tag{33}$$

for $x \in R$ and $0 < \alpha \leq 1$.

The local fractional Fourier coefficients of Eq. (33) can be computed as

$$\begin{cases} a_0 = \dfrac{1}{T^\alpha}\Gamma(1+\alpha)\,{}_0I_T f(t), \\ a_k = \left(\dfrac{2}{T}\right)^\alpha \Gamma(1+\alpha)\,{}_0I_T\left\{f(t)\sin_\alpha\left(k^\alpha t^\alpha\right)\right\}, \\ b_k = \left(\dfrac{2}{T}\right)^\alpha \Gamma(1+\alpha)\,{}_0I_T\left\{f(t)\cos_\alpha\left(k^\alpha t^\alpha\right)\right\}. \end{cases} \tag{34}$$

The following results are valid (Yang, 2011).

Theorem 6 (Yang, 2011).

Let $f(t)$ be 2π -periodic, bounded and local fractional integral on $[-\pi, \pi]$. Suppose that $f(t) \sim \frac{a_0}{2} + \sum_{k=1}^{\infty}\left(a_k \cos_\alpha (kt)^\alpha + b_k \sin_\alpha (kt)^\alpha\right)$, we get

$$\frac{1}{\pi^\alpha}\int_{-\pi}^{\pi} f^2(t)(dt)^\alpha = \frac{a_0^2}{2} + \sum_{k=0}^{\infty}\left(a_k^2 + b_k^2\right). \tag{35}$$

Proof. See (Yang, 2011).

Theorem 7 (Convergence theorem for local fractional Fourier series) (Yang, 2011).

Let $f(t)$ be 2π -periodic and local fractional integral on $[-\pi, \pi]$. If $f(x) \in C_\alpha(a,b)$, then the local fractional series of $f(t)$ converges to $f(t)$ at $t \in [-\pi, \pi]$, and

$$f(t) = a_0 + \sum_{k=1}^{\infty}\left(a_k \cos_\alpha (kt)^\alpha + b_k \sin_\alpha (kt)^\alpha\right), \tag{36}$$

where

$$a_n = \frac{1}{\pi^\alpha}\int_{-\pi}^{\pi} f(x)\cos_\alpha (nx)^\alpha (dt)^\alpha \tag{37}$$

and

$$b_n = \frac{1}{\pi^\alpha}\int_{-\pi}^{\pi} f(x)\sin_\alpha (nx)^\alpha (dt)^\alpha . \tag{38}$$

Proof. See (Yang, 2011).

Local Fractional Fourier Transform

Definition 5 (Yang, 2011; Yang, 2014b, Zhao, 2013b, Zhang, 2013; Wang, 2014).

Let $f(x) \in C_\alpha(-\infty,\infty)$. The local fractional Fourier transform of $f(x)$ of order α is defined as:

$$F_\alpha\{f(x)\} = f_\omega^{F,\alpha}(\omega) = \frac{1}{\Gamma(1+\alpha)}\int_{-\infty}^{\infty} E_\alpha\left(-i^\alpha \omega^\alpha x^\alpha\right) f(x)(dx)^\alpha . \tag{39}$$

Let $f(x) \in C_\alpha(-\infty,\infty)$. If $F_\alpha\{f(x)\} \equiv f_\omega^{F,\alpha}(\omega)$, its inversion formula of Eq.(39) reads as follows:

$$f(x) = F_\alpha^{-1}\left(f_\omega^{F,\alpha}(\omega)\right) := \frac{1}{(2\pi)^\alpha}\int_{-\infty}^{\infty} E_\alpha\left(i^\alpha \omega^\alpha x^\alpha\right) f_\omega^{F,\alpha}(\omega)(d\omega)^\alpha . \tag{40}$$

The following results are valid (Yang, 2011).

Theorem 8 (Yang, 2011).

Let $F_\alpha\{f(x)\} = f_\omega^{F,\alpha}(\omega)$ and $F_\alpha\{g(x)\} = g_\omega^{F,\alpha}(\omega)$, and let a,b be two constants. if $f(x), g(x) \in C_\alpha(-\infty,\infty)$, then we obtain

$$F_\alpha\{af(x)+bg(x)\} = aF_\alpha\{f(x)\} + bF_\alpha\{g(x)\}. \tag{41}$$

Proof. See (Yang, 2011).

Theorem 9 (Yang, 2011).

Let $F_\alpha\{f(x)\} = f_\omega^{F,\alpha}(\omega)$. If $\lim_{|x|\to\infty} f(x) = 0$ and $f(x) \in C_\alpha(-\infty,\infty)$, then we have

$$F_\alpha\left\{f^{(\alpha)}(x)\right\} = i^\alpha \omega^\alpha F_\alpha\left\{f(x)\right\}. \tag{42}$$

Proof. See (Yang, 2011).

Local Fractional Laplace Transform

Definition 6 (Yang, 2011; Liu et al., 2013; Zhao, 2014b).

Let $f(x) \in C_\alpha(-\infty, \infty)$. The local fractional Laplace transform of $f(x)$ of order α is defined as:

$$L_\alpha\left\{f(x)\right\} = f_s^{L,\alpha}(s) = \frac{1}{\Gamma(1+\alpha)} \int_0^\infty E_\alpha\left(-s^\alpha x^\alpha\right) f(x)(dx)^\alpha. \tag{43}$$

Let $f(x) \in C_\alpha(-\infty, \infty)$. If $F_\alpha\left\{f(x)\right\} \equiv f_\omega^{F,\alpha}(\omega)$, the inverse formula of Eq.(43) is expressed as follows:

$$f(x) = L_\alpha^{-1}\left\{f_s^{L,\alpha}(s)\right\} = \frac{1}{(2\pi)^\alpha} \int_{\beta-i\infty}^{\beta+i\infty} E_\alpha\left(s^\alpha x^\alpha\right) f_s^{L,\alpha}(s)(ds)^\alpha, \tag{44}$$

where $f(x)$ is local fractional continuous, $s^\alpha = \beta^\alpha + i^\alpha \infty^\alpha$ and $\mathrm{Re}(s) = \beta > 0$.

The following results are valid (Yang, 2011).

Theorem 10 (Yang, 2011).

Let $L_\alpha\left\{f(x)\right\} = f_s^{L,\alpha}(s)$ and $L_\alpha\left\{g(x)\right\} = g_s^{L,\alpha}(s)$, and let a, b be two constants. If $f(x), g(x) \in C_\alpha(-\infty, \infty)$, then we obtain

$$F_\alpha\left\{af(x) + bg(x)\right\} = aF_\alpha\left\{f(x)\right\} + bF_\alpha\left\{g(x)\right\}. \tag{45}$$

Proof. See (Yang, 2011).

Theorem 11 (Yang, 2011).

Let $L_\alpha\left\{f(x)\right\} = f_s^{L,\alpha}(s)$. If $\lim_{|x|\to\infty} f(x) = 0$ and $f(x) \in C_\alpha(-\infty, \infty)$, then we have

$$L_\alpha\left\{f^{(\alpha)}(x)\right\} = s^\alpha L_\alpha\left\{f(x)\right\}. \tag{46}$$

Proof. See (Yang, 2011).

Applications

In this section, we present the applications of local fractional intergral transforms to local fractional ODEs and PDEs.

Local Fractional Fourier Series for Solving Local Fractional ODE and PDE

We give now some examples for the LFODE and LFPDE by using the local fractional Fourier series.

Example 1. We now present the local fractional ODE in the following form:

$$\frac{d^{2\alpha}y}{d^{2\alpha}x}+y=\frac{x^{\alpha}}{\Gamma(1+\alpha)},\ x\in(0,\pi) \tag{47}$$

subject to initial value conditions

$$y(0)=0,\ \ y(\pi)=0. \tag{48}$$

In view of Eqs.(36-38), the function $x^{\alpha}/\Gamma(1+\alpha)$ can be written as

$$\frac{x^{\alpha}}{\Gamma(1+\alpha)}=\sum_{n=1}^{\infty}b_n\sin_{\alpha}\left(n^{\alpha}x^{\alpha}\right),\ x\in(0,\ \pi), \tag{49}$$

where

$$b_n=\frac{1}{\pi^{\alpha}}\int_0^{2\pi}\frac{x^{\alpha}}{\Gamma(1+\alpha)}\sin_{\alpha}\left(n^{\alpha}x^{\alpha}\right)(dx)^{\alpha}=-\frac{1}{n^{\alpha}}\frac{(2\pi)^{\alpha}}{\Gamma(1+\alpha)}. \tag{50}$$

We also write $y(x)$ as

$$y(x)=\sum_{n=1}^{\infty}B_n\sin_{\alpha}\left(n^{\alpha}x^{\alpha}\right),\ x\in(0,\ \pi), \tag{51}$$

which leads to

$$\frac{d^{2\alpha}y}{d^{2\alpha}x}=-\sum_{n=1}^{\infty}B_n n^{2\alpha}\sin_{\alpha}\left(n^{\alpha}x^{\alpha}\right),\ x\in(0,\ \pi). \tag{52}$$

Submitting Eqs.(49-52) into Eq.(47) , we obtain

$$-\sum_{n=1}^{\infty} B_n \pi^{2\alpha} \sin_\alpha\left(n^\alpha x^\alpha\right)+\sum_{n=1}^{\infty} B_n \sin_\alpha\left(n^\alpha x^\alpha\right)=-\sum_{n=1}^{\infty} \frac{1}{n^\alpha} \frac{(2\pi)^\alpha}{\Gamma(1+\alpha)} \sin_\alpha\left(n^\alpha x^\alpha\right). \quad (53)$$

Matching the coefficients for the same terms in two series, from Eq.(53) we conclude:

$$\left(1-n^{2\alpha}\right) B_n=-\frac{1}{n^\alpha} \frac{(2\pi)^\alpha}{\Gamma(1+\alpha)}, \quad \text{n=1, 2, 3,....} \quad (54)$$

In view of Eq.(54) , we have the following formula

$$B_n=\frac{(2\pi)^\alpha}{\Gamma(1+\alpha) n^\alpha\left(n^{2\alpha}-1\right)} \quad (55)$$

so that

$$y(x)=\sum_{n=1}^{\infty} \frac{(2\pi)^\alpha}{\Gamma(1+\alpha) n^\alpha\left(\pi^{2\alpha}-1\right)} \sin_\alpha\left(n^\alpha \pi^\alpha\right). \quad (56)$$

Example 2. The local fractional Laplace equation is given by

$$\frac{\partial^{2\alpha} u(x, y)}{\partial y^{2\alpha}}+\frac{\partial^{2\alpha} u(x, y)}{\partial x^{2\alpha}}=0 \quad (57)$$

with the fractal boundary conditions

$$u(x, 0)=E_\alpha\left(x^\alpha\right), \quad (58)$$

$$u(0, y)=u(L, y)=0, \quad (59)$$

$$\frac{\partial^\alpha u(L, 0)}{\partial x^\alpha}=E_\alpha\left(x^\alpha\right), \quad (60)$$

$$\frac{\partial^\alpha u(L, y)}{\partial x^\alpha}=0. \quad (61)$$

Now we can structure particular solutions in the following form

$$u(x, y)=XY \quad (62)$$

and we have

$$X^{(2\alpha)}Y=-XY^{(2\alpha)}. \tag{63}$$

Separating the variables, it yields

$$\frac{X^{(2\alpha)}}{Y}=-\frac{X^{(2\alpha)}}{X}=-\lambda^{2\alpha}. \tag{64}$$

Therefore, we have the following formulas

$$X^{(2\alpha)}+\lambda^{2\alpha}X=0, \tag{65}$$

$$Y^{(2\alpha)}-\lambda^{2\alpha}Y=0 \tag{66}$$

with the boundary conditions

$$u(0)=u^{(\alpha)}(L)=0. \tag{67}$$

Hence, the non-differentiable solution of Eq. (65) becomes:

$$u(x)=C_1\cos_\alpha\lambda^\alpha x^\alpha+C_2\sin_\alpha\lambda^\alpha x^\alpha\ \left(C_1=\cos t,C_2=\cos t\right). \tag{68}$$

For $x=0$ and $x=L$ we deduce that

$$u(0)=C_1=0, \tag{69}$$

and

$$u(l)=u(x)\big|_{x=L}=C_2\sin_\alpha\lambda^\alpha L^\alpha=0. \tag{70}$$

Assuming that $C_2\neq 0$, we observe that

$$\lambda_n^{\ \alpha}L^\alpha=n^\alpha\pi^\alpha, \tag{71}$$

where n is an integer.

Therefore, we get the following formulas

$$\lambda_n^\alpha=\left(\frac{n\pi}{L}\right)^\alpha,\left(n\in Z^+\cup 0\right), \tag{72}$$

$$X_n(x) = \sin_\alpha \lambda_n^\alpha x^\alpha = \sin_\alpha n^\alpha \left(\frac{\pi x}{L}\right)^\alpha = 0, \left(n \in Z^+ \cup 0\right). \tag{73}$$

For $\lambda^\alpha = \lambda_n^\alpha$ Eq. (66) yields

$$Y_n(y) = A_n \cos_\alpha \lambda_n^\alpha y^\alpha + B_n \sin_\alpha \lambda_n^\alpha y^\alpha, \left(n \in Z^+ \cup 0\right) \tag{74}$$

and therefore

$$u_n(x,y) = \left(A_n \cos_\alpha \left(\frac{\pi x}{L}\right)^\alpha y^\alpha + B_n \sin_\alpha \left(\frac{\pi x}{L}\right)^\alpha y^\alpha\right) \sin_\alpha n^\alpha \left(\frac{\pi x}{L}\right)^\alpha, \left(n \in Z^+ \cup 0\right). \tag{75}$$

By applying the local fractional Fourier series, we conclude that

$$u(x,y) = \sum_{n=1}^{\infty} u_n(x,y) = \sum_{n=1}^{\infty} \left(A_n \cos_\alpha \left(\frac{\pi x}{L}\right)^\alpha y^\alpha + B_n \sin_\alpha \left(\frac{\pi x}{L}\right)^\alpha y^\alpha\right) \sin_\alpha n^\alpha \left(\frac{\pi x}{L}\right)^\alpha. \tag{76}$$

We now consider that

$$\begin{aligned} u(x,0) &= \sum_{n=1}^{\infty} u_n(x,0) = \sum_{n=1}^{\infty} A_n \sin_\alpha n \left(\frac{\pi x}{L}\right)^\alpha \\ &= E_\alpha\left(x^\alpha\right) \end{aligned} \tag{77}$$

and

$$\begin{aligned} &\frac{\partial^\alpha u(L,0)}{\partial x^\alpha} = \\ &\sum_{n=1}^{\infty} \left(-A_n \left(\frac{\pi x}{L}\right)^\alpha \sin_\alpha \left(\frac{\pi x}{L}\right)^\alpha y^\alpha + B_n \left(\frac{\pi x}{L}\right)^\alpha \cos_\alpha \left(\frac{\pi x}{L}\right)^\alpha y^\alpha\right) \sin_\alpha n^\alpha \left(\frac{\pi x}{L}\right)^\alpha \Big|_{y=0} \\ &= \sum_{n=1}^{\infty} B_n \left(\frac{\pi x}{L}\right)^\alpha \sin_\alpha n^\alpha \left(\frac{\pi x}{L}\right)^\alpha = E_\alpha\left(x^\alpha\right) \end{aligned} \tag{78}$$

By applying the local fractional Fourier series, we conclude that the local fractional Fourier series yields the its coefficients, which read

$$\Gamma(1+\alpha)\, {}_0I_L \left\{\sin_\alpha^2 n^\alpha \left(\frac{\pi x}{L}\right)^\alpha\right\} = \frac{L^\alpha}{2} \left(n \in Z^+ \cup 0\right), \tag{79}$$

$$A_n = \frac{2\Gamma(1+\alpha)}{L^\alpha} {}_0I_L \left(E_\alpha(x^\alpha) \sin_\alpha \left[\left(\frac{n\pi}{L} \right)^\alpha x^\alpha \right] \right) (n \in Z^+ \cup 0), \tag{80}$$

$$B_n = \frac{2\Gamma(1+\alpha)}{a^\alpha \lambda_n^\alpha L^\alpha} {}_0I_L \left(E_\alpha(x^\alpha) \sin_\alpha \left[\left(\frac{n\pi}{L} \right)^\alpha x^\alpha \right] \right) (n \in Z^+ \cup 0). \tag{81}$$

In view of Eqs. (68) and (76), we have the following form

$$u_n(x,y) = \left\{ A_n \cos_\alpha \left[\left(\frac{n\pi}{L} \right)^\alpha y^\alpha \right] + B_n \sin_\alpha \left[\left(\frac{n\pi}{L} \right)^\alpha y^\alpha \right] \right\} \sin_\alpha \left[\left(\frac{n\pi}{L} \right)^\alpha x^\alpha \right], \tag{82}$$

where

$$\begin{aligned} A_n &= \frac{2\Gamma(1+\alpha)}{L^\alpha} {}_0I_L \left(E_\alpha(x^\alpha) \sin_\alpha \left[\left(\frac{n\pi}{L} \right)^\alpha x^\alpha \right] \right) \\ &= \frac{2\Gamma(1+\alpha)}{L^\alpha} \frac{E_\alpha(x^\alpha) \left[\sin_\alpha \left(\left(\frac{n\pi}{L} \right)^\alpha x^\alpha \right) - \left(\frac{n\pi}{L} \right)^\alpha \cos_\alpha \left(\left(\frac{n\pi}{L} \right)^\alpha x^\alpha \right) \right] + \left(\frac{n\pi}{L} \right)^\alpha}{1 + \left(\frac{n\pi}{L} \right)^{2\alpha}} \end{aligned} \tag{83}$$

and

$$\begin{aligned} B_n &= \frac{(n\pi)^\alpha \Gamma(1+\alpha)}{2} {}_0I_L \left(E_\alpha(x^\alpha) \sin_\alpha \left[\left(\frac{n\pi}{L} \right)^\alpha x^\alpha \right] \right) \\ &= \frac{(n\pi)^\alpha \Gamma(1+\alpha)}{2} \frac{E_\alpha(x^\alpha) \left[\sin_\alpha \left(\left(\frac{n\pi}{L} \right)^\alpha x^\alpha \right) - \left(\frac{n\pi}{L} \right)^\alpha \cos_\alpha \left(\left(\frac{n\pi}{L} \right)^\alpha x^\alpha \right) \right] + \left(\frac{n\pi}{L} \right)^\alpha}{1 + \left(\frac{n\pi}{L} \right)^{2\alpha}} \\ &\qquad (n \in Z^+ \cup 0) \end{aligned} \tag{84}$$

Finally, we obtain

$$u_n(x,y)=\left(A_n\cos_\alpha\left(\frac{n\pi}{L}\right)^\alpha y^\alpha+B_n\sin_\alpha\left(\frac{n\pi}{L}\right)^\alpha y^\alpha\right)\sin_\alpha\left(\frac{n\pi}{L}\right)^\alpha x^\alpha$$

$$=\frac{\Gamma(1+\alpha)E_\alpha(x^\alpha)\left[\sin_\alpha\left(\left(\frac{n\pi}{L}\right)^\alpha x^\alpha\right)-\left(\frac{n\pi}{L}\right)^\alpha\cos_\alpha\left(\left(\frac{n\pi}{L}\right)^\alpha x^\alpha\right)\right]+\left(\frac{n\pi}{L}\right)^\alpha}{1+\left(\frac{n\pi}{L}\right)^{2\alpha}}. \tag{85}$$

$$\sin_\alpha\left(\left(\frac{n\pi}{L}\right)^\alpha x^\alpha\right)\left[\frac{2}{L^\alpha}\cos_\alpha\left(\frac{n\pi}{L}\right)^\alpha+\frac{(n\pi)^\alpha}{2}\sin_\alpha\left(\frac{n\pi}{L}\right)^\alpha y^\alpha\right],\qquad \left(n\in Z^+\cup 0\right)$$

In view of Eq.(85), we have the local fractional Fourier series solution as

$$u(x,y)$$

$$=\sum_{n=1}^{\infty}\frac{E_\alpha(x^\alpha)\left[\sin_\alpha\left(\left(\frac{n\pi}{L}\right)^\alpha x^\alpha\right)-\left(\frac{n\pi}{L}\right)^\alpha\cos_\alpha\left(\left(\frac{n\pi}{L}\right)^\alpha x^\alpha\right)\right]+\left(\frac{n\pi}{L}\right)^\alpha}{1+\left(\frac{n\pi}{L}\right)^{2\alpha}} \tag{87}$$

$$\Gamma(1+\alpha)\sin_\alpha\left(\left(\frac{n\pi}{L}\right)^\alpha x^\alpha\right)\left[\frac{2}{L^\alpha}\cos_\alpha\left(\frac{n\pi}{L}\right)^\alpha+\frac{(n\pi)^\alpha}{2}\sin_\alpha\left(\frac{n\pi}{L}\right)^\alpha y^\alpha\right]$$

Local Fractional Fourier Transform for Solving Local Fractional ODE and PDE

Below we show some examples for the LFODE and LFPDE by using the local fractional Fourier transforms.

Example 3. Let us consider the following local fractional ordinary differential equation as

$$\frac{d^\alpha u(t)}{dt^\alpha}+u(t)=\delta_\alpha(t),\ t>0, \tag{88}$$

where

$${}_{-\infty}I_{\infty}^{(\alpha)}\delta_\alpha(t)=1 \tag{89}$$

and

$$u(0)=1. \tag{90}$$

In view of Eqs.(42) and (88), we get

$$u_{\omega}^{F,\alpha}(\omega)=\frac{1}{i^{\alpha}\omega^{\alpha}+1}. \tag{91}$$

Taking the inverse version of the local fractional Fourier transform, from Eq.(91) we obtain

$$u(t)=E_{\alpha}\left(-t^{\alpha}\right). \tag{92}$$

The chart of Eq.(92) for $\alpha=\ln 2/\ln 3$ is shown in Figure 1.

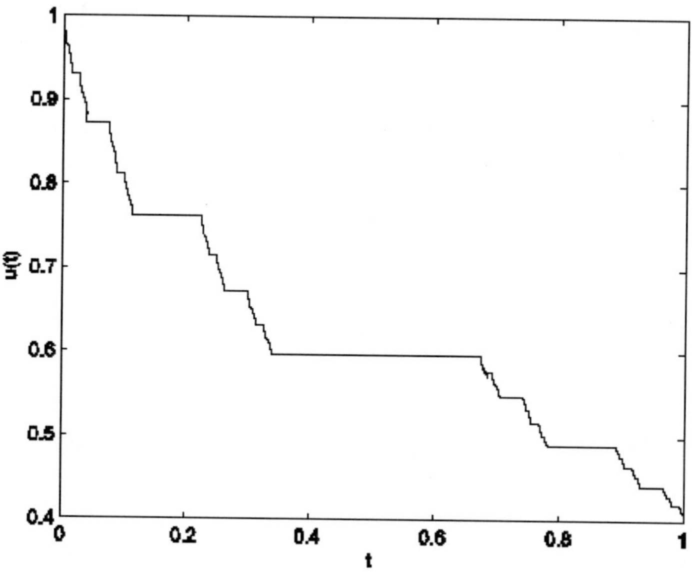

Figure 1. Graph of $u(t)$ for $\alpha=\ln 2/\ln 3$.

Example 4. The local fractional partial differential equation is given in the form

$$\frac{\partial^{2\alpha}}{\partial t^{\alpha}}u(x,t)+\frac{\partial^{2\alpha}}{\partial x^{\alpha}}u(x,t)=0,\ t>0, \tag{93}$$

subject to the initial value condition

$$u(x,0)=E_{\alpha}\left(-x^{\alpha}\right), \tag{94}$$

where

$${}_{-\infty}I_{\infty}^{(\alpha)}\delta_{\alpha}(t)=1. \tag{95}$$

Here, from Eqs.(42) and (93) we have

$$\frac{\partial^{2\alpha}}{\partial t^{\alpha}} u_{\omega}^{F,\alpha}(\omega,t) + i^{\alpha}\omega^{\alpha} u_{\omega}^{F,\alpha}(\omega,t) = 0, \tag{96}$$

and

$$u_{\omega}^{F,\alpha}(\omega,0) = \frac{1}{1+i^{\alpha}\omega^{\alpha}}. \tag{97}$$

Making use of Eqs.(96) and (97), we get

$$u_{\omega}^{F,\alpha}(\omega,t) = \frac{1}{1+i^{\alpha}\omega^{\alpha}} E_{\alpha}\left(-i^{\alpha}\omega^{\alpha}t^{\alpha}\right). \tag{98}$$

Taking the inverse version of the local fractional Fourier transform, from Eq.(98) we obtain

$$u(x,t) = E_{\alpha}\left(-x^{\alpha}\right) E_{\alpha}\left(t^{\alpha}\right). \tag{99}$$

The chart of Eq.(99) for $\alpha = \ln 2 / \ln 3$ is shown in Figure 2.

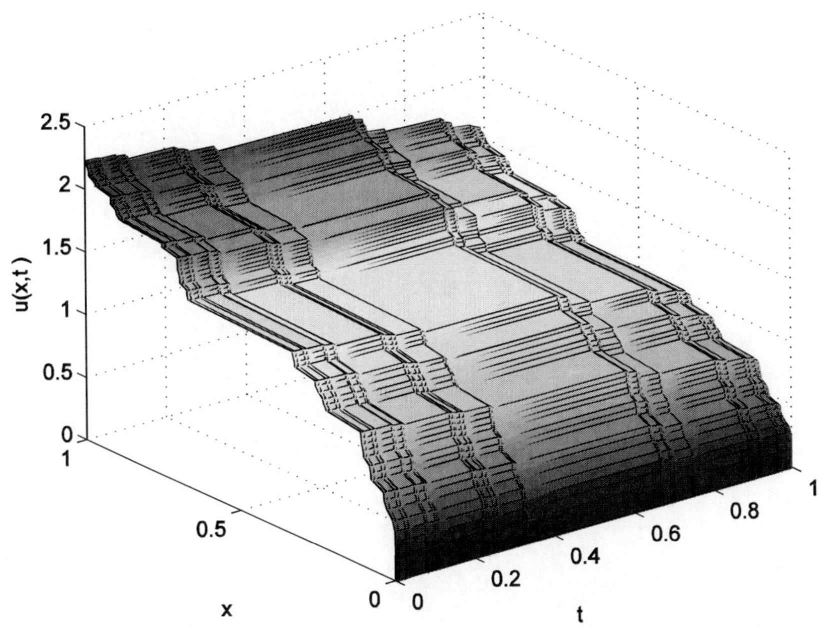

Figure 2. Graph of $u(x,t)$ for $\alpha = \ln 2 / \ln 3$.

Local Fractional Laplace Transform for Solving Local Fractional ODE and PDE

Below we invesigate some examples for the LFODE and LFPDE by using the local fractional Laplace transforms.

Example 5. We now consider the local fractional differential equation

$$\frac{d^\alpha u(t)}{dt^\alpha}-u(t)=1,\ t>0 \tag{100}$$

subject to the initial value

$$u(0)=1. \tag{101}$$

In view of Eq.(46), from Eq.(100) we have the following form

$$s^\alpha u_s^{L,\alpha}(s)-x(0)-u_s^{L,\alpha}(s)=\frac{1}{s^\alpha}. \tag{102}$$

From Eq.(102), we obtain

$$u_s^{L,\alpha}(s)=\frac{1}{s^\alpha-1}+\frac{1}{s^\alpha(s^\alpha-1)}. \tag{103}$$

Therefore, making use of Eq.(44), the solution of Eq.(100) can be written as

$$u(t)=2E_\alpha(t^\alpha)-1. \tag{104}$$

The chart of Eq.(104) for $\alpha=\ln 2/\ln 3$ is shown in Figure 3.

Example 6. Let us consider the following inhomogeneous local fractional partial differential equation as

$$\frac{\partial^{2\alpha}}{\partial t^\alpha \partial x^\alpha}u(x,t)=-\sin_\alpha(t^\alpha),\ t>0. \tag{105}$$

The initial boundary conditions are as follows

$$u(x,0)=\frac{x^\alpha}{\Gamma(1+\alpha)}, \tag{106}$$

$$u(0,t)=0. \tag{107}$$

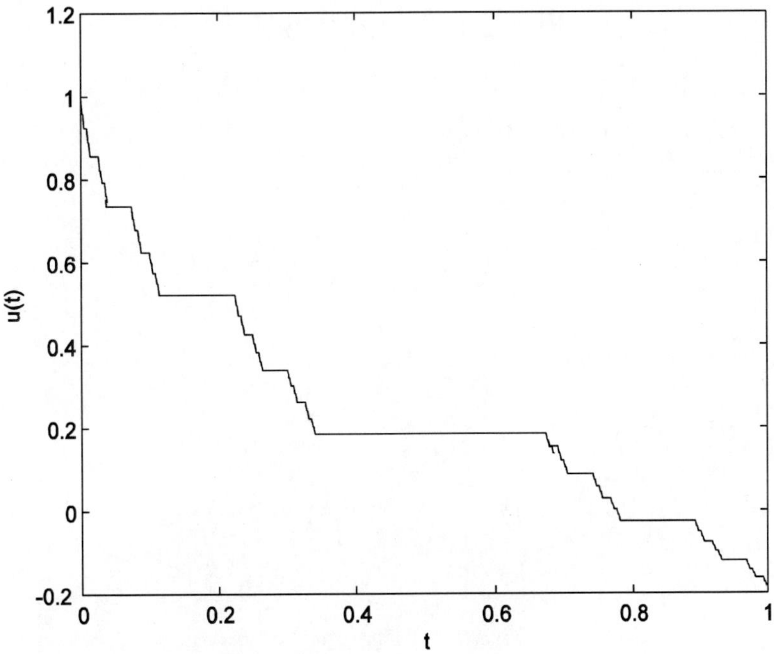

Figure 3. Graph of $u(t)$ for $\alpha = \ln 2/\ln 3$.

Making use of Eq.(46), we get

$$\frac{d^{\alpha} u_s^{L,\alpha}(x,s)}{dx^{\alpha}} = \frac{s^{\alpha}}{s^{2\alpha}+1}, \tag{108}$$

$$u_s^{L,\alpha}(0,s)=0. \tag{109}$$

Hence, we have

$$u_s^{L,\alpha}(x,s) = \frac{x^{\alpha}}{\Gamma(1+\alpha)} \frac{s^{\alpha}}{s^{2\alpha}+1} + A, \tag{110}$$

where A is a constant.

In view of Eq.(109), Eq.(110) becomes

$$u_s^{L,\alpha}(x,s) = \frac{x^{\alpha}}{\Gamma(1+\alpha)} \frac{s^{\alpha}}{s^{2\alpha}+1}. \tag{111}$$

Therefore, the non-differentiable solution of Eq.(105) reads as follows:

$$u(x,t)=\frac{x^{\alpha}}{\Gamma(1+\alpha)}\cos_{\alpha}\left(t^{\alpha}\right). \tag{112}$$

The chart of Eq.(112) for $\alpha=\ln 2/\ln 3$ is shown in Figure 4.

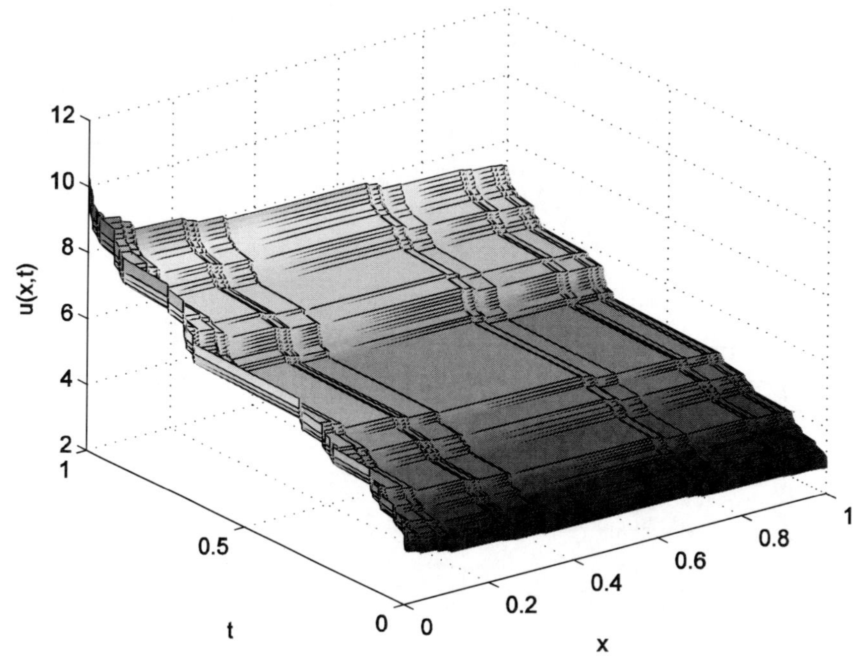

Figure 4. Graph of $u(x,t)$ for $\alpha=\ln 2/\ln 3$.

Conclusion

In this work we investigated the LFITs within the local fractional Fourier series, and local fractional Fourier and Laplace transforms. We used the LFITs to solve the local fractional ODEs and PDEs. The non-differentiable solutions were obtained. The results show that they are the powerful tools to solve LFDEs.

References

A. A. Kilbas, H. M. Srivastava, and J. J. Trujillo, Theory and Applications of Fractional Differential Equations, Elsevier, Amsterdam, The Netherlands, 2006.

Adda, F. B., Cresson, J., (2001). About non-differentiable functions. *Journal of Mathematical Analysis and Applications* 263(2), 721-737.

Babakhani, A., Gejji, V. D., (2002). On calculus of local fractional derivatives. *Journal of Mathematical Analysis and Applications* 270(1), 66-79.

Baleanu, D., Machado, J. A. T., Cattani, C., Baleanu, M.C., Yang, X. J., (2014). Local fractional variational iteration and decomposition methods for wave equation on Cantor sets within local fractional operators. *Abstract and Applied Analysis* 2014, 535048.

Bracewell, R. N., (1986). The Fourier transform and its applications. New York: McGraw-Hill.

Brockwell, P. J., Davis, R. A., (2009). Time Series: Theory and Methods. New York: Springer.

Brown, J. W., Churchill, R. V., (1993). Fourier series and boundary value problems. New York: McGraw-Hill.

Carpinteri, A., Cornetti, P., (2002). A fractional calculus approach to the description of stress and strain localization in fractal media. *Chaos, Solitons & Fractals* 13(1), 85-94.

Carpinteri, A., Cornetti, P., Kolwankar, K. M., (2004). Calculation of the tensile and flexural strength of disordered materials using fractional calculus. *Chaos, Solitons & Fractals* 21(3), 623-632.

Chen, Y., Yan, Y., Zhang, K., (2010). On the local fractional derivative. Journal of Mathematical Analysis and Applications, 362(1), 17-33.

Cherubini, U., Della Lunga, G., Mulinacci, S., Rossi, P., (2010). Fourier Transform Methods in Finance, Wiley.

D. Baleanu, J. A. Tenreiro Machado, A. C. J. Luo, Fractional Dynamics and Control, Springer, New York, 2011.

D. Baleanu, K. Diethelm, E. Scalas and J.J. Trujillo, Fractional calculus models and numerical methods, Series on Complexity, Nonlinearity and Chaos, World Scientific, Boston, 2012.

Davies, B., (2002). Integral Transforms and Their Applications. New York: Springer.

Dehghan, M., Manafian, J., Saadatmandi, A., (2010). Solving nonlinear fractional partial differential equations using the homotopy analysis method.*Numerical Methods for Partial Differential Equations* 26(2), 448-479.

F. Mainardi, Fractional Calculus and Waves in Linear Viscoelasticity, Imperial College Press, London, UK, 2010.

Golmankhaneh, A. K., Golmankhaneh, A. K., Baleanu, D., (2011). On nonlinear fractional Klein–Gordon equation. *Signal Processing* 91(3), 446-451.

Grubbström, R. W. (1967). On the application of the Laplace transform to certain economic problems, *Management Science* 13(7), 558-567.

Hao, Y. J., Srivastava, H. M., Jafari, H. Yang, X. J., (2013). Helmholtz and diffusion equations associated with local fractional derivative operators involving the Cantorian and Cantor type cylindrical coordinates. *Advances in Mathematical Physics*, 2013,754248.

Hristov, J., (2010). Heat-balance integral to fractional (half-time) heat diffusion sub-model. Thermal Science, 14(2), 291-316.

Hu, M. S., Agarwal, R. P., Yang, X. J., (2012). Local fractional Fourier series with application to wave equation in fractal vibrating string. *Abstract and Applied Analysis* 2012, 567401.

I. Podlubny, Fractional Differential Equations, Academic Press, New York, NY, USA, 1999.

J. Klafter, S. C. Lim, and R. Metzler, Fractional Dynamics in Physics: Recent Advances, World Scientific, Singapore, 2012.

Jafari, H., Seifi, S., (2009). Solving a system of nonlinear fractional partial differential equations using homotopy analysis method. *Communications in Nonlinear Science and Numerical Simulation* 14(5), 1962-1969.

Kolwankar, K. M., Gangal, A. D., (1998). Local fractional Fokker-Planck equation. *Physical Review Letters* 80(2), 214.

Komornik, V., Loreti, P., (2005). Fourier Series in Control Theory. New York: Springer.

Lathi, B. P., (1998). Signal Processing and Linear Systems. Berkeley: Cambridge Press.

Liao, M. K., Yang, X. J., Yan, Q., (2013). A new viewpoint to Fourier analysis in fractal space, G. A. Anastassiou, O. Duman (Eds), Advances in Applied Mathematics and Approximation Theory, Chapter 26, pp 397-409, Springer, New York.

Liu, C. F., Kong, S. S., Yuan, S. J., (2013). Reconstructive schemes for variational iteration method within Yang-Laplace transform with application to fractal heat conduction problem. *Thermal Science* 17(3), 715-721.

Ma, X. J., Srivastava, H. M., Baleanu, D., Yang, X. J., A new Neumann series method for solving a family of local fractional Fredholm and Volterra integral equations. *Mathematical Problems in Engineering* 2013, 325121.

Meerschaert, M. M., Tadjeran, C., (2004). Finite difference approximations for fractional advection–dispersion flow equations. *Journal of Computational and Applied Mathematics* 172(1), 65-77.

Momani, S., (2006). Non-perturbative analytical solutions of the space-and time-fractional Burgers equations. *Chaos, Solitons & Fractals* 28(4), 930-937.

Momani, S., Odibat, Z., (2006). Analytical solution of a time-fractional Navier–Stokes equation by Adomian decomposition method. *Applied Mathematics and Computation* 177(2), 488-494.

N. Laskin, Fractional Schrödinger equation, Physical Review E, vol.66, no.5, 056108, 2002.

R. L. Magin, Fractional Calculus and Bioengineering, Begell House, New York, NY, USA, 2006.

Schiff, J. L., (1999). The Laplace transform: Theory and Applications. New York: Springer.

Schnars, U., Jueptner, W., (2005). Digital Holography. Berlin: Springer.

Su, W. H., Baleanu, D., Yang, X. J., Jafari, H., (2013). Damped wave equation and dissipative wave equation in fractal strings within the local fractional variational iteration method. *Fixed Point Theory and Applications* 2013(1), 1-11.

Tranter, C. J., (1966). Integral Transforms in Mathematical Physics. New York: Methuen.

V. E. Tarasov, Fractional Dynamics: Applications of Fractional Calculus to Dynamics of Particles, Fields and Media, Springer, Berlin, 2011.

Wang, S. Q., Yang, Y. J., Jassim, H. K., (2014). Local fractional function decomposition method for solving inhomogeneous wave equations with local fractional derivative. *Abstract and Applied Analysis* 2014, 176395.

West, B., Bologna, M., Grigolini, P., (2003). Physics of Fractal Operators, Springer, New York.

Yang, A. M., Chen, Z. S., Srivastava, H. M., Yang, X. -J., (2013g). Application of the local fractional series expansion method and the variational iteration method to the Helmholtz equation involving local fractional derivative operators, *Abstract and Applied Analysis* 2013, 259125.

Yang, X. J., (2012). Advanced Local Fractional Calculus and Its Applications. World Science, New York, NY, USA.

Yang, X. J., Baleanu, D., (2013). Fractal heat conduction problem solved by local fractional variation iteration method. *Thermal Science* 17(2), 625-628.

Yang, X. -J., Baleanu, D., Machado, J. A. T., (2013e). Mathematical aspects of Heisenberg uncertainty principle within local fractional Fourier analysis, *Boundary Value Problems* 2013:131.

Yang, X. J., (2011). Local Fractional Functional Analysis and Its Applications, Asian Academic Publisher Limited, HongKong.

Yang, X.-J., Baleanu, D., He, J. H., (2013c). Transport equations in fractal porous media within fractional complex transform method. *Proceedings of the Romanian Academy, Series A* 14(4), 287-292.

Yang, X.-J., Baleanu, D., Khan, Y., Mohyud-Din, S.T., (2014a). Local fractional variational iteration method for diffusion and wave equations on Cantor sets. *Romanian Journal of Physics* 59(1-2), 36-48.

Yang, X.-J., Baleanu, D., Machado, J. A. T., (2013b). Systems of Navier-Stokes equations on Cantor sets. *Mathematical Problems in Engineering* 2013, 769724.

Yang, X.-J., Baleanu, D., Machado, J. A. T., (2014b). Application of the local fractional Fourier series to fractal signals, In: J. A. Tenreiro Machado, D. Baleanu, Luo, Albert C. J. , Discontinuity and Complexity in Nonlinear Physical Systems, Springer, New York.

Yang, X.-J., Baleanu, D., Zhong, W. P., (2013d). Approximate solutions for diffusion equations on Cantor space-time. *Proceedings of the Romanian Academy, Series A* 14(2), 127-133.

Yang, X.-J., Srivastava, H. M., He, J. -H., Baleanu, D., (2013a). Cantor-type cylindrical-coordinate method for differential equations with local fractional derivatives. *Physics Letters A* 377(28–30), 1696–1700.

Yang, Y.-J., Baleanu, D., Yang, X. -J., (2013f). Analysis of fractal wave equations by local fractional Fourier series method, *Advances in Mathematical Physics* 2013, 632309.

Zayed, A. I., (2002). A class of fractional integral transforms: a generalization of the fractional Fourier transform. *Signal Processing, IEEE Transactions on* 50(3), 619-627.

Zhang, Y., Yang, A., Yang, X. J., (2013). 1-D heat conduction in a fractal medium: A solution by the local fractional Fourier series method. *Thermal Science* 17(3), 953-956.

Zhao, Y., Baleanu, D., Baleanu, M. C., Cheng, D. F., Yang, X.-J., (2013b). Mappings for special functions on Cantor sets and special integral transforms via local fractional operators. *Abstract and Applied Analysis* 2013, 316978.

Zhao, Y., Baleanu, D., Cattani, C., Cheng, D. F., Yang, X. -J., (2013a). Maxwell's equations on Cantor sets: A local fractional approach. *Advances in High Energy Physics* 2013, 686371.

Zhao, Z., Li, C., (2012). Fractional difference/finite element approximations for the time–space fractional telegraph equation. *Applied Mathematics and Computation* 219(6), 2975–2988.

Zhou, Y., Jiao, F., (2010). Nonlocal Cauchy problem for fractional evolution equations. *Nonlinear Analysis: Real World Applications* 11(5), 4465-4475.

In: Fractional Calculus: Theory
Editors: R. Abi Zeid Daou and X. Moreau
ISBN: 978-1-63463-002-3

Chapter 5

EXTENDED BOREL TRANSFORM AND FRACTIONAL CALCULUS

Akira Asada
Sinsyu University, Japan

Abstract

We have introduced an integral transform of $\mathcal{R}$;
$\mathcal{R}[f(s)](x) = \int_{-\infty}^{\infty} \frac{x^s}{\Gamma(1+s)} f(s)ds$ that intertwines fractional order differentiation with the parallel transform of a variable:

$$\frac{d^a}{dx^a}\mathcal{R}[f(s)](x) = \mathcal{R}[f(s+a)](x).$$

It is recognized that the suitable choice of domains for $\mathcal{R}$ are spaces of discrete support distributions or ultra distributions. In these cases, $\mathcal{R}$ is decomposed as $\mathcal{B} \circ \mathcal{N}$, where $\mathcal{B}$ is the extended Borel transform and $\mathcal{N}$ is closely related to the Mellin transform.

In this paper, we explain these relations, mainly focusing on the interpretation of an extended Borel transform that is the extension of the Borel transform to functions expressed by power series of $\log x$. Then we will discuss the applications of $\mathcal{R}$ to fractional differential equations.

Keywords: Integral transform $\mathcal{R}$, Integral transform $\mathcal{N}$, Extended Borel transform, Discrete delta potential, Finite exponential type function, Fractional differential equation

1. Introduction

Today fractional calculus is applied to several areas of science and technology. But there remain many mathematical problems around fractional calculus. For example, there are two definitions of fractional order differentiation: Riemann-Liouville fractional differentiation and Caputo (or Riesz-Caputo) differentiation, which are markedly different ([1],[2],[12]). We have remarked that these two kinds of definitions coincide if we use the domain of fractional calculus to be the space of operators of Mikusinski ([17] [4],[19]). In this case, $\{\frac{d^a}{dx^a} | a \in \mathbb{R}\}$ becomes a 1-parameter group. Its generating operator is $\log(\frac{d}{dx})$ and

$\log(\frac{d}{dx})f(x) = -(\gamma f(x) + \int_0^x \log(x-t)\frac{df_+(t)}{dt}dt)$. Here γ is the Euler constant and $\frac{df_+(t)}{dt} = \begin{cases} \frac{df(t)}{dt}, t > 0, \\ f(0)\delta, t = 0, \end{cases}$ where δ is the Dirac function.

Introducing an integral transformation $\mathcal{R}$:

$$\mathcal{R}[f(s)](x) = \int_{-\infty}^{\infty} \frac{x^s}{\Gamma(1+s)} f(s)ds,$$

we have

$$\frac{d^a}{dx^a}\mathcal{R}[f(s)](x) = \mathcal{R}[\tau_a f(s)](x), \quad \tau_a f(s) = f(s+a),$$
$$\log(\frac{d}{dx})\mathcal{R}[f(s)](x) = \mathcal{R}[\frac{df(s)}{ds}](x),$$

where differentiation in the second formula is taken in the sense of distribution ([6],[7],[8] cf.[5]). Therefore, the 1-parameter group $\{\tau_a | a \in \mathbb{R}\}$ acting on the space of functions on $\mathbb{R}$ is mapped to the 1-parameter group $\{\frac{d^a}{dx^a} | a \in \mathbb{R}\}$ acting on the space of functions on $\mathbb{R}_+ = \{x | x > 0\}$ and its generating operator $\frac{d}{ds}$ is mapped to $\log(\frac{d}{dx})$ ([4],[19]).

The transform $\mathcal{R}$ is closely related to a bilateral Laplace transform $\mathcal{L}$; $\mathcal{L}[f(s)](t) = \int_{-\infty}^{\infty} e^{st} f(s)ds$. With the variable change $t = \log x$, we have

$$\mathcal{L}[f(s)](\log x) = \mathcal{N}[f(s)](x) = \int_{-\infty}^{\infty} x^s f(s)ds,$$
$$\mathcal{R}[f(s)](x) = \mathcal{N}[\frac{f(s)}{\Gamma(1+s)}](x).$$

Integral transform $\mathcal{N}$ is useful, especially if we take discrete support distributions or a suitable class of generalized functions that are not distributions in the sense of Schwartz, $\mathcal{R}$ is decomposed as

$$\mathcal{R} = \mathcal{B} \circ \mathcal{N},$$

where $\mathcal{B}$ is the extended Borel transform that is explained in this paper (cf.[8]).

This paper is organized as follows: In §2, we review fractional calculus and the logarithm of differentiation $\log(\frac{d}{dx})$. $\mathcal{R}$ and related transforms $\mathcal{N}$ and $\mathcal{R}_+$ are explained in §3. §4 deals with the inverse and adjoint of $\mathcal{R}$ and $\mathcal{N}$. Then extended Borel transform $\mathcal{B}$ and the relation between $\mathcal{R}$ and $\mathcal{B}$ are discussed in §5 and §6. These chapters constitute the main body of the paper. Applications of $\mathcal{B}$ and $\mathcal{R}$ to fractional differential equations with constant coefficients, Euler type fractional differential equations, and fractional analogs of regular singularity are exposed in §7, §8, and §9 (cf.[9],[11],[15],[16],[18]). In addition, there is another way to extend a Borel transform ([3]). Adopting this extension to the formula $\mathcal{R} = \mathcal{B} \circ \mathcal{N}$, we can extend the domain of $\mathcal{R}$. In the Appendix, we provide such extension without using an extension of the Borel transform discussed in §5.

2. Review of Fractional Calculus

Let $f(x)$ be a function defined on $\{x|x \geq c\}$. Then we use

$$I_c^a f(x) = \frac{1}{\Gamma(a)} \int_c^x (x-t)^{a-1} f(t)dt, \quad a > o \tag{1}$$

as the a-th order indefinite integral of f form c. For simplicity, we denote I^a instead of I_0^a.

Note. I_c^a is also defined for complex a provided $\Re a > 0$. Then I_c^a is analytic in a.

The Riemann-Liouville and Caputo fractional differentiations are defined by

$$\frac{d^{n-a}}{dx^{n-a}} f(x) = \frac{d^n}{dx^n}(I^a f(x)), \quad \frac{d^{n-a}}{dx^{n-a}} f(x) = I^a(\frac{d^n f(t)}{dt^n})(x),$$

respectively. These two definitions are different. But if we use the space of operators of Mikusinski as the domain of fractional calculus, and use distribution, then these definitions coincide. As a consequence, constant functions are not in the domain of fractional calculus. Constant function 1 is replaced by Heaviside function Y; $Y(x) = \begin{cases} 1, \ x \geq 1. \\ 0, \ x < 0 \end{cases}$. For a function f on $\mathbb{R}$, we set $f_+(x) = f(x)Y(x)$. Then in order to use Mikusinski's operator as the domain of fractional differentiation, we replace f by f_+. As a result, $\frac{d^a}{dx^a} f(x) = 0$ implies $f(x) = 0$ in this case.

The Riemann-Liouville and Caputo fractional differentiations do not satisfy

$$\frac{d^a}{dx^a} \cdot \frac{d^b}{dx^b} = \frac{d^{a+b}}{dx^{a+b}}$$

in general. But this equality holds if we choose a suitable domain, such as the space of operators of Mikusinski. Under such a selection of the domain of fractional differentiation, $\{\frac{d^a}{dx^a}|a \in \mathbb{R}\}$, $\frac{d^{-a}}{dx^{-a}} = I^a$, becomes a 1-parameter group. The generating operator $\log(\frac{d}{dx}) = \frac{d}{da}(\frac{d^a}{dx^a})|_{a=0}$ is given by

$$\log(\frac{d}{dx}) f(x) = -(\gamma f(x) + \int_0^x \log(x-t) \frac{df_+(t)}{dt} dt), \tag{2}$$

where $\frac{df_+(t)}{dt} = \frac{df(t)}{dt} + f(0)\delta$ ([4],[19]). Since $\log(\frac{d}{dx})$ is the generating operator of $\{\frac{d^a}{dx^a}|a \in \mathbb{R}\}$, we have

$$e^{a \log(\frac{d}{dx})} = \frac{d^a}{dx^a}. \tag{3}$$

Moreover, if both c and $c-a$ are not negative integers, then

$$\frac{d^a}{dx^a} x^c = \frac{\Gamma(1+c)}{\Gamma(1+c-a)} x^{c-a}. \tag{4}$$

Note. Strictly speaking, we need to assume $\Re a > 0$ to hold (4). But since this right hand side is analytic in a, c, and $.x$, we have (4) for arbitrary a by analytic continuation. We also note that if a is not a natural number, then $\frac{d^a}{dx^a}x^{-n}$ does not exist when we define $\frac{d^a}{dx^a}x^{-n}$ according to (4).

Since $\frac{\partial^m}{\partial c^m}x^c = (\log x)^m x^c$, we have

$$\frac{d^a}{dx^a}((\log x)^m)x^c) = \sum_{k=0}^{m} \frac{m!}{k!(m-k)!}\frac{\partial^k}{\partial c^k}(\frac{\Gamma(1+c)}{\Gamma(1+c-a)}(\log x)^{m-k}x^{c-a}.$$

According to these formulas, we can define the fractional order differentiation of $\sum_m \sum_n b_{n,m}(\log x)^m x^{c_n}$ if it satisfies a suitable convergence condition. However, since $\frac{d^{1/2}}{dx^{1/2}}x^{-1/2} = 0$, but $\frac{d}{dx}x^{-1/2} \neq 0$, we do not have $\frac{d^a}{dx^a}\frac{d^b}{dx^b} = \frac{d^{a+b}}{dx^{a+b}}$ in general. This ambiguity occurs because $\Gamma(s)$ is an entire function with 0-points at $s = 0, -1, \ldots$. Hence we have:

Lemma 1. *If a is a real number, then the following holds.*

1. $(\frac{d^a}{dx^a})^n = \frac{d^{na}}{dx^{na}}$, if a is an integer or an irrational number.
2. $(\frac{d^a}{dx^a})^n = \frac{d^{na}}{dx^{na}}$ does not hold if a is a non integral rational number.

3. Integral Transform $\mathcal{R}$

Definition 1. *We define an integral transform $\mathcal{R}$ by*

$$\mathcal{R}[f(s)](x) = \int_{-\infty}^{\infty} \frac{x^s}{\Gamma(1+s)} f(s)ds. \tag{5}$$

We also introduce an integral transform $\mathcal{N}$:

$$\mathcal{N}[f(s)](x) = \int_{-\infty}^{\infty} x^s f(s)ds. \tag{6}$$

As for the ranges of $\mathcal{R}$ and $\mathcal{N}$, it seems appropriate to take the space of analytic functions on $\mathbb{C} \setminus \{x|x \geq 0\}$ or the space of many-valued functions on $\mathbb{C}^\times = \mathbb{C} \setminus \{0\}$.

Since $\frac{1}{\Gamma(1-s)} = \frac{\sin(\pi s)\Gamma(s)}{\pi}$, searching for an appropriate domain of $\mathcal{R}$ seems difficult. Experience suggests that using the space of distributions or generalized functions as domains of $\mathcal{R}$ is useful ([7],[8]).

If f satisfies a suitable condition, such as

$$f(s) = O(e^{A|s|^\mu}),\ A > 0, \mu > 1,\ s \to -\infty,$$
$$f(s) = O(e^{Bs}),\ B > 0,\ s \to \infty,$$

then by Fubini's Theorem, we have

$$\frac{d^a}{dx^a}\mathcal{R}[f(s)](x) = \int_{-\infty}^{\infty} \frac{d^a}{dx^a} x^s \frac{f(s)}{\Gamma(1+s)} ds. \tag{7}$$

Theorem 1 ([6]). *Let $\tau_a f(s)$ be $f(s+a)$. Then if (7) holds, we have*

$$\frac{d^a}{dx^a}\mathcal{R}[f(s)](x) = \mathcal{R}[\tau_a f(s)](x). \tag{8}$$

Proof. With (7) and (4), we have

$$\begin{aligned}\frac{d^a}{dx^a}\mathcal{R}[f(s)](x) &= \int_{-\infty}^{\infty} x^{s-a} \frac{\Gamma(1+s)}{\Gamma(1+s-a)} \frac{f(s)}{\Gamma(1+s)} ds \\ &= \int_{-\infty}^{\infty} \frac{x^{s-a}}{\Gamma(1+s-a)} f(s) ds.\end{aligned}$$

Hence we have (8) through the variable change $s - a = t$.

Corollary. *If* (8) *holds, then*

$$\log(\frac{d}{dx})\mathcal{R}[f(s)](x) = \mathcal{R}[\frac{df(s)}{ds}](x). \tag{9}$$

Note. In the above discussion, we assume a is a real number. If a is a complex number $b + ic$, then we have

$$\frac{d^a}{dx^a}\mathcal{R}[f(s)](x) = \int_{-\infty-ic}^{\infty-ic} \frac{x^s}{\Gamma(1+s)} f(s) ds.$$

But if $|f(s)| \leq Ae^{-B|s|^2}$ on $\{s||\Im s| \leq |c|\}$, (8) holds according to Cauchy's integral formula.

In (9), $\frac{df(s)}{ds}$ is taken in the sense of distribution. For example, if $f(s)$ is continuous on $s \geq c$, differentiable on $s > c$, and 0 on $s < c$, then we have

$$\begin{aligned}&\frac{I - \frac{d^{-h}}{dx^{-h}}}{h}\mathcal{R}[f(s)](x) \\ =&\mathcal{R}[\frac{f(s) - f(s-h)}{h}](x) = (\int_{c+h}^{\infty} + \int_{c}^{c+h}) \frac{x^s}{\Gamma(1+s)} \frac{f(s) - f(s-h)}{h} ds \\ =& \int_{c+h}^{\infty} \frac{x^s}{\Gamma(1+s)} \frac{f(s) - f(s-h)}{h} ds + \frac{1}{h}\int_{c}^{c+h} \frac{x^s}{\Gamma(1+s)} f(s) ds.\end{aligned}$$

Hence we have

$$\lim_{h\to 0} \frac{I - \frac{d^{-h}}{dx^{-h}}}{h}\mathcal{R}[f(s)](x) = \mathcal{R}[f'(s)](x) + \frac{x^c}{\Gamma(1+c)} f(c),$$

where $f'(s) = \dfrac{df(s)}{ds}$ defined on $s > c$. Therefore in (1), it must be

$$\frac{df(s)}{ds} = \begin{cases} \dfrac{df(s)}{ds}, & s > c, \\ f(c)\delta, & s = 0. \end{cases} \tag{10}$$

Later we use the formula

$$\mathcal{R}[\delta_a](x) = \frac{1}{\Gamma(1+a)} x^a.$$

Then we have $\mathcal{R}[\delta_{-n}] = 0$. Hence we have

$$\frac{d^a}{dx^a}\mathcal{R}[\delta_{-n}] = 0 \neq \mathcal{R}[\delta_{-n-a}] = \frac{x^{-n-a}}{\Gamma(1-n-a)},$$

if a is not an integer. Therefore (8) may not hold when $f(s)$ is a distribution.

On the other hand, since

$$\int_0^\infty \frac{x^s}{\Gamma(1+s)} g(x)dx = -\int_0^\infty \frac{x^{s+1}}{\Gamma(s+2)} g'(x)dx,$$

if $s > -1$, we get

$$\lim_{s \downarrow -1} \int_0^\infty \frac{x^s}{\Gamma(1+s)} g(x)dx = g(0).$$

Hence to define alternative integral transform $\mathcal{R}_+$,

$$\mathcal{R}_+[f(s)](x) = \int_{-\infty}^\infty \frac{x_+^s}{\Gamma(1+s)} f(s)ds, \tag{11}$$

we have

$$\mathcal{R}_+[\delta_{-1}] = \delta, \quad \mathcal{R}_+[\delta_{-n-1}] = \delta^{(n)}. \tag{12}$$

(8) holds for $\mathcal{R}_+$; $\dfrac{d^a}{dx^a}\mathcal{R}_+[f(s)](x) = \mathcal{R}_+[\tau_a f(s)](x)$, and (12) shows that (8) still holds for $f(s) = \delta_a$.

There are no problems in the higher dimensional extension of $\mathcal{R}$;

$$\begin{aligned} &\mathcal{R}[f(s_1, \ldots, s_n)](x_1, \ldots, x_n) \\ &= \int_{\mathbb{R}^n} \prod_{i=1}^n \frac{x_i^{s_i}}{\Gamma(1+s_i)} f(s_1, \ldots, s_n) ds_1 \cdots ds_n. \end{aligned}$$

But the situation is a little different when we consider $\mathcal{R}$ on an infinite dimensional space $\ell(\aleph)$. Here, elements of $\ell(\aleph)$ are a series $(s_1, s_2, \ldots)$ with suitable conditions. An integral on $\ell(\aleph)$ is defined by

$$\int_{\ell(\aleph)} f(s_1, s_2, \ldots) d\ell(\aleph) = \lim_{n \to \infty} \int_{\mathbb{R}^n} f(s_1, \ldots, s_n; s_{n+1}, \ldots) ds_1 \cdots ds_n.$$

We assume the existence of this limit is a strong constraint on f. But we will not discuss this problem in depth.

To define $\mathcal{R}$ for functions on $\ell(\aleph)$, we need a convergence of the infinite product $\prod_{n=1}^{\infty} \frac{x_n^{s_n}}{\Gamma(1+s_n)}$. Since

$$\log(\prod_{n=1}^{\infty} \frac{x_n^{s_n}}{\Gamma(1+s_n)}) = \sum_{n=1}^{\infty}(s_n \log x_n - \log(\Gamma(1+s_n))),$$

and $\log(\Gamma(1+s)) = -\gamma s + \sum_{n=2}^{\infty} \frac{(-1)^n \zeta(n)}{n} s^n$, this infinite product converges if $\ell(\aleph)$ is contained in ℓ^1 as a set and $(\log x_1, \log x_2, \ldots) \in \ell(\aleph)^\dagger$.

4. Inverse and Adjoint of $\mathcal{R}$

To describe the inverse of $\mathcal{R}$, it is convenient to consider $\mathcal{R}$ as an integral transform from single valued functions on $\mathbb{R}$ to many-valued (analytic) functions on $\mathbb{C}^\times$. Let $g(x) = \mathcal{R}[f]$. We can regard $\tilde{g}(t) = g(e^{it})$ to be a function on $\mathbb{R}$. $\tilde{g}$ may not be a periodic function, because $g(e^{it})$ and $g(e^{i(t+2pi)})$ may take different values. Then by using Fourier's inversion formula, we obtain:

Proposition 1. *If* $\tilde{g} \in L^2(\mathbb{R}) \cap L^1(\mathbb{R})$, *then*

$$f(s) = \frac{\Gamma(1+s)}{2\pi} \int_{-\infty}^{\infty} e^{-its} \tilde{g}(t) dt. \tag{13}$$

For the rest, we use the notation

$$\mathcal{R}^{-1}[g](s) = \frac{\Gamma(1+s)}{2\pi} \int_{-\infty}^{\infty} e^{-its} \tilde{g}(t) dt.$$

Similarly, $\mathcal{N}^{-1}$ is given by

$$\mathcal{N}^{-1}[g](s) = \frac{1}{2\pi} \int_{-\infty}^{\infty} e^{-its} \tilde{g}(t) dt.$$

We can apply (13) for more general functions and distributions. Such an extension is necessary. Because if g is a finitely many-valued function, then $\tilde{g}$ is a periodic function. Hence if $\mathcal{R}^{-1}[g]$ is defined, it should be a distribution or an ultra distribution but not a function. For example, we have

$$\mathcal{R}[\sum_{n=0}^{\infty} c^n \delta_n](x) = e^{cx},$$

where δ_n is the Dirac function with support $s = n$. Therefore, it must be

$$\mathcal{R}^{-1}[e^{cx}] = \sum_{n=0}^{\infty} c_n \delta_n.$$

Since $\sum_n c^n \delta_n$ is not a distribution in the sense of Schwartz, we need an ultra distribution in the discussion of $\mathcal{R}^{-1}$.

Note. $\mathcal{R}^{-1}[g](s)$ may have a pole of order 1 at $s = -n$, given that n a positive integer, while $\mathcal{N}^{-1}[g](s)$ does not have poles. Therefore, strictly saying, $\mathcal{R}^{-1}$ is not well defined but $\mathcal{N}^{-1}$ is well defined.

If $f(s)$ and $g(x)$ satisfy suitable conditions, e, g, both f and g have compact supports, then we have

$$\begin{aligned}\int_0^\infty \mathcal{R}_+[f(s)]g(x)dx &= \int_0^\infty (\int_{-\infty}^\infty \frac{x^s}{\Gamma(1+s)} f(s)ds)g(x)dx \\ &= \int_{-\infty}^\infty \frac{1}{\Gamma(1+s)}(\int_0^\infty x^s g(x)dx)f(s)ds.\end{aligned}$$

Since functions that have compact supports are dense in $L^2(0,\infty)$ and $L^2(-\infty,\infty)$, by using the Mellin transform

$$\mathcal{M}[h(x)](s) = \int_0^\infty x^{s-1}h(x)dx$$

of $h(x)$, we obtain:

Proposition 2. *As an operator from $L^2(0,\infty)$ to $L^2(-\infty,\infty)$, the adjoint operator $\mathcal{R}_+^\dagger$ of $\mathcal{R}_+$ is given by*

$$\mathcal{R}_+^\dagger[g(x)](s) = \frac{1}{\Gamma(1+s)}\mathcal{M}[xg(x)](s). \tag{14}$$

Similarly, denoting $\mathcal{N}_+[f] = \int_{-\infty}^\infty x_+^s f(s)ds$ and regarding $\mathcal{N}_+$ as an operator from $L^2(0,\infty)$ to $L^2(-\infty,\infty)$, we have

$$\mathcal{N}_+^\dagger[g(x)](s) = \mathcal{M}[xg(x)](s). \tag{15}$$

Since $\mathcal{R}[f] = \mathcal{R}_+[f]$ and $\mathcal{N}[f] = \mathcal{N}_+[f]$, if $f(x) = 0, x < 0$, we can use (14) and (15) when we restrict $\mathcal{R}$ and $\mathcal{N}$ to be operators of $L^2(0,\infty)$.

Using (14) and (15), to define $\mathcal{R}[T](f)$ by $T(\mathcal{R}^\dagger[f])$, *etc.*, if $\Re a > 0$, we have

$$\mathcal{R}[\delta_a](x) = \frac{x^a}{\Gamma(1+a)}, \quad \mathcal{N}[\delta_a] = x^a. \tag{16}$$

Through analytic continuation on a, we regard (16) as being valid for all $a \in \mathbb{C}$.

Alternatively, using $\lim_{t\to\infty} \dfrac{1}{\sqrt{\pi t}} e^{-t(s-a)^2} = \delta_a$ in the sense of distribution, (16) is justified as follows: Since $x^s e^{-t(s-a)^2} = e^{s\log x - t(s-a)^2}$ and

$$-t(s-a)^2 + s\log x = -t(s-c-\frac{\log x}{2t})^2 + (a\log x - \frac{(\log x)^2}{4t}),$$

we have

$$\mathcal{N}[\frac{e^{-t(s-a)}}{\sqrt{\pi t}}](x) = x^a e^{-\frac{(\log x)^2}{4t}} (= -x^{a-\frac{\log x}{4t}}).$$

Hence $\lim_{t\to\infty}\mathcal{N}[\frac{e^{-t(s-a)^2}}{\sqrt{\pi t}}] = x^a$ on $\mathbb{C}^\times$. Note that this justification works for arbitrary $a \in \mathbb{C}$.

Let $\delta_a^{(n)}$ be $\frac{d^n}{dx^n}\delta_a$. Then since $\delta_a(x) = \delta(x-a)$, we have

$$\delta_a^{(n)} = (-1)^n \frac{\partial^n}{\partial a^n}\delta_a.$$

Hence we obtain:

Lemma 2. *The following holds:*

$$\mathcal{R}[\delta_a^{(n)}](x) = (-1)^n \frac{\partial^n}{\partial c^n}\left(\frac{x^c}{\Gamma(1+c)}\right)|_{c=a}, \tag{17}$$

$$\mathcal{N}[\delta_a^{(n)}](x) = (-1)^n (\log x)^n x^a. \tag{18}$$

With this lemma, we can define $\mathcal{R}$ and $\mathcal{N}$ on the space of discrete delta potentials, or more generally, on the space of discrete support distributions and suitable spaces of ultra distributions (analytic functionals cf.[14],[21],[23],[24]) (cf.[8]).

5. Extended Borel Transform

Let $\mathcal{O}$ be the algebra of germs of holomorphic functions at the origin with an ordinary product, and let $\mathrm{Exp}(\mathbb{C})$ be the algebra of finite exponential type functions with the $\sharp$ product. Here, $\sharp$ product $f\sharp g$ of f, g is defined by

$$f(z)\sharp g(z) = \frac{d}{dz}\int_0^z f(z-\zeta)g(\zeta)d\zeta.$$

The Borel transform $\mathcal{B}$ is an algebra isomorphism from $\mathcal{O}$ to $\mathrm{Exp}(\mathbb{C})$ defined by

$$\mathcal{B}[\phi(\zeta)](z) = \frac{1}{2\pi i}\oint e^{\frac{z}{\zeta}}\frac{f(\zeta)}{\zeta}d\zeta. \tag{19}$$

If $\phi(\zeta) = \sum_{n=0}^\infty c_n\zeta^n$, then $\mathcal{B}[\phi(\zeta)] = \sum_{n=0}^\infty \frac{c_n}{n!}z^n$. We define convergences of a series $\{\phi_n\}$ of $\mathcal{O}$ and $\{f_n\}$ of $\mathrm{Exp}(\mathbb{C})$ by

1. $\lim_{n\to\infty}\phi_n = \phi$, if there is an $\epsilon > 0$ such that each ϕ_n is holomorphic on $D_\epsilon = \{z||z| < \epsilon\}$ and $\{\phi_n\}$ converges uniformly in a wider sense to ϕ on D_ϵ.

2. $\lim_{n\to\infty} f_n = f$, if there are $A > 0$ and $B > 0$ such that $|f_n(z)| < Ae^{|B|z|}$ for all n and $\{f_n\}$ converges to f on $\mathbb{C}$ in a wider sense.

Then $\mathcal{B}$ is continuous by this topology.

Originally, $\mathcal{B}$ is defined only for germs of holomorphic functions at the origin. But the right hand side of (19) is defined for any analytic function $\phi = \sum_{n=-\infty}^\infty c_n z^n$ as having isolated singularity at the origin. Then we have

$$\mathcal{B}[\phi](z) = \mathcal{B}[\phi_+](z), \quad \phi_+(z) = \sum_{n=0}^\infty c_n z^n. \tag{20}$$

In this extended sense, we have

$$\frac{d^n}{dz^n}\mathcal{B}[\phi(\zeta)](z) = \mathcal{B}[\zeta^{-n}\phi(\zeta)](z). \tag{21}$$

Note. If either f or g has singularity at the origin, we do not have $\mathcal{B}[fg] = \mathcal{B}[f]\sharp\mathcal{B}[g]$ in general. For example, we have

$$1 = \mathcal{B}[\zeta^{-n}\zeta^n] \neq \mathcal{B}[\zeta^{-n}]\sharp\mathcal{B}[\zeta^n] = 0.$$

For a germ of holomorphic f function at the origin of $\mathbb{C}^n$, we can define its Borel transform $\mathcal{B}[f]$ by

$$\mathcal{B}[f(\zeta)](z) = \frac{1}{(2\pi i)^n}\int_{|\zeta_1|=\epsilon_1,\ldots,|\zeta_n|=\epsilon_n} e^{\frac{z_1}{\zeta_1}+\cdots+\frac{z_n}{\zeta_n}}\frac{f(\zeta_1,\ldots,\zeta_n)}{\zeta_1\cdots\zeta_n}d\zeta_1\cdots d\zeta_n.$$

But when f has singularity at the origin, owing to the non-uniqueness of the Laurent expansion of f at the origin, we need to take care on the magnitude of $\epsilon_1,\ldots,\epsilon_n$. For example, since

$$\frac{x}{x-y} = \begin{cases} 1+\sum_{n=1}^{\infty} x^{-n}y^n, & |x|>|y|, \\ -\sum_{n=1}^{\infty} x^n y^{-n}, & |x|<|y|, \end{cases}$$

we have

$$\frac{-1}{4\pi^2}\int_{\xi=\epsilon_1}\int_{\eta=\epsilon_2} e^{\frac{x}{\xi}+\frac{y}{\eta}}\frac{\xi}{\xi\eta(\xi+\eta)}d\xi d\eta = \begin{cases} 1, & \epsilon_1>\epsilon_2|, \\ 0, & \epsilon_1<\epsilon_2. \end{cases}$$

In general, if $f(z)$ has singularities on Y, $0\in Y$, then to denote

$$H_{n,*}(U(0)\setminus Y;\mathbb{Z}) = \lim_{\epsilon\to 0}\{H^n(U(\epsilon)\setminus Y;\mathbb{Z}) : r^{\epsilon}_{\epsilon'}\},$$

where $U(\epsilon)=\{z|\|z\|<\epsilon\}$ and $r^{\epsilon}_{\epsilon'}$ is the restriction homomorphism, let $\gamma_1,\ldots,\gamma_m$ be the generators of $H_{n,*}(U(0)\setminus Y;\mathbb{Z})$. Then we must use the notation

$$\mathcal{B}_{\gamma_j}[f(\zeta)](z) = \frac{1}{(2\pi i)^n}\int_{\gamma_j} e^{\frac{z_1}{\zeta_1}+\cdots+\frac{z_n}{\zeta_n}}\frac{f(\zeta_1,\ldots,\zeta_n)}{\zeta_1\cdots\zeta_n}d\zeta_1\cdots d\zeta_n.$$

The Borel transform of f depends on the outcome of γ_j. However, we will not discuss this phenomenon in the course of this work.

In the above extended sense, the Borel transform has a non-trivial kernel. But as a transform defined on $\mathcal{O}$, the Borel transform has an inverse. The inverse Borel transform is given by

$$\mathcal{B}^{-1}[f(t)](x) = \int_0^{\infty} e^{-t}f(xt)dt. \tag{22}$$

$\mathcal{B}^{-1}$ can be defined for some $f\notin \mathrm{Exp}(\mathbb{C})$. For example, we have

$$\mathcal{B}^{-1}[t^a](x) = \Gamma(1+a)x^a, \quad \mathcal{B}^{-1}[\log t](x) = \log x - \gamma.$$

Strictly speaking, these formulas hold on $\{x|\Re x > 0\}$ and as for the first formula, we use analytic continuation on a, if $\Re a \leq 0$.

Definition 2. *We extend the Borel transform to z^a, $a \in \mathbb{C}$ and $\log z$ by*

$$\mathcal{B}[\zeta^a](z) = \frac{z^a}{\Gamma(1+a)}, \quad \mathcal{B}[\log\zeta](z) = \log z + \gamma. \tag{23}$$

Note. We have

$$\lim_{\epsilon\downarrow 0} \mathcal{B}[(\zeta+\epsilon)^a] = \frac{z^a}{\Gamma(1+a)}, \quad \lim_{\epsilon\downarrow 0} \mathcal{B}[\log(\zeta+\epsilon)] = \log z + \gamma,$$

uniformly in a wider sense on $\{z|\Re z > 0\}$. This suggests the natural domain of $\mathcal{B}[\phi]$ is the right half plane $\{z|\Re z > 0\}$. In the next section, we present an alternative definition of the Borel transform according to this suggestion. It is denoted by $\mathcal{B}_+$.

In the rest of this section, we do not follow this suggestion, and instead hold that $\mathcal{B}[\phi](z)$ is defined on $\mathbb{C} \setminus \{x|x \in \mathbb{R}, x \leq 0\}$ or on $\mathbb{C}^\times = \mathbb{C} \setminus \{0\}$. In the latter case, we regard $\mathcal{B}[\phi]$ as a many-valued function.

According to Definition 2, we have

$$\mathcal{B}[\zeta^a\zeta^b](z) = \mathcal{B}[\zeta^a](z)\sharp\mathcal{B}[\zeta^b](z), \tag{24}$$

provided both a and b are not negative integers. We also have $\mathcal{B}[\log\zeta](z) = \frac{\partial}{\partial a}\mathcal{B}[\zeta^a](z)|_{a=0}$. Hence we define $\mathcal{B}[(\log\zeta)^n\zeta^a](z)$ by

$$\mathcal{B}[(\log\zeta)^n\zeta^a](z) = \frac{\partial^n}{\partial a^n}\mathcal{B}[\zeta^a](z). \tag{25}$$

Lemma 3. *Let $f^{\sharp n}$ be $\overbrace{f\sharp\cdots\sharp f}^{n}$. Then we have*

$$\mathcal{B}[(\log\zeta)^n](z) = (\mathcal{B}[\log\zeta](z))^{\sharp n}. \tag{26}$$

Proof. According to the above definitions, we have

$$\begin{aligned}\mathcal{B}[(\log\zeta)^2](z) =& \frac{\partial^2}{\partial a\partial b}\mathcal{B}[\zeta^a\zeta^b](z)|_{a=b=0} \\ =& \frac{\partial^2}{\partial a\partial b}(\mathcal{B}[\zeta^a](z)\sharp\mathcal{B}[\zeta^b](z))|_{a=b=0} \\ =& \mathcal{B}[\log\zeta](z)\sharp\mathcal{B}[\log\zeta](z).\end{aligned}$$

In the same vein, we have $\mathcal{B}[(\log\zeta)^n](z) = \mathcal{B}[(\log\zeta)^n](z)\sharp\mathcal{B}[\log\zeta](z)$. Hence we have the lemma.

Example. Since $\zeta^a = e^{a\log\zeta}$, we obtain

$$\begin{aligned}\frac{z^a}{\Gamma(1+a)} =& \sum_{n=0}^{\infty}\frac{\mathcal{B}[a^n(\log\zeta)^n](z)}{n!} \\ =& \sum_{n=0}^{\infty}\frac{a^n(\log z+\gamma)^{\sharp n}}{n!} = e^{\gamma a}(\sum_{n=0}^{\infty}\frac{(\log z)^{\sharp n}}{n!}).\end{aligned}$$

Hence we get

$$\sum_{n=0}^{\infty}\frac{a^n(\log z)^{\sharp n}}{n!}=\frac{e^{-\gamma a}}{\Gamma(1+a)}z^a. \tag{27}$$

The *extended Borel transform* is defined on suitable algebra A spanned by $\{z^a; a\in\mathbb{C}\}$ and $\log z$. Then, assuming multiplications by z^a and $\log z$ are possible in A, we have:

Proposition 3. *The following holds:*

$$\frac{d^a}{dz^a}\mathcal{B}[f(\zeta)](z)=\mathcal{B}[\zeta^{-a}\phi(\zeta)](z), \tag{28}$$

$$\log(\frac{d}{dz})\mathcal{B}[\phi(\zeta)](z)=-\mathcal{B}[\log\zeta\phi(\zeta)](z). \tag{29}$$

Example. With (28) and (29), we have

$$\log(\frac{d}{dx})\mathcal{B}[\zeta^{-a}](x)=-\mathcal{B}[\log\zeta\cdot\zeta^{-a}](x)=-\frac{d^a}{dx^a}\mathcal{B}[\log\zeta](x).$$

We have $\mathcal{B}[(1-\lambda\zeta)^{-1}](z)=e^{\lambda z}$. Similarly, by using the Mittag-Leffler function $E_\rho(z)=\sum_{n=0}^{\infty}\dfrac{z^n}{\Gamma(1+n/\rho)}$ ([10]), we have

$$\mathcal{B}[(1-\lambda\zeta^a)^{-1}](z)=E_{1/a}(\lambda z^a). \tag{30}$$

Since $\dfrac{\partial^n}{\partial\lambda^n}\dfrac{1}{1-\lambda\zeta^a}=(n-1)!\dfrac{\zeta^{an}}{(1-\lambda\zeta^a)^{n+1}}$, through (39) we obtain

$$\mathcal{B}[(1-\lambda\zeta^a)^{-(n+1)}](z)=\frac{1}{(n-1)!}\frac{d^{-na}}{dz^{-na}}(\frac{\partial^n}{\partial\lambda^n}E_{1/a}(\lambda z^a)). \tag{31}$$

These are used in the study of fractional differential equations with constant coefficients (§7).

Note. We can define $\mathcal{B}[(\log\zeta)^{-n}]$ as follows: First we note

$$-\int_a^{\infty}x^s ds=\frac{x^a}{\log x},\ |x|<1,\quad \int_{-\infty}^{a}x^s ds=\frac{x^a}{\log x},\ |x|>1.$$

Hence if $\int_a^{\infty}\mathcal{B}[\zeta^a]=\mathcal{B}[\int_a^{\infty}\zeta^s ds]$, we may conclude

$$\mathcal{B}[\frac{\zeta^a}{\log\zeta}]=-\int_a^{\infty}\frac{x^s}{\Gamma(1+s)}ds.$$

Here we do not need to restrict $|x|<1$, because the integral of the right hand side converges for all x. Especially, if we have

$$\mathcal{B}[\frac{1}{\log\zeta}](x)=-\int_0^{\infty}\frac{x^s}{\Gamma(1+s)}ds=-\mathcal{R}[Y(s)](x).$$

In fact, we have

$$\log(\frac{d}{dx})\mathcal{B}[\frac{1}{\log\zeta}](x) = -\mathcal{B}[1](x) = -1,$$

$$\log(\frac{d}{dx})(-\mathcal{R}[Y(s)](x) = -\mathcal{R}[\delta](x) = 1.$$

As a result, it should be

$$\mathcal{B}[(\log\zeta)^{-n}](x) = (-1)^n\Big(\int_0^\infty \frac{x^s}{\Gamma(1+s)}ds\Big)^{\sharp n}.$$

This answer may be rewritten as

$$\mathcal{B}[(\log\zeta)^{-n}](x) = (-1)^n\frac{1}{(n-1)!}\mathcal{R}[s_+^{n-1}](x).$$

Whether we can rewrite these answers as a Laurent series of $\log x$ or not is a problem.

6. $\mathcal{R}$ and the Extended Borel Transform

Theorem 2 ([8]). *Let T be a discrete support distribution. Then we have*

$$\mathcal{R}[T] = \mathcal{B}\circ\mathcal{N}[T]. \tag{32}$$

Proof. Discrete support distributions are sums of $\delta_a^{(n)}$, $a\in\mathbb{C}, n\in\mathbb{N}$. With (15) and (23), we have

$$\mathcal{R}[\delta_a] = \mathcal{B}[\zeta^a] = \mathcal{B}\circ\mathcal{N}[\delta_a].$$

Then, since $\frac{\partial^n}{\partial a^n}\delta_a = (-1)^n\delta_a^{(n)}$, we obtain

$$\mathcal{R}[\delta_a^{(n)}] = (-1)^n\frac{\partial^n}{\partial a^n}\mathcal{B}[\zeta^a] = \mathcal{B}\circ\mathcal{N}[\delta_a^{(n)}].$$

Hence we have a theorem.

Note. Let F_a, $a>0$, the space of the holomorphic function on $\{z||z|\le a\}$. Then a generalized function (analytic functionals) T on F_a takes the form $T=\sum_{n=0}^\infty c_n\delta^{(n)}$, $|c_n|\le Ca^{-n}$. Therefore, regarding $\mathcal{R}$ and $\mathcal{N}$ as transforms on the space of generalized functions on F_a, Theorem 2 is still valid.

With Theorem 2 and Lemma 3, we obtain:

Lemma 4. *Let $\frac{1}{\Gamma(1+z)} = \sum_{n=0}^\infty \alpha_n z^n$. Then we have*

$$\mathcal{B}[(\log\zeta)^n](z) = \sum_{k=0}^n \frac{n!}{k!}\alpha_{n-k}(\log z)^k. \tag{33}$$

Theorem 3. *Let $f(z)=\sum_{n=0}^\infty c_nz^n$ and let $f(z)^\sharp$ be $\sum_{n=0}^\infty c_n(\log z)^{\sharp n}$. Then if $f(z)$ is a finite exponential function, $f(z)^\sharp$ is well defined and takes the form $F(\log z)$. $F(z)$ is a* *finite exponential type function.*

Proof. By assumption, $f(z-\gamma)$ is a finite exponential type function. Therefore, to show the theorem, it is sufficient to show
$\sum_{n=0}^{\infty} c_n\mathcal{B}[(\log\zeta)^n](z)$ is a finite exponential type function.

Since we have $f(z-\gamma)^\sharp = \sum_{n=0}^{\infty} c_n(\log z-\gamma)^n$ and $\mathcal{B}[(\log\zeta - \gamma)^n](z) = (\log z)^{\sharp n}$, to show the theorem, it is sufficient to show $\sum_{n=0}^{\infty} c_n\mathcal{B}[(\log\zeta)^n](z)$ is a finite exponential type function.

Since f is a finite exponential type function, there exist positive numbers M and C such that $n!|c_n| \le MC^n$. Hence with (33), we get

$$\sum_{n=0}^{\infty} c_n\mathcal{B}[(\log\zeta)^n](z) = \sum_{k=0}^{\infty}(\sum_{n\ge k} c_n n!\alpha_{n-k})(\log z)^k.$$

Since $\dfrac{1}{\Gamma(1+x)}$ is an entire function,

$$\sum_{n\ge k}|c_n n!\alpha_{n-k}| \le \sum_{n\ge k} MC^n|\alpha_{n-k}| \le C^k B,$$

for some $B>0$. Therefore we have a theorem.

Note. In [8], an alternative proof of this theorem is provided.

Of course, there are many classes of functions upon which an extended Borel transform is defined. For example, if $\Re a > 0$ and $f(z) = \sum_{n\ge m}^{\infty} c_n z^{na}$, $m \in \mathbb{Z}$, and $|c_n| \le MC^n$ for some positive M and C, then $\mathcal{B}[f(\zeta)](z)$ is defined on $\mathbb{C}^\times$ as a many-valued function.

Since $z^{na} = e^{na\log z}$, formally, we can write

$$\begin{aligned}\sum_{n=0}^{\infty} c_n z^{na} &= \sum_{n=0}^{\infty} c_n(\sum_{m=0}^{\infty}\frac{(na)^m}{m!}(\log z)^m)\\ &= \sum_{m=0}^{\infty}\frac{a^m}{m!}(\sum_{n=0}^{\infty} c_n n^m)(\log z)^m.\end{aligned}$$

This rewriting is justified if $\{c_n | n \in \mathbb{N}\}$ satisfies the suitable conditions. For example, since $\exp(e^z)$ is an entire function, if $\sum_{n=0}^{\infty} c_n z^n$ is a finite exponential type function, this is possible. On the other hand, such rewriting is impossible if $\sum_{n=0}^{\infty} c_n z^n$ has only a finite convergence radius. But for some special f, we can rewrite $f(z)$ as a power series of $\log z$.

Example. Let $f(z)$ be $\sum_{n=0}^{\infty} z^n$ and set

$$f(z,s) = 1 + \sum_{n=1}^{\infty} n^{-s}z^s.$$

Then we have

$$
\begin{aligned}
1+\sum_{n=1}^{\infty} n^{-s}z^{n} &= 1+\sum_{n=1}^{\infty}(\sum_{m=0}^{\infty}\frac{n^{-s}n^{m}}{m!}(\log z)^{m}) \\
&= 1+\sum_{m=0}^{\infty}(\sum_{n=1}^{\infty}\frac{n^{m-s}}{m!})(\log z)^{m} \\
&= 1+\sum_{m=0}^{\infty}\frac{\zeta(s-m)}{m!}(\log z)^{m}.
\end{aligned}
$$

Hence through analytic continuation with respect to s, we conclude

$$\sum_{n=0}^{\infty} z^{n} = 1-\sum_{m=0}^{\infty}\frac{B_{m+1}}{(m+1)!}(\log z)^{m}. \tag{34}$$

Here, B_m is the m-th Bernoulli number. Therefore $(1-z)^{-1}$ is expressed by the infinite sum of even powers of $\log z$ (cf.[8]).

Note. Since $(\log x)^n = \mathcal{N}[\delta^{(n)}]$ and $x^n = \mathcal{N}[\delta_n]$, to rewrite a power series $f(x) = \sum_n c_n z^n$ as a power series of $\log z$ is essentially the same thing as computing the Taylor expansion of a discrete delta potential $\sum_n a_n \delta_n$.

$\mathcal{N}$ has inverse $\mathcal{N}^{-1}$ on the space of discrete support distributions and is suitable for its completion. Then by (32), we have:

Theorem 4. *On suitable algebra of power series of z^a and $\log z$, we have*

$$\mathcal{B} = \mathcal{R}\circ\mathcal{N}^{-1}. \tag{35}$$

Definition 3 ([8]). *We define an alternative extension of the Borel transform $\mathcal{B}_+$ as*

$$\mathcal{B}_+ = \mathcal{R}_+\circ\mathcal{N}^{-1}. \tag{36}$$

$\mathcal{B}_+[\phi]$ is a distribution on $\{z|\Re z \geq 0\}$. It satisfies

$$\mathcal{B}_+[fg] = \mathcal{B}_+[f]\sharp\mathcal{B}_+[g],$$

although f or g has singularity at the origin. On $\{z\in\mathbb{C}|\Re z>0\}$, we can regard it as a function, and we have

$$\mathcal{B}[\phi]|\{z\in\mathbb{C}|\Re z>0\} = \mathcal{B}_+[\phi]|\{z\in\mathbb{C}|\Re z>0\}.$$

With this and (12), we have:

Proposition 4. *The following holds:*

1. $\mathcal{B}_+[\zeta^{-n}] = \delta^{(n)}$, $n\in\mathbb{N}$.

2. We say that $\frac{d^a}{dz^a}\mathcal{B}_+[\phi] = \mathcal{B}[\zeta^{-a}\phi]$ and

$$\frac{d^a}{dz^a}(\frac{d^b}{dz^b}\mathcal{B}_+[\phi]) = \frac{d^{a+b}}{dz^{a+b}}\mathcal{B}_+[\phi]$$

holds without exception on a, b.

Note. By definition, we have $\mathcal{B}_+[(1-\lambda\xi)^{-1}](x) = e^{\lambda x}|_+$. Hence we see that

$$\frac{d}{dx}\mathcal{B}_+[(1-\lambda\xi)^{-1}](x) = \lambda(\delta + \mathcal{B}_+[(1-\lambda\xi)^{-1}](x)).$$

It is not a solution for the equation $\frac{dy}{dx} = \lambda y$, but it is a fundamental solution.

7. Applications to Fractional Differential Equations

Let $\Re a > 0$ and set $y = \mathcal{B}[u]$. Then the equation

$$\frac{d^a y}{dz^a} = \lambda y, \tag{37}$$

is solved if $\mathcal{B}[(\zeta^{-a} - \lambda)u(\zeta)] = 0$. Therefore u is a sum of

$$u_n(z) = u_{n;a,\lambda}(z) = \mathcal{B}[\frac{\zeta^{a-n}}{1-\lambda\zeta^a}](z), n \in \mathbb{N}. \tag{38}$$

By using (30), we have

$$u_n(z) = \frac{d^n}{dz^n}(I^a(E_{1/a}(\lambda z^a))).$$

More accurately, u_n may not satisfy (37) and $u_1, u_2, \ldots$ is not independent in general ([7]).

Proposition 5. *The following holds:*

1. If a is a positive integer, then $u_1(z), \ldots, u_a(z)$ is independent, and other $u_n(z)$ depend on $u_1(z), \ldots, u_a(z)$.

2. If $a = \frac{q}{p}$, $(p, q) = 1$ is a rational number but not an integer, then $u_1(z), \ldots, u_q(z)$ is independent and $u_r(z)$, $r > q$ does not satisfy (37).

3. If a is an irrational number, then $u_1(z), u_2(z), \ldots$ are all independent and satisfy (37).

Proof. Since $\Re a > 0$, we have

$$\frac{\zeta^{a-n}}{1-\lambda\zeta^a} = \sum_{m=0}^{\infty} \lambda^m \zeta^{(m+1)a-n}. \tag{39}$$

If a is an integer, we set $n = pa + q$. Then

$$\sum_{m=0}^{\infty} \lambda^m \zeta^{(m+1)a-n} = \sum_{m=0}^{\infty} \lambda^m \zeta^{(m+1-p)a-q}.$$

Since $\mathcal{B}[\zeta^{(m+1-p)a-q}] = 0$ if $m + 1 - p \leq 0$, we get

$$\mathcal{B}[\sum_{m=0}^{\infty} \lambda^m \zeta^{(m+1-p)a-q}] = \lambda^{p-1}\mathcal{B}[\sum_{k=0}^{\infty} \zeta^{ka-q}].$$

Hence we have 1.

If $n > q$, the expansion (39) contains terms z^{-s} and z^{s-a}, where s is a positive integer. Since u_n does not contain the term z^{-s}, $\frac{d^a}{dz^a}u_n(z)$ does not contain the term z^{s-a}. Hence u_n does not satisfy (37). Hence we have 2.

Since $\mathcal{B}[\zeta^{am-n}](z) = \dfrac{z^{am-n}}{\Gamma(1 + am - n)} \neq 0$ for all $n \in \mathbb{N}$, because a is an irrational number, we have 3.

Note. Alternatively, if we use $\mathcal{B}_+$, then we have

$$\frac{d^a}{dz^a}\mathcal{B}_+[\frac{\zeta^{a-n}}{(1-\lambda\zeta^a}](z) = \lambda\mathcal{B}_+[\frac{\zeta^{a-n}}{1-\lambda\zeta^a}](z) + \delta^{(n-1)}. \tag{40}$$

Contrary to the case used, $\mathcal{B}$, (40) holds whenever $\Re a > 0$, and we need no further assumptions on a.

To get the same results by using $\mathcal{R}$, we consider the equation

$$\tau_a T = \lambda T + \delta_{-n}. \tag{41}$$

Here T is a distribution (or a generalized function). In the category of discrete delta potentials, a solution of (41) is

$$T = \sum_{m=0}^{\infty} \lambda^m \delta_{ma-n}. \tag{42}$$

From (42), we can recover Proposition 5 (cf.[7]).

Proposition 5 shows only those $u_{n;a,\lambda}$, $1 \leq n < \Re a + 1$ are stable under the variation of a as a solution of (37). Especially, if $\Re a < 2$, only $u_{1;a,\lambda} = \frac{d}{dz}I^a E_{1/a}(\lambda z^a)$ is the stable solution of (37) under the variation of a. As a consequence, solutions of the fractional evolution equation

$$\frac{\partial^a}{\partial t^a}U(t,x) = D_x U(t,x), \tag{43}$$

where $|a - 1|$ is small, should take the forms

$$U(t,x) = \sum_n c_n(\frac{d^a}{dt^a}I^a E_{1/a}(\lambda_n t^a))\phi_n(x) \tag{44}$$

if D admits the spectral decomposition $Df(x) = \sum_n \lambda_n(\phi_n(x), f(x))$.

Therefore to study the asymptotic properties of $E_a(x)$ with respect to a, it will be useful to the study of asymptotic properties of equation (43) with respect to a when a is close to 1. Of course, there are alternative ways to study the asymptotic properties of (43). For example, if $U(t,x) = \mathcal{R}_s[V(s,x)](t,x)$ satisfies (43), then it must be that $V(s+a,x) = D_xV(s,x)$. We set $a = 1+\epsilon$, and $V = \sum_{n=0}^{\infty} \epsilon^n V_n$, where $U_0 = \mathcal{R}[V_0]$ satisfies

$$\frac{\partial}{\partial t}U(t,x) = D_xU(t,x). \tag{45}$$

Then it should be true that

$$\frac{\partial}{\partial s}V_0(s+1,x) + D_xV_1(s,x) = 0.$$

Hence, if D has the Green operator G, and if $\mathcal{R}_s[V(s,x)]$ satisfies (43), then it must be that

$$V(s,x) = V_0(s,x) - \epsilon G_x(\frac{\partial}{\partial s}V_0(s+1,x)) + O(\epsilon^2). \tag{46}$$

Here $U(t,x) = \mathcal{R}[V_0(s,x)]$ is a solution for (45).

We also expect to research how to define the Mittag-Leffler function of operators such as a Laplacian. Such an endeavor should be useful to the study of fractional evolution equations.

Note. Since we do not have $\frac{d^a}{dx^a}\frac{d^b}{dx^b} = \frac{d^{a+b}}{dx^{a+b}}$ in general, we need to distinguish $(\frac{d^a}{dx^a})^2$ and $\frac{d^{2a}}{dx^{2a}}$, *etc.*, in the study of fractional differential equations. For example, solutions of the equation $\frac{d^{1/2}y}{dx^{1/2}} = \pm y$ are different from e^x. This is because $I^{1/2}E_2(\pm\sqrt{t})$ is not differentiable at the origin (see (40)).

By the same reasoning, solutions for the equation

$$(\frac{d^a}{dx^a} - \lambda)^m y = 0 \tag{47}$$

are given by a linear combination of

$$u_{n,k;a,\lambda}(x) = \mathcal{B}[\frac{\xi^{ka-n}}{(1-\lambda\xi^a)^k}](x), \quad 1 \le k \le m. \tag{48}$$

While for the equation

$$\sum_{k=0}^{m} \frac{m!}{k!(m-k)!}\lambda^m \frac{d^{(m-k)a}}{dx^{(m-k)a}}y = 0,$$

we cannot obtain the same conclusion.

In general, the equation

$$(\frac{d^{a_0}}{dx^{a_0}} + c_1\frac{d^{a_1}}{dx^{a_1}} + \cdots + c_m)y = 0, \tag{49}$$
$$\Re a_0 > \Re a_1 \ge \ldots \ge \Re a_{m-1} \ge 0,$$

has solutions expressed as linear combinations of

$$\mathcal{B}[\frac{\xi^{a_0-n}}{1+c_1\xi^{a_0-a_1}+\cdots+c_n\xi^{a_0}}](x).$$

We can define this Borel transform, because by the assumption on $a_0, a_1, \ldots, a_{m-1}$, we know that $|c_1\xi^{a_0-a_1}+\cdots+c_n\xi^{a_0}|<1$ if $|\xi|$ is sufficiently small.

To treat the fractional differential equation $P(\frac{d^a}{dx^a})y=0$ with constant coefficients similarly to an ordinary differential equation with constant coefficients, we need to assume

$$P(\frac{d^a}{dx^a})=P(x)|_{X=\frac{d^a}{dx^a}}=(\frac{d^a}{dx^a})^n+c_1(\frac{d^a}{dx^a})^{n-1}+\cdots+c_n,$$

$$P(X)=X^n+c_1X^{n-1}+\cdots+c_n=\prod_{k=1}^{j}(X-\lambda_k)^{m_j}.$$

Then the equation $P(\frac{d^a}{dx^a})y=0$ has solutions of the form $u_{n,k;a,\lambda}(x)$.

The same problem appears for fractional differential equations with variable coefficients. Since we have $\frac{d^a}{dx^a}x^{a-1}=0$, we know that

$$(\frac{d^a}{dx^a})^n x^{a-1}=0,$$

while $\frac{d^{na}}{dx^{na}}x^{a-1}=\frac{\Gamma(a)}{\Gamma((1-n)a)}x^{(1-n)a-1}\neq 0$, if $(1-n)a$ is not a negative integer. In the same vein, we know that

$$(\frac{d^a}{dx^a})^n x^{a-k}=0 \tag{50}$$

if $na-k$ is not a negative integer. Hence we can construct a solution of the form $y(x)=\sum_{n=1}^{\infty}b_nx^{na-k}$ for the equation

$$((\frac{d^a}{dx^a})^n+g_1(x^a)(\frac{d^a}{dx^a})^{n-1}+\cdots+g_n(x^a))y=0, \tag{51}$$

$$g_j(x^a)=\sum_{n=0}^{\infty}c_{j,n}x^{an}, \quad j=1,\ldots,n. \tag{52}$$

If a is an irrational number, an arbitrary $k\in\mathbb{N}$ is allowed in this construction. We are still looking for the meaning of solutions with $k\geq 2$.

Note. The above method can be used for the equation $(\frac{d^a}{dx^a})^2y+x^ay=0$. But we cannot use the above method for the equation $\frac{d^{2a}}{dx^{2a}}y+x^ay=0$.

8. Fractional Euler Type Equations

We say that equation K:

$$Ky = \sum_{k=1}^{n}\sum_{j=1}^{m_k} b_{k,j}x^{(a_k)j}(\frac{d^{a_k}}{dx^{a_k}})^j y, \tag{53}$$

$$m_1\Re a_1 \geq m_2\Re a_2 \geq \cdots \geq m_n\Re a_n$$

is a Euler type fractional equation. By definition, we have

$$Kx^q = (\sum_{k=1}^{n} b_{k,j}\prod_{j=1}^{m_k}\frac{\Gamma(1+q-(j-1)a_k)}{\Gamma(1+q-ja_k)})x^q. \tag{54}$$

Here $\frac{\Gamma(1+q-(j-1)a_k)}{\Gamma(1+q-ja_k)}$ means $\lim_{\epsilon\to 0}\frac{\Gamma(1+q+\epsilon-(j-1)a_k)}{\Gamma(1+q+\epsilon-ja_k)}$ and we set $\prod_{j=1}^{m_k}\frac{\Gamma(1+q-(j-1)a_k)}{\Gamma(1+q-ja_k)} = 0$ if one of these factors is equal to 0.

Note. To use $\frac{d^a}{dx^a}x^a$ instead of $x^a\frac{d^a}{dx^a}$, or to use $(x^a\frac{d^a}{dx^a})^m$ instead of $x^{ma}(\frac{d^a}{dx^a})^m$, *etc.*, we obtain alternative definitions of Euler type fractional equations. The differences between these operators are understood via the fractional analog of canonical commutation relation ([4],[5],[20], cf.[21],[22]). But we do not use these alternative definitions in the rest of the work.

Definition 4. *We say that*

$$\chi_K(q) = \sum_{k=1}^{n} b_{k,j}\prod_{j=1}^{m}\frac{\Gamma(1+q-(j-1)a_k)}{\Gamma(1+q-ja_k)} \tag{55}$$

are the characteristic functions of K.

If λ is a constant, then we have

$$\chi_{L+\lambda}(q) = \chi_K(q) + \lambda \tag{56}$$

by definition. Hence, if $\chi_K(q)$ is not a rational function, by Great Picard's Theorem, the equation $Ky = \lambda y$ has infinitely many independent solutions for almost all λ with exceptions at two λ's. Finding a suitable boundary condition for the equation

$$Ky = \lambda y$$

is a problem.

Example. Let $K = K(a) = x\frac{d}{dx} + ax^\alpha\frac{d^\alpha}{dx^\alpha}$, $a \neq 0$, then $\chi_K(q) = q + a\frac{\Gamma(1+q)}{\Gamma(1+q-\alpha)}$. Hence $Kx^q = 0$ if

$$\frac{\Gamma(q)}{\Gamma(1+q-\alpha)} = -\frac{1}{a}.$$

Since this left hand side has essential singularity at $q = \infty$, except for two values of a, there exist infinitely many q such that $K(a)x^q = 0$.

We note that if $n < \Re\alpha < n+1$, then the constant coefficients equation $\frac{dy}{dx} + a\frac{d^\alpha y}{dx^\alpha} = 0$ is changed to the Volterra type integral equation:

$$y(x) + \frac{a}{\Gamma(\alpha+n)} \int_0^x (x-t)^{n-\alpha} y(t)dt = C_0 + C_1 x + \cdots + C_n x^n.$$

Hence it has $(n+1)$ independent solutions.

Proposition 6. *If $\chi_K(q_0) = 0$ with order m, then*

$$K(x^{q_0}(\log x)^\ell) = 0, \quad 0 \le \ell \le m-1. \tag{57}$$

Proof. Since we have

$$K(x^q(\log x)^\ell) = \sum_{j=0}^{\ell} \frac{\ell!}{j!(\ell-j)!}(\frac{d^j}{dx^j}\chi_K(q))x^q(\log x)^{\ell-j},$$

we have a proposition according to the definition of χ_K and the assumption on q_0.

9. Fractional Analog of Regular Singularity

Let $\Re a > 0$ and consider the fractional differential operator

$$L = g_0(x)(\frac{d^a}{dx^a})^n + g_1(x)(\frac{d^a}{dx^a})^{n-1} + \cdots + g_n(x), \tag{58}$$

$$g_m(x) = \sum_{k=0}^{\infty} b_{m,k} x^{(k+n-m)a}, \; m = 0, \ldots, n, \; b_{n,0} = 1.$$

By definition, we can write

$$g_m(x) = x^{a(n-m)} h_m(x^a), \quad h_m(x) = \sum_{k=0}^{\infty} h_{m,k} x^n.$$

We assume each $h_m(x)$ is holomorphic at the origin.

Definition 5. *We say L has a regular singularity at $x = 0$.*

We set $L_0 = b_{n,0}x^{na}(\frac{d^a}{dx^a})^n + b_{n-1,0}x^{(n-1)a}(\frac{d^a}{dx^a})^{n-1} + \cdots + b_{0,0}$. Then we define the characteristic function $chi_L(q)$ as

$$\chi_L(q) = \chi_{L_0}(q).$$

In the rest of the work, we assume a is real and an irrational number. Then by Lemma 1, we can set

$$L = g_0(x)\frac{d^{na}}{dx^{na}} + g_1(x)\frac{d^{(n-1)a}}{dx^{(n-1)a}} + \cdots + g_n(x), \tag{59}$$

$$\chi_L(q) = \sum_{k=1}^{n} b_n \frac{\Gamma(1+q)}{\Gamma(1+q-ka)}. \tag{60}$$

We search for a solution to the equation $Ly = 0$, having the form $x^q + \sum_k c_k x^{q+ka}$. Since

$$x^{ka}\frac{d^{ka}}{dx^{ka}}x^{q+ma} = \frac{\Gamma(1+q+ma)}{\Gamma(1+q+(m-k)a)}x^{q+ma},$$

to set $Ly = \sum_{k=0}^{\infty} d_k x^{q+ka}$, we have to accept $d_0 = \chi_L(q)$. Hence q should satisfy $\chi_L(q) = 0$.

As for $d_k; k \geq 1$, we set

$$d_{i,0}(q) = \sum_{k=0}^{n} b_{k,i}\frac{\Gamma(1+q)}{\Gamma(1+q-(n-k)a)}, \quad d_{i,j}(q) = d_{i,0}(q+ja).$$

Then we have

$$d_k = \sum_{i+j=k} c_j d_{i,j} = c_k d_{0,k} + c_{k-1}d_{1,k-1} + \cdots + d_{k,0}.$$

Hence if $c_1, \ldots, c_{k-1}$ is determined to satisfy $d_j = 0, j = 1, \ldots, k-1$ and $d_{0,k} \neq 0$, then we have $d_k = 0$ if we take

$$c_k = -\frac{c_{k-1}d_{1,k-1} + \cdots + d_{k,0}}{d_{0,k}}.$$

Since $d_{0,k}(q) = \chi_L(q+ka)$, we have:

Lemma 5. *If $s \in \mathbb{C}$ satisfies*

$$\chi_L(s) = 0, \quad \chi_L(s+ka) \neq 0, \ k = 1, 2, \ldots, \tag{61}$$

then there exists a series $y = x^q + \sum_{k=1}^{\infty} c_k y^{s+ka}$ that satisfies $Ly = 0$ formally.

Theorem 5. *The formal solution y converges if $|x|$ is small.*

Proof. To show the theorem, it is sufficient to set $f(x) = \sum_{k=0}^{\infty} c_n x^n$. Then $f(x)$ converges on $|x| < \epsilon$ for some $\epsilon > 0$.

Since a is a positive real number, we have

$$\frac{\Gamma(1+s+ja)}{\Gamma(1+s+(j-k)a)} = (1+s+ja)^{ka}(1+O(\frac{1}{j}))$$

by Stirling's formula. Hence

$$d_{k,j} \sim b_{0,k}(1+s+ja)^{na} + b_{1,k}(1+s+ka)^{(n-1)a} + \cdots,$$

because each $h_m(x)$ is holomorphic at the origin, and therefore there exists $\mu > 0$ such that

$$|b_{m,k}| < C\mu^k, \quad 0 \le m \le n.$$

Since we may assume $\{|b_{i,j}||0 \le k \le n, j = 1, 2, \ldots\}$ to be a bounded set by the variable change $x = \alpha t$, if necessary, we can conclude that there exists $A > 0$ such that

$$|\frac{d_{k,j}}{d_{0,k+j}}| \le A, \quad 1 \le k \le n, \; j = 1, 2, \ldots.$$

We take R such that $R > 1 + A$. Then we have $1 + R + \cdots + R^{n-1} \le R^{n-1}\dfrac{R}{R-1}$. Hence if we have $|c_k| \le MR^n$, $1 \le k \le n-1$, for some $M > 0$, then we have

$$|c_n| \le MR^n.$$

It is clear $f(x)$ converges if $|x| < \dfrac{A}{\mu}$. Therefore we have the theorem.

Since $y = \sum_{n=0}^{\infty} c_n x^{q+an}$, it is a many-valued function unless q and a are both integers.

Example. Let L be a fractional analog of Bessel operator

$$L = x^{2a}\frac{d^{2a}}{dx^{2a}} + x^a\frac{d^a}{dx^a} + (x^{2a} - \nu^{2a}).$$

Then $\chi_L(q) = \dfrac{\Gamma(1+q)}{\Gamma(1+q-2a)} + \dfrac{\Gamma(1+q)}{\Gamma(1+q-a)} - \nu^{2a}$. If there exists s such that $\chi_L(s) = 0$, $\chi_L(s+2na) \neq 0$, $n = 1, 2, \ldots$, then to determine c_{2n} by

$$c_{2n}\chi_L(s+2na) + c_{2n-2} = 0, \quad c_0 = 1,$$

$y = x^s + \sum_{n=1}^{\infty} c_{2n}x^{2na}$ satisfies $Ly = 0$. Since

$$c_{2n} = c_{2n}(s) = (-1)^n \prod_{k=1}^{n} \frac{1}{\chi_L(s+2ka)},$$

y is a many-valued function on $\mathbb{C}^\times$ if a is not an integer. Of course, y may not be a single valued function when a is an integer, because s may not be an integer in this case.

If s is a multiple solution of the equation $\chi_L(q) = 0$ of order m, then we can construct the solution y of the equation $Ly = 0$ having the form

$$y(x) = (x^q(\log x)^k f(x^a), \quad 1 \le k \le m-1$$

in the same way. But we will omit these details in this work.

Appendix. $\mathcal{R}[e^{cs}]$ as a Generalized Function

We have extended the domain of $\mathcal{B}$ to different directions, as explained in §5 ([3]). To apply this extension and Theorem 2, we obtain an extension of the domain of $\mathcal{R}$, as understood from discussions in §6. In this Appendix, we give a direct approach for such an extension without using an extension of the Borel transform given in [3]. With Theorem 4, discussions in this Appendix may recover the results yielded in [3].

We set H_3, the set of entire functions $h(s)$ such that

$$|h(s)| \leq Ae^{B|s|^3} \quad \text{for some } A > 0, B > 0,$$

and set

$$\mathsf{F}_4 = \{f(s)e^{-as^4} | f(s) \in \mathsf{H}_3,\ c > 0\} \tag{62}$$

(cf.[13]). With the map $e^{-as^4} \to \log a$, we can identify $\{e^{-as^4}|a > 0\}$ and $\mathbb{R}$. Hence as a set, we have

$$\mathsf{F}_4 \cong \mathsf{H}_3 \times \mathbb{R}. \tag{63}$$

If $f_1(s) = h_1(s)e^{-c_1 s^4}$, $f_2(s) = h_2(s)e^{-c_2 s^4}$, and $|h_1(s)| < A_1 e^{B_1|s|^3}$, $|h_2(s)| < A_2 e^{B_2|s|^3}$, then we have

$$|a_1 f_1(s) + a_2 f_2(s)| \leq (|a_1| + |a_2|)\max\{A_1, A_2\}e^{\max\{B_1,B_2\}|s|^3}.$$

Therefore, F_4 is a vector space.

Definition 6. *We say a series* $\{f_n | f_n \in \mathsf{H}_3\}$ *converges to* f *if there exists* $A > 0,\ B > 0$ *such that*

$$|f_n(s)| \leq Ae^{B|s|^3}$$

for all n*, and* $\{f_n\}$ *converges uniformly in a wider sense to* f *on* $\mathbb{C}$.

H_3 is considered complete by this topology. That is, if $\lim_n f_n = f$, then $f \in \mathsf{H}_3$. We give the F_4 product space a topology of H_3 and $\mathbb{R}$. For later convenience, we denote $\dfrac{1}{\Gamma(1+s)}\mathsf{F}_4$ with $\tilde{\mathsf{F}}_4$.

Let $\mathcal{M}^{-1}(\mathsf{F}_4)$ be $\{p(x)|p(x) = \mathcal{M}^{-1}[q(s)](x), q(s) \in \mathsf{F}_4\}$ and set

$$\mathsf{G}_4 = \{g(x)|xg(x) \in \mathcal{M}^{-1}(\mathsf{F}_4)\}. \tag{64}$$

The topology of G_4 is the induced topology from F_4. Then G_4 becomes a topological vector space.

Definition 7. Let $T : \tilde{\mathsf{F}}_4 \to \mathbb{C}$ be a linear functional on $\tilde{\mathsf{F}}_4$. Then we define $\mathcal{R}[T] : \mathsf{G}_4 \to \mathbb{C}$ by

$$\mathcal{R}[T](g(x)) = T(\mathcal{R}^\dagger[g]). \tag{65}$$

For a function $u(s)$ on $\mathbb{R}$, we define a linear functional T_u of F_4, *etc.*, by

$$T_u(f) = \int_{-\infty}^{\infty} u(s)f(s)ds.$$

We regard u to be a generalized function T_u, which is defined on F_4, *etc.* That is, we define $\mathcal{R}[u]$ by $\mathcal{R}[T_u]$.

Note. In the rest of the work, we often use the notation

$$(u, f) = \int_{-\infty}^{\infty} u(s) f(s) ds.$$

By using this notation, we have

$$\mathcal{R}[u](g) = \mathcal{R}[T_u](g) = (u, \mathcal{R}^\dagger[g]).$$

According to this definition, we define $\mathcal{R}[e^{cs}] : \mathsf{G}_4 \to \mathbb{C}$ by

$$\begin{aligned}\mathcal{R}[e^{cs}](g(x)) &= (e^{cs}, \mathcal{R}^\dagger[g(x)]) \\ &= \int_{-\infty}^{\infty} e^{cs} \frac{1}{\Gamma(1+s)} \mathcal{M}[xg(x)](s) ds. \end{aligned} \tag{66}$$

Since $\mathcal{R}^\dagger \in \tilde{\mathsf{F}}_4$, the integral of the right hand side of (66) converges. Hence $\mathcal{R}[e^{cs}]$ is defined as a generalized function defined on G_4.

If $t > 0$, we can define $\mathcal{R}[e^{as}e^{-ts^2}]$ as a function, and $(\mathcal{R}[e^{as}e^{-ts^2}](x), g(x)), g(x) \in \mathsf{G}_4$ is defined for any $g \in \mathsf{G}_4$. By Lebesgue's convergence theorem, we have:

Lemma 6. As generalized functions on G_4, we have

$$\lim_{t \downarrow 0} \mathcal{R}[e^{-as}e^{-ts^2}] = \mathcal{R}[e^{as}]. \tag{67}$$

We know that $\int_{-\infty}^{\infty} \tau_a f(s) g(s) ds = \int_{-\infty}^{\infty} f(s) \tau_{-a} g(s) ds$. Hence, assuming if $g(x) \in \mathsf{G}_4$, then $T[\tau_{-a} g]$ exists, we define the fractional order differentiation of a generalized function T based on G_4 as follows.

Definition 8. Let T be a generalized function defined on F_4, while $\mathcal{R}[T]$ is also defined. Then we define $\dfrac{d^{\mathcal{R}}[T]}{dx^a}$ by

$$(\frac{d^a}{dx^a} \mathcal{R}[T])[g(x)] = T[\tau_{-a} \mathcal{R}^\dagger[g(x)](s)]. \tag{68}$$

According to this definition, we have

$$(\frac{d^a}{dx^a} \mathcal{R}[e^{cs}])[g(x)] = (e^{cs}, \tau_{-a} \mathcal{R}^\dagger[g(x)](s)) = (\tau_a e^{cs}, \mathcal{R}^\dagger[g(x)](s)).$$

Hence $\dfrac{d^a}{dx^a} \mathcal{R}[e^{cs}]$ exists and by Lemma 1,

$$\lim_{t \downarrow 0} \frac{d^a}{dx^a} \mathcal{R}[e^{cs}e^{-ts^2}] = \frac{d^a}{dx^a} \mathcal{R}[e^{cs}] = e^{ac} \mathcal{R}[e^{cs}]. \tag{69}$$

Example. The equation $\frac{d^a y}{dx^a} = \lambda y$ has eigen generalized functions:

$$y = \mathcal{R}[e^{cs}], \quad e^{ca} = \lambda. \tag{70}$$

Note. We have already shown that the equation $\frac{d^a y}{dx^a} = \lambda y$ has a solution

$$u_1(x) = \frac{d}{dx}\big(I^a(E_{1/a}(\lambda x^a)\big).$$

This $u_1(x)$ differs from the aforementioned $\mathcal{R}[e^{cs}]$ based on the reason explained below.

If $\mathcal{R}[f(s)]$ is defined as a generalized function on G_4 and satisfies

$$\frac{d^a}{dx^a}\mathcal{R}[f(s)] = \lambda \mathcal{R}[f(s)],$$

then $f(s+a)$ should be $\lambda f(s)$. Therefore we have

$$f(s) = e^{\nu s}h(s), \quad e^{\nu a} = \lambda, \quad h(s+a) = h(s). \tag{71}$$

Since a periodic function is bounded on $\mathbb{R}$, we can define $\mathcal{R}[e^{cs}h(s)]$ as a generalized function on G_3 by

$$\mathcal{R}[e^{cs}h(s)](g(x)) = (e^{cs}h(s), \mathcal{R}^\dagger[g(x)](s)).$$

But this means the equation $\frac{dU}{dx} = \lambda U$ has infinitely many independent solutions. Therefore $\mathcal{R}[e^{cs}]$ is neither a function nor a distribution in the sense of Schwartz.

Fixing $\mu = \log\lambda$, c is written as $\frac{\mu}{a} + \frac{2n}{a}\pi i$, $n \in \mathbb{Z}$;

$\mu = \begin{cases} \log\lambda, & \lambda > 0, \\ \log|\lambda| + \pi i, & \lambda < 0, \end{cases}$. Hence equation (37) has infinitely many independent eigen (generalized) functions, considering $\frac{d^\alpha}{dx^\alpha}$ is defined on the space of operators of Mikusinski. As we have shown in §7, (37) has infinitely many independent eigenfunctions if α is an irrational number. We ask whether there exists a relation between these two observations.

Note. Similarly, we define $\log(\frac{d}{dx})T$ by $(\log(\frac{d}{dx})T)[f(s)] = -T[\frac{df(s)}{ds}]$. Then we obtain

$$\log(\frac{d}{dx})\mathcal{R}[e^{cs}] = \mathcal{R}[\frac{d}{ds}e^{cs}] = c\mathcal{R}[e^{cs}]. \tag{72}$$

Therefore a solution for the equation $\log(\frac{d}{dx})T = \lambda T$ is given by $T = \mathcal{R}[e^{\lambda s}]$.

If $h(s) \in \mathsf{H}_3$, then $\int_{-\infty}^{\infty} h(s)f(s)ds$, $f \in \tilde{\mathsf{F}}_4$ exists. Hence we can define $\mathcal{R}[h(s)]$ as a generalized function on G_3.

In general, to set

$$\mathsf{H}_n = \{h(s) | |h(s)| \le Ae^{B|s|^n}, \text{ for some } A > 0, B > 0\}$$

and to set $\mathsf{F}_n = \{h(s)e^{-cs^n} | h \in \mathsf{H}_{n-1}, c > 0\}$, we can define $\mathcal{R}[h(s)]$,$h(s) \in \mathsf{H}_{4n-1}$ as a generalized function on

$$\mathsf{G}_{4n} = \{g(x)|xg(x) \in \mathcal{M}^{-1}(\mathsf{F}_{4n}\}.$$

This satisfies

$$\lim_{t \downarrow 0} \mathcal{R}[h(s)e^{-ts^{4n}}] = \mathcal{R}[h(s)],$$

where $\mathcal{R}[h(s)e^{-ts^{4n}}]$ exists as a function if $t > 0$. Therefore, we define

$$\frac{d^a}{dx^a}\mathcal{R}[h(s)] = \mathcal{R}[\tau_a h(s)], \quad \log(\frac{d}{dx})\mathcal{R}[h(s)] = \mathcal{R}[\frac{dh(s)}{ds}].$$

References

[1] Agrawal, Om. Baleanu, D., Fractional formulation within Caputo's derivatives, *Czech. J. Phys.* 56(2006), 1087-1092,

[2] Almeida, R. Malianowska, A. Torres, D., A fractional calculus of variations for multiple integrals with application to vibrating strings, *J. Math. Phys.* 51(2010), 033503, 12pp.

[3] Asada, A., Borel transformation in non-analytic category, *J. Fac. Sci. Shinshu Univ.* 12(1977), 1-36.

[4] Asada, A., Fractional calculus and infinite order differential operators, *Yokohama Math. J.* 55(2010), 129-147.

[5] Asada, A., Fractional Calculus and Gamma Function, Geometric Theory of Dynamical Systems and Related Topics (*eds*. Iwai, T. Tanimura, S. Yamaguchi, Y.). *Publ. RIMS* 1692(2010), 17-38.

[6] Asada, A., Lie algebra generated by logarithm of differentiation and logarithm. *Balkan J. Geom. and Its Appl.* 16(2011), 1-11.

[7] Asada, A., Integral transform arose from fractional calculus and discrete delta potential. *Contemporary topics in Mathematics and Statistics with Applications*, ed. Adhikari, M.R., Chap. 3. Kolkata 2013.

[8] Asada A. *Domain of integral transform arose from fractional calculus and extended Borel transform*, to appear.

[9] Bubscri, D. Dattoli, G. Quattrromini, M. *Relativistic equations with fractional and pseudo-differential equations*, arXiv. 1101.5066v1.

[10] Erdéli, A. Magnus, W. Oberbettinger, F. Tricomi, F. G., Higher Transcendental. *Functions, Chap.* 18. New York 1981.

[11] Ertürk, V. S. Monami, S., Solving systems of fractional differential equations using differential transformation method, *J. Comp. and Appl. Math.* 215(2008), 142-151.

[12] Fedelico, G. Torres, D., Fractional Noether's theorem in the Riesz-Caputo sense, *Appl. Math. Comp.* 217(2010), 1023-1033.

[13] Fujii, K., Beyond the Gaussian. *I. SIGMA* 7(2011), 022, 12pp, II. arXiv 1103.4428v1.

[14] Gelfand, I. M. Shilov, G. E., *Generalized Functions*, II. New York, 1964.

[15] Kilbas, A. A. Srivastra, H. M. Trujillo, J. J., Theory and Applications of Fractional Differential Equations, *North-Holland Math. Studies* 204, Amsterdam 2006.

[16] Li, M. F. Ren, J. R. Zhu, T., *Series expansion in fractional calculus and fractional differential equations*, arXiv, 09010.4819v1. *Fractional vector calculus and fractional special functions*, arXiv. 1001.2829.

[17] Mikusinski, J., Operational Calculus, *Pergamon*, 1959.

[18] Morita, T. Sato, K., Solutions of fractional differential equations in terms of distribution theory, *Interdisc. Inf. Sci.* 12(2006), 71-83.

[19] Nakanishi, N., Logarithm type functions of the differential operator, *Yokohama J. Math.* 55(2010), 149-163.

[20] Purishchay, M. S., Deformed Heisenberg Algebras, Fractional Spin Fields, and Supersymmetry without Fermions, *Ann. Phys.* 245(1996), 339-360.

[21] Chaichian, M. Mnatsakanova, M. Tureanu, A. Vernov, Y., Test functions space in noncommutative quantum field theory, *JHEP* 0809(2008), 125.

[22] Fischer, A. Szabo, R. J., Duality covariant quantum field theory on noncommutative Minkowski space, *JHEP* 0902(2009), 031.

[23] Soloviev, M. A., Star product algebras of test functions, *Theor. Math. Phys.* 153(2007), 1351.

[24] Zahn, J., Divergence in quantum field theory on the noncommutative two-dimensional Minkowski space with Grosse-Wulkenhaar potential, *Ann. Henri Poincaré* 12(2011), 777-804.

In: Fractional Calculus: Theory
Editors: R. Abi Zeid Daou and X. Moreau

ISBN: 978-1-63463-002-3

Chapter 6

INTRODUCTION TO STABILITY THEORY OF LINEAR FRACTIONAL DIFFERENCE EQUATIONS

Jan Čermák and Tomáš Kisela*
Institute of Mathematics, Brno University of Technology
Brno, Czech Republic

Abstract

One of the important aspects of the current development of fractional calculus is its expansion to various discrete settings. Foundations of discrete fractional calculus were set in 1960s and 1970s, but its systematic study started only recently. This research is carried out especially in two main directions: discrete fractional calculus as a special branch of discrete analysis, and as a tool of numerical investigations of underlying continuous fractional problems.

In this chapter, we will summarize and develop some recent results achieved in the field of discrete fractional calculus, with a special attention devoted to fractional difference equations and essentials of their qualitative theory. First we will recall some basic definitions and properties of discrete fractional calculus, which are useful in the following investigations. Then we will introduce the notion of linear fractional difference equation considered on a uniform mesh of points, investigate the well-posedness of the corresponding initial value problem and discuss the structure of the solution space. Also, an explicit form of the general solution will be described via discrete analogues of the Mittag-Leffler type functions. As the main results, basic qualitative properties of some test equations will be stated, with emphasis on their stability and asymptotics. Finally, several notes will discuss connections of these results to numerical analysis of underlying fractional differential equations and we will also outline their possible extensions to other (non-uniform) discrete settings.

Keywords: fractional h-calculus, fractional difference equation, stability, asymptotic behaviour

---

*E-mail address: kisela@fme.vutbr.cz (Corresponding author)

1. Introduction

The well-known history of continuous fractional calculus started with some preliminary notes on its basic introductions more than three hundred years ago. During this period, many outstanding mathematicians contributed to the discussion on key definitions of fractional integrals and fractional derivatives, their motivations, properties and consequences (for a historical survey on this topic as well as its theoretical foundations we refer, e.g., to [27, 34, 37, 39]). For a real function $f(t)$ having appropriate properties, a generally accepted definition of the fractional integral of real order $\gamma > 0$ is

$$ {}_aI^{\gamma}f(t) = \int_a^t \frac{(t-\tau)^{\gamma-1}}{\Gamma(\gamma)} f(\tau)\,\mathrm{d}\tau\,, \qquad t > a, \quad a \in \mathbb{R}\,, \tag{1.1} $$

where $\Gamma(\cdot)$ is the Euler Gamma function (sometimes the notation ${}_a\mathrm{D}^{-\gamma} \equiv {}_aI^{\gamma}$ is used instead). Regarding the definition of the fractional derivative of real order $\alpha > 0$, two approaches are standardly used. The Riemann-Liouville fractional derivative is given by composition of integer-order derivative and fractional integral, namely

$$ {}_a\mathrm{D}^{\alpha}f(t) = \frac{\mathrm{d}^{\lceil\alpha\rceil}}{\mathrm{d}t^{\lceil\alpha\rceil}}\, {}_aI^{\lceil\alpha\rceil-\alpha}f(t)\,, \qquad t > a, \quad a \in \mathbb{R}\,, \tag{1.2} $$

where $\lceil\alpha\rceil = \min(m \in \mathbb{Z}^+ : m \geq \alpha)$ is the ceiling function. In many applications of fractional calculus, the Caputo fractional derivative

$$ {}_a^{\mathrm{C}}\mathrm{D}^{\alpha}f(t) = {}_aI^{\lceil\alpha\rceil-\alpha}\frac{\mathrm{d}^{\lceil\alpha\rceil}f(t)}{\mathrm{d}t^{\lceil\alpha\rceil}}\,, \qquad t > a, \quad a \in \mathbb{R} \tag{1.3} $$

is used. Obviously, for $\alpha \in \mathbb{Z}^+$ both definitions become the standard integer-order derivative (fractional integral of zero order is considered to be the function itself). Throughout this chapter, we utilize all these definitions and notation (if the lower bound a is omitted then $a = 0$).

Contrary to its continuous counterpart, the history of discrete fractional calculus, dealing with sums and differences of non-integer orders, is not so long and rich. Its foundations trace back to 1960s when R. P. Agarwal and W. A. Al-Salam published their pioneering works [1, 2] introducing fractional calculus in the framework of so-called q-calculus, i.e., calculus on the discrete mesh of points forming a geometrical sequence. The essentials of fractional calculus on discrete sets consisting of a uniform mesh of points were set by J. B. Díaz and T. S. Osler. In their paper [16], the notion of fractional difference on the set of integers was introduced, for a real sequence $f(n)$, via generalization of the forward difference operator as

$$ {}^{\mathrm{DO}}\Delta^{\alpha}f(n) = \sum_{j=0}^{\infty}(-1)^j\binom{\alpha}{j}f(n+\alpha-j)\,, \qquad n \in \mathbb{Z}\,, $$

where the order α is arbitrary real. The problem with this definition is that it requires values of $f(n)$ at non-integer values of its argument. For this reason, a similar definition based on the backward difference operator, which removes this domain issue, was proposed by C. W. J. Granger and R. Joyeux [23] and by J.R.M. Hosking [26]. Their introduction was

$$^{\mathrm{GJH}}\nabla^{\alpha} f(n) = \sum_{j=0}^{\infty} (-1)^j \binom{\alpha}{j} f(n-j) \,, \qquad n \in \mathbb{Z} \,.$$

Although the fractional differences based on this definition have many interesting properties, they require the sequence $f(n)$ to be defined for all $n \in \mathbb{Z}$. More precisely, there is no lower bound in this introduction corresponding to the parameter a involved in the definitions (1.1)-(1.3) of continuous fractional operators. This discrepancy was overcome by H. L. Gray and N. F. Zhang in [24]. Here the authors introduced a new definition utilizing a finite sum, namely

$$^{\mathrm{GZ}}_{\;a}\nabla^{\alpha} f(n) = \nabla^{\lceil \alpha \rceil} \sum_{j=a}^{n} (-1)^{n-j} \binom{\alpha - \lceil \alpha \rceil}{n-j} f(j) \,, \qquad n \geq a, \quad a \in \mathbb{Z} \,,$$

where α is a positive real and $\nabla^{\lceil \alpha \rceil}$ means the standard backward difference of $\lceil \alpha \rceil$ order.

In recent years, many papers on other related introductions and properties of basic notions of discrete fractional calculus have appeared. In particular, more general discrete settings, such as a mesh of points involving arithmetic and geometrical sequences as special cases (see, e.g., [15]), have been considered. For some recent papers on properties of fractional sums and differences we refer, e.g., to [6, 25, 35, 36].

Despite the recent developments of discrete fractional calculus, the systematic investigations of fractional difference equations (i.e., difference equations involving differences of non-integer orders) is just at its beginning. Some introductory results on linear fractional difference equations can be found, e.g., in [7, 21, 22, 32] or [10, 12]. The main goal of this chapter is to present basic qualitative results on linear fractional difference equations originating from the recent papers [11, 13, 14, 31] as well as their other new extensions.

The basic framework for our following investigations is formed by the so-called h-calculus considered on the domain $h\mathbb{Z} = \{t_n = nh : n \in \mathbb{Z}, \ h > 0\}$. Compared to the standard difference case (with $h = 1$), extensions to h-calculus are mostly only technical, but enable interesting discussions of various properties with respect to changing h. As it is customary, we define the backward h-difference

$$\nabla_h f(t_n) = \frac{f(t_n) - f(t_{n-1})}{h} \tag{1.4}$$

implying its h-antidifference (sum) as

$$_a\nabla_h^{-1} f(t_n) = \sum_{k=\frac{a}{h}+1}^{n} h\, f(t_k) \,, \qquad a \in h\mathbb{Z} \,. \tag{1.5}$$

Differences and sums of higher orders are then introduced recursively for $m \in \mathbb{Z}^+$ as

$$\nabla_h^m \equiv \nabla_h \nabla_h^{m-1}$$

and

$$ {}_a\nabla_h^{-m} \equiv {}_a\nabla_h^{-1} \, {}_a\nabla_h^{-m+1} \,, $$

respectively.

The reason why we prefer to discuss the backward difference operators instead of the forward ones is two-fold. First, as mentioned above, the forward fractional operators require sequences prescribed also at non-integer values of their arguments, which may be inconvenient (see, e.g., [24] and [10]). Secondly, keeping in mind possible applications to numerical analysis of fractional differential equations, it is well-known from the integer-order case that difference schemes utilizing backward operators are preferred especially due to their better stability properties.

The organization of this chapter is as follows. In Section 2, we recall definitions and properties of basic fractional operators on the discrete set $h\mathbb{Z}$, including some illustrative examples. Section 3 discusses the essentials of basic theory of linear fractional difference equations. In particular, we show the well-posedness of corresponding initial value problems and describe the structure of the solution space. Section 4 is devoted to stability analysis of single-term linear fractional difference equations involving the Riemann-Liouville or the Capute difference operator. We derive here necessary and sufficient stability conditions for both types, including the asymptotic decay rate of the solutions. In Section 5, a technique of extending these stability results to multi-term fractional difference equations is proposed. Its main idea is illustrated by the discrete analogue of the Basset equation. Section 6 concludes this chapter with several remarks and outlines.

2. Basic Properties of Fractional Difference Operators

In this section, we discuss the essentials of the so-called fractional h-calculus, i.e., fractional calculus on the set $h\mathbb{Z}$. A survey of its basic notions and properties can be found, e.g., in [25, 35].

First we notice that the kernel of fractional integral operator in continuous calculus is formed by the power function $(t-\tau)^\beta/\Gamma(\beta+1)$, $\beta \in \mathbb{R}$. In h-calculus, this role is played by the so-called h-Pochhammer symbol

$$ \frac{(t_n - t_k)_h^{(\beta)}}{\Gamma(\beta+1)} = \frac{h^\beta \Gamma(n-k+\beta)}{\Gamma(\beta+1)\Gamma(n-k)} \,, \qquad n \geq k \,, $$

which was utilized already in [24]. Throughout this chapter, we often employ its expression via the binomial numbers formula

$$ \frac{(t_n - t_k)_h^{(\beta)}}{\Gamma(\beta+1)} = \frac{(-1)^{n-k-1}}{h^{-\beta}} \binom{-\beta-1}{n-k-1} \,. \tag{2.6} $$

Instead of formal introductions of basic notions of fractional h-calculus (replacing appropriate symbols in (1.1)-(1.3) by (1.4)-(2.6)), we prefer to present their mathematical justification, which arises from the Cauchy formula for repeated integration (see, e.g., [37, 39]).

Proposition 2.1 (The discrete Cauchy formula). *Let $m \in \mathbb{Z}^+$ and $a \in h\mathbb{Z}$. Then, for a real sequence $f(t_n)$, it holds*

$$ {}_a\nabla_h^{-m} f(t_n) = \sum_{k=\frac{a}{h}+1}^{n} h \frac{(t_n - t_{k-1})_h^{(m-1)}}{\Gamma(m)} f(t_k) . \tag{2.7} $$

Proof. This assertion can be proved by the induction principle. If $m = 1$ then (2.7) obviously holds. Let $m \geq 2$ and assume that (2.7) holds with m replaced by $m-1$, i.e.

$$ {}_a\nabla_h^{-m+1} f(t_n) = \sum_{k=\frac{a}{h}+1}^{n} h \frac{(t_n - t_{k-1})_h^{(m-2)}}{\Gamma(m-1)} f(t_k) . $$

Using the definition of m-th multiple sum we can write

$$ \begin{aligned} {}_a\nabla_h^{-m} f(t_n) &= \sum_{k=\frac{a}{h}+1}^{n} h \, {}_a\nabla_h^{-m+1} f(t_k) = \sum_{k=\frac{a}{h}+1}^{n} h \sum_{j=\frac{a}{h}+1}^{k} h \frac{(t_k - t_{j-1})_h^{(m-2)}}{\Gamma(m-1)} f(t_j) \\ &= \sum_{j=\frac{a}{h}+1}^{n} \frac{f(t_j)}{h^{-m}} \sum_{k=j}^{n} (-1)^{k-j} \binom{-m+1}{k-j} = \sum_{j=\frac{a}{h}+1}^{n} \frac{f(t_j)}{h^{-m}} \sum_{k=0}^{n-j} (-1)^{k} \binom{-m+1}{k} \\ &= \sum_{j=\frac{a}{h}+1}^{n} \frac{(-1)^{n-j}}{h^{-m}} \binom{-m}{n-j} f(t_j) = \sum_{j=\frac{a}{h}+1}^{n} h \frac{(t_n - t_{j-1})_h^{(m-1)}}{\Gamma(m)} f(t_j) , \end{aligned} $$

where the basic properties of binomial numbers have been utilized. □

Clearly, we can replace the integer m in (2.7) with a positive real γ. Thus, the formula (2.7) is a corner stone for the following introduction.

Definition 2.1. *Let $\gamma \in \mathbb{R}_0^+$ and $a \in h\mathbb{Z}$. Then, for a real sequence $f(t_n)$, we define the fractional sum of order $\gamma \in \mathbb{R}^+$ (with the lower bound a) as*

$$ {}_a\nabla_h^{-\gamma} f(t_n) = \sum_{k=\frac{a}{h}+1}^{n} h \frac{(t_n - t_{k-1})_h^{(\gamma-1)}}{\Gamma(\gamma)} f(t_k) , \qquad t_n > a \tag{2.8} $$

and for $\gamma = 0$ we put ${}_a\nabla_h^0 f(t_n) = f(t_n)$.

The h-Pochhammer symbol has several interesting properties. In particular, it holds a discrete analogue of the power rule (see, e.g., [39]), which is the fundamental relation of the continuous fractional calculus.

Lemma 2.1. *Let $\gamma \in \mathbb{R}^+$, $\beta \in \mathbb{R} \setminus \mathbb{Z}^-$ and $a \in h\mathbb{Z}$. Then*

$$ {}_a\nabla_h^{-\gamma} \frac{(t_n - a)_h^{(\beta)}}{\Gamma(\beta+1)} = \frac{(t_n - a)_h^{(\beta+\gamma)}}{\Gamma(\beta+\gamma+1)} . $$

Proof. Using (2.6) we can write

$$ {}_a\nabla_h^{-\gamma}\frac{(-1)^{n-\frac{a}{h}-1}}{h^{-\beta}}\binom{-\beta-1}{n-\frac{a}{h}-1}=\sum_{k=\frac{a}{h}+1}^{n}\frac{(-1)^{n-k}}{h^{-\gamma}}\binom{-\gamma}{n-k}\frac{(-1)^{n-\frac{a}{h}-1}}{h^{-\beta}}\binom{-\beta-1}{k-\frac{a}{h}-1} $$

$$ =\frac{(-1)^{n-\frac{a}{h}-1}}{h^{-\gamma-\beta}}\sum_{k=\frac{a}{h}+1}^{n}\binom{-\gamma}{n-k}\binom{-\beta-1}{k-\frac{a}{h}-1}=\frac{(-1)^{n-\frac{a}{h}-1}}{h^{-\gamma-\beta}}\binom{-\beta-\gamma-1}{n-\frac{a}{h}-1}, $$

where we have used the Vandermonde identity [38, the formula (6:5:5)] on the last row. □

Example 2.1. *Consider $0<h<1$ and the sequence $f(t_n)$ given by*

$$ f(t_n)=\begin{cases} t_n-\frac{1}{2}, & 0\le n\le\lceil\frac{1}{h}\rceil,\\ -t_n+\frac{3}{2}, & n>\lceil\frac{1}{h}\rceil \end{cases} $$

and calculate its fractional sum of order $\gamma>0$. Using Lemma 2.1 we get

$$ {}_0\nabla_h^{-\gamma}f(t_n)={}_0\nabla_h^{-\gamma}\frac{(t_n)_h^{(1)}}{\Gamma(2)}-\frac{1}{2}{}_0\nabla_h^{-\gamma}\frac{(t_n)_h^{(0)}}{\Gamma(1)}=\frac{(t_n)_h^{(\gamma+1)}}{\Gamma(\gamma+2)}-\frac{1}{2}\frac{(t_n)_h^{(\gamma)}}{\Gamma(\gamma+1)} $$

for $0\le n\le\lceil\frac{1}{h}\rceil$. Similar procedures as those used in the proof of Lemma 2.1 yield

$$ {}_0\nabla_h^{-\gamma}f(t_n)=\sum_{k=1}^{\lceil\frac{1}{h}\rceil}h\frac{(t_n-t_{k-1})_h^{(\gamma-1)}}{\Gamma(\gamma)}\left(t_k-\frac{1}{2}\right)+\sum_{k=\lceil\frac{1}{h}\rceil+1}^{n}h\frac{(t_n-t_{k-1})_h^{(\gamma-1)}}{\Gamma(\gamma)}\left(-t_k+\frac{3}{2}\right) $$

$$ =\sum_{k=1}^{\lceil\frac{1}{h}\rceil}h\frac{(t_n-t_{k-1})_h^{(\gamma-1)}}{\Gamma(\gamma)}(2t_k-2)+\sum_{k=1}^{n}h\frac{(t_n-t_{k-1})_h^{(\gamma-1)}}{\Gamma(\gamma)}\left(-t_k+\frac{3}{2}\right) $$

$$ =2\sum_{k=1}^{\lceil\frac{1}{h}\rceil}h\frac{(t_n-t_{k-1})_h^{(\gamma-1)}}{\Gamma(\gamma)}\frac{(t_k)_h^{(1)}}{\Gamma(2)}-2\sum_{k=1}^{\lceil\frac{1}{h}\rceil}h\frac{(t_n-t_{k-1})_h^{(\gamma-1)}}{\Gamma(\gamma)}\frac{(t_k)_h^{(0)}}{\Gamma(1)} $$

$$ -\frac{(t_n)_h^{(\gamma+1)}}{\Gamma(\gamma+2)}+\frac{3}{2}\frac{(t_n)_h^{(\gamma)}}{\Gamma(\gamma+1)} $$

for $n>\lceil\frac{1}{h}\rceil$.

It might be interesting to compare this result with its continuous counterpart. Consider the function

$$ F(t)=\begin{cases} t-\frac{1}{2}, & 0\le t\le 1,\\ -t+\frac{3}{2}, & t>1. \end{cases} $$

Utilizing the ideas presented in [39] we obtain

$$ {}_0\mathrm{D}^{-\gamma}F(t)=\begin{cases} \frac{t^{\gamma+1}}{\Gamma(\gamma+2)}-\frac{1}{2}\frac{t^{\gamma}}{\Gamma(\gamma+1)}, & 0\le t\le 1,\\ 2\frac{t^{\gamma+1}B_{1/t}(\gamma,2)}{\Gamma(\gamma)}-2\frac{t^{\gamma}B_{1/t}(\gamma,1)}{\Gamma(\gamma)}-\frac{t^{\gamma+1}}{\Gamma(\gamma+2)}+\frac{3}{2}\frac{t^{\gamma}}{\Gamma(\gamma+1)}, & t>1, \end{cases} $$

where $B_z(\cdot,\cdot)$ denotes the incomplete Beta function (see, e.g., [38]), which illustrates a formal agreement between the continuous and discrete case. Moreover, we can observe similarities in resulting sequences and functions (see Figure 1 depicting ${}_0\nabla_h^{-\gamma}f(t_n)$ and ${}_0\mathrm{D}^{-\gamma}F(t)$ for several values of γ) as well as a "continuous transition" between the original sequence and its first sum with respect to γ changing from 0 to 1.

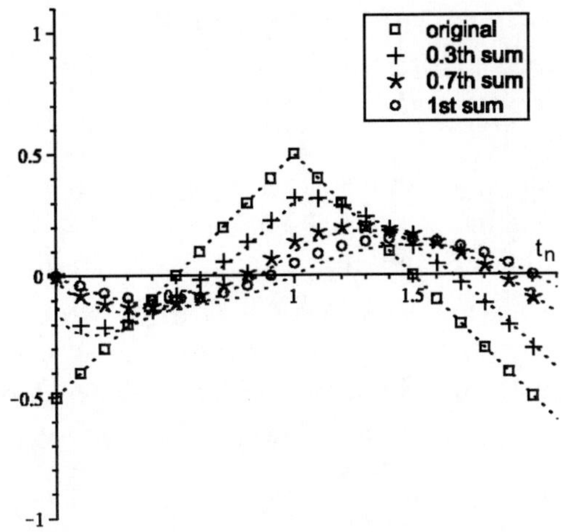

Figure 1. Fractional sums of $f(t_n)$ for $h = 0.1$. The dotted lines represent the corresponding fractional integrals of $F(t)$.

Now, we proceed to the introduction of fractional differences. As it was recalled in Section 1, this notion admits more approaches. Following the continuous patterns (1.2) and (1.3), we introduce two types of fractional differences.

Definition 2.2. *Let $\alpha \in \mathbb{R}^+$ and $a \in h\mathbb{Z}$. Then, for a real sequence $f(t_n)$, we define the Riemann-Liouville fractional difference of order α (with the lower bound a) as*

$$ {}_a\nabla_h^{\alpha} f(t_n) = {}_a\nabla_h^{\lceil\alpha\rceil}\, {}_a\nabla_h^{-(\lceil\alpha\rceil-\alpha)} f(t_n) \,. \tag{2.9}$$

Definition 2.3. *Let $\alpha \in \mathbb{R}^+$ and $a \in h\mathbb{Z}$. Then, for a real sequence $f(t_n)$, we define the Caputo fractional difference of order α (with the lower bound a) as*

$$ {}_a^{\mathrm{C}}\nabla_h^{\alpha} f(t_n) = {}_a\nabla_h^{-(\lceil\alpha\rceil-\alpha)}\, {}_a\nabla_h^{\lceil\alpha\rceil} f(t_n) \,. \tag{2.10}$$

Further, we formulate the properties analogous to that of Lemma 2.1.

Corollary 2.1. *Let $\alpha \in \mathbb{R}^+$, $\beta \in \mathbb{R} \setminus \mathbb{Z}^-$ and $a \in h\mathbb{Z}$. Then*

(i) $$ {}_a\nabla_h^{\alpha} \frac{(t_n-a)_h^{(\beta)}}{\Gamma(\beta+1)} = \begin{cases} \frac{(t_n-a)_h^{(\beta-\alpha)}}{\Gamma(\beta-\alpha+1)}, & \beta-\alpha \notin \{-1,\dots,-\lceil\alpha\rceil\}, \\ 0, & \beta-\alpha \in \{-1,\dots,-\lceil\alpha\rceil\}. \end{cases}$$

(ii) $$ {}_a^{\mathrm{C}}\nabla_h^{\alpha} \frac{(t_n-a)_h^{(\beta)}}{\Gamma(\beta+1)} = \begin{cases} \frac{(t_n-a)_h^{(\beta-\alpha)}}{\Gamma(\beta-\alpha+1)}, & \beta \notin \{0,\dots,\lceil\alpha\rceil-1\}, \\ 0, & \beta \in \{0,\dots,\lceil\alpha\rceil-1\}. \end{cases}$$

Proof. (i) Lemma 2.1 implies

$$ {}_a\nabla_h^{\alpha} \frac{(t_n-a)_h^{(\beta)}}{\Gamma(\beta+1)} = \nabla_h^{\lceil\alpha\rceil} \left({}_a\nabla_h^{-(\lceil\alpha\rceil-\alpha)} \frac{(t_n-a)_h^{(\beta)}}{\Gamma(\beta+1)} \right) = \nabla_h^{\lceil\alpha\rceil} \frac{(t_n-a)_h^{(\beta+\lceil\alpha\rceil-\alpha)}}{\Gamma(\beta+\lceil\alpha\rceil-\alpha+1)} \,.$$

Then the statement follows from (2.6) with respect to the basic properties of binomial numbers (for more details see, e.g., [10]).
(ii) The proof procedure is analogous. □

Now, we are in a position to discuss the composition rules. This matter was already studied in [15], where the authors assumed the function $f(t_n)$ to be defined on the whole $h\mathbb{Z}$ and extended the validity of (2.8) to $t_n \le a$. Consequently, the derived composition rules have a simple form ${}_a\nabla_h^{\alpha}{}_a\nabla_h^{\beta} f(t_n) = {}_a\nabla_h^{\alpha+\beta} f(t_n)$ for all $\alpha, \beta \in \mathbb{R}$.

Taking into account Definitions 2.1 and 2.2 (i.e., without extension of the domain of $f(t_n)$) we get the following composition rules which correspond to the continuous pattern. We note that similar results were derived in [25] for the case $h = 1$.

Lemma 2.2. *Let $\alpha \in \mathbb{R}$, $\beta \in \mathbb{R}^+$. Then it holds*

$$ {}_a\nabla_h^{\alpha}{}_a\nabla_h^{-\beta} f(t_n) = {}_a\nabla_h^{\alpha-\beta} f(t_n)\,, $$

$$ {}_a\nabla_h^{\alpha}{}_a\nabla_h^{\beta} f(t_n) = {}_a\nabla_h^{\alpha+\beta} f(t_n) - \sum_{j=1}^{\lceil\beta\rceil} \frac{(t_n-a)_h^{(-\alpha-j)}}{\Gamma(-\alpha-j+1)} \left(\nabla_h^{\beta-j} f(t_n)\big|_{t_n=a}\right). $$

Proof. The proof is only a technical matter (for more details we refer, e.g., to [29]). □

Using this result it is possible to specify the relation between the Riemann-Liouville and Caputo fractional difference in the form

$$ {}_a^{\mathrm{C}}\nabla_h^{\alpha} f(t_n) = {}_a\nabla_h^{\alpha} f(t_n) - \sum_{j=0}^{\lceil\alpha\rceil-1} \frac{(t_n-a)_h^{(j-\alpha)}}{\Gamma(j-\alpha+1)} \left(\nabla_h^{j} f(t_n)\big|_{t_n=a}\right). \tag{2.11} $$

This relation is a discrete analogue of [27, the formula (2.4.6)], which is the formula well-known from the continuous fractional calculus. We can see that both the operators coincide either for $a \to -\infty$, which is a rather theoretical condition, or in the case when the sequence $f(t_n)$ and all its integer-order differences up to the order $\lceil\alpha\rceil - 1$ are zero at point $t_n = a$.

Example 2.2. *Let $f(t_n)$ be the same sequence as in Example 2.1 and consider its Riemann-Liouville and Caputo differences. Calculations concerning evaluation of ${}_0\nabla_h^{\alpha} f(t_n)$ and ${}_0^{\mathrm{C}}\nabla_h^{\alpha} f(t_n)$ are analogous to those made in Example 2.1 and therefore are omitted. We plot the results in Figures 2 and 3. Analogously to the case of fractional sums, we can observe a connection with the corresponding continuous operators as well as a "continuous transition" from the original sequence to its first difference as α varies from 0 to 1. Notice also that the differences are not defined at the intial point $a = 0$. In particular, it enables the Riemann-Liouville fractional difference to follow its continuous counterpart even though it is unbounded in a neighbourhood of the origin.*

We have already emphasized that the previous definitions of discrete fractional operators based on the h-Pochhammer symbol reveal interesting analogues between continuous and discrete fractional calculus. However, in the sequel we are going to utilize its alternative introductions, which are more convenient for computational reasons.

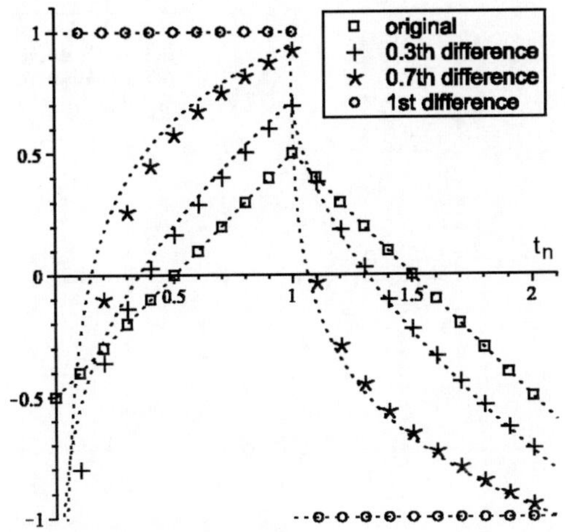

Figure 2. Riemann-Liouville fractional differences of $f(t_n)$ for $h = 0.1$. The dotted lines represent the corresponding fractional derivatives of $F(t)$.

Let ${}_a\nabla_h^\alpha f(t_n)$ be given by (2.9) and $0 < \alpha < 1$. Then, the direct calculations show that

$$\begin{aligned} {}_a\nabla_h^\alpha f(t_n) &= \frac{1}{h}\left({}_a\nabla_h^{\alpha-1} f(t_n) - {}_a\nabla_h^{\alpha-1} f(t_{n-1})\right) \\ &= \sum_{k=\frac{a}{h}+1}^{n} \frac{(-1)^{n-k}}{h^\alpha}\binom{\alpha-1}{n-k} f(t_k) - \sum_{k=\frac{a}{h}+1}^{n-1} \frac{(-1)^{n-1-k}}{h^\alpha}\binom{\alpha-1}{n-1-k} f(t_k) \\ &= \frac{f(t_n)}{h^\alpha} + \sum_{k=\frac{a}{h}+1}^{n-1} \frac{(-1)^{n-k}}{h^\alpha}\left(\binom{\alpha-1}{n-k} + \binom{\alpha-1}{n-1-k}\right) f(t_k), \end{aligned}$$

which, after employing the well-known binomial formula [38, the formula (6:5:3)], yields the equivalent expression of the Riemann-Liouville fractional difference in the form

$$ {}_a\nabla_h^\alpha f(t_n) = \sum_{k=\frac{a}{h}+1}^{n} \frac{(-1)^{n-k}}{h^\alpha}\binom{\alpha}{n-k} f(t_k). \tag{2.12}$$

Moreover, using the same line of arguments we can verify the validity of (2.12) for all $\alpha > 0$, where we put $\binom{\xi}{m} = 0$ for positive integers $\xi < m$. Analogously, as the Caputo fractional difference (2.10) regards, we arrive for all $\alpha > 0$ at

$$ {}_a^C\nabla_h^\alpha f(t_n) = \sum_{k=1}^{n} \frac{(-1)^{n-k}}{h^\alpha}\binom{\alpha}{n-k} f(t_k) - \sum_{j=0}^{\lceil\alpha\rceil-1} \frac{(-1)^{n-1}}{h^{\alpha-j}}\binom{\alpha-j-1}{n-1} \nabla_h^j f(t_n)\Big|_{t_n=a}. \tag{2.13}$$

Notice that if we admit negative values of α in (2.12), we obtain the expression for the fractional sum (2.8) rewritten with respect to (2.6), which allows a unique notation of fractional sums and differences.

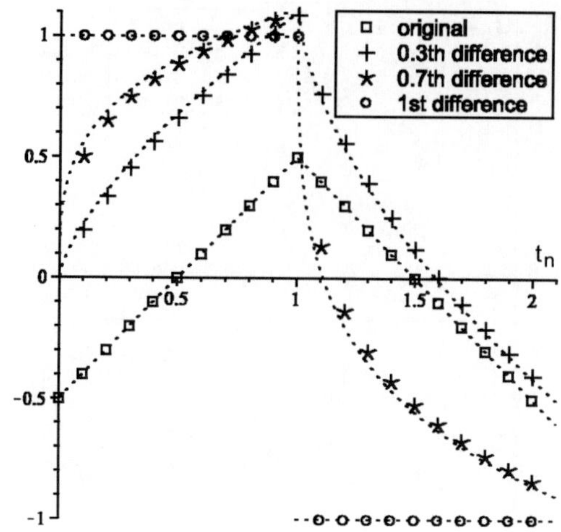

Figure 3. Caputo fractional differences of $f(t_n)$ for $h = 0.1$. The dotted lines represent the corresponding fractional derivatives of $F(t)$.

We recall that even if the formulae (2.12) and (2.13) admit n being an arbitrary integer, it is necessary to respect the domains of these operators implied by Definitions 2.2 and 2.3, respectively. In particular, if $f(t_n)$ is defined for $n \geq a$ than both ${}_a\nabla_h^\alpha f(t_n)$ and ${}_a^{\mathrm{C}}\nabla_h^\alpha f(t_n)$ are defined for $n \geq a + \lceil \alpha \rceil$.

To simplify our notation, throughout the remaining parts of this chapter we put $a = 0$ and write shortly ${}_0\nabla_h^\alpha \equiv \nabla_h^\alpha$ and ${}_0^{\mathrm{C}}\nabla_h^\alpha \equiv {}^{\mathrm{C}}\nabla_h^\alpha$.

3. Linear Fractional Difference Equations

This section provides an introductory study to the theory of linear fractional difference equations. A general form of such equations is (with respect to the type of involved fractional differences) either

$$\sum_{j=0}^{m} p_j(t_n)\, \nabla_h^{\alpha_j} y(t_n) = f(t_n)\,, \qquad n > \lceil \alpha_0 \rceil\,,$$

or

$$\sum_{j=0}^{m} p_j(t_n)\, {}^{\mathrm{C}}\nabla_h^{\alpha_j} y(t_n) = f(t_n)\,, \qquad n > \lceil \alpha_0 \rceil\,,$$

where $\alpha_0 > \alpha_1 > \cdots > \alpha_m = 0$ are real scalars and $f(t_n)$, $p_j(t_n)$ $(j = 0, \ldots, m)$ are real sequences.

Following the continuous case (see, e.g., [18]), we call these equations single-term if $m = 1$, otherwise we call them multi-term. For the sake of simplicity, we are going to study especially the single-term linear fractional difference equations (with a nonconstant as well as a constant coefficient).

In this section, our aim is to investigate some basic properties of these equations such as the well-possedness of corresponding initial value problems or the form of general solutions and their related representation formulae. Basic stability and asymptotic properties of these equations (including the two-term ones) are described in Sections 4 and 5.

The results of this section (derived originally for a special multi-term equation) are the subject of the paper [12].

3.1. Single-Term Fractional Equations and Structure of Their Solution Space

We start with the initial value problems involving single-term fractional difference equations with a nonconstant coefficient, namely

$$\nabla_h^{\alpha} y(t_n) + p(t_n)\, y(t_n) = f(t_n)\,, \qquad n > \lceil\alpha\rceil\,, \tag{3.14}$$

$$\nabla_h^{\alpha-j} y(t_n)\big|_{n=\lceil\alpha\rceil} = y_{\alpha-j}\,, \qquad j = 1, 2, \ldots, \lceil\alpha\rceil \tag{3.15}$$

and

$${}^{\mathrm{C}}\nabla_h^{\alpha} y(t_n) + p(t_n)\, y(t_n) = f(t_n)\,, \qquad n > \lceil\alpha\rceil\,, \tag{3.16}$$

$$\nabla_h^{\lceil\alpha\rceil-j} y(t_n)\big|_{n=\lceil\alpha\rceil} = y_{\lceil\alpha\rceil-j}\,, \qquad j = 1, 2, \ldots, \lceil\alpha\rceil\,, \tag{3.17}$$

where $\alpha \in \mathbb{R}^+$, $f(t_n), p(t_n)$ are arbitrary real sequences and $y_{\alpha-j}, y_{\lceil\alpha\rceil-j}$ $(j = 1, \ldots, \lceil\alpha\rceil)$ are arbitrary real scalars.

First, we discuss the well-posedness of these initial value problems. If α is a positive integer, then (3.14), (3.15) as well as (3.16), (3.17) correspond to the standard initial value problem studied in classical difference calculus. If α is not an integer, then applying (2.12) and (2.13) to both the equations, respectively, we can observe that (3.14) and (3.16) can be written in the form

$$\left(\frac{1}{h^{\alpha}} + p(t_n)\right) y(t_n) + \sum_{k=1}^{n-1} \frac{(-1)^{n-k}}{h^{\alpha}} \binom{\alpha}{n-k} y(t_k) = g(t_n)\,, \qquad n > \lceil\alpha\rceil \tag{3.18}$$

with $g(t_n)$ depending on the type of utilized fractional difference (in particular, $g(t_n) = f(t_n)$ for the Riemann-Liouville fractional difference). Such difference equations are usually referred to as the equations of Volterra type. If this equation has two different solutions, then their values differ at least at one of the starting points $t_1, t_2, \ldots, t_{\lceil\alpha\rceil}$. In particular, if $1+p(t_n)h^{\alpha} \neq 0$ for all $n > \lceil\alpha\rceil$, then arbitrary values of $y(t_1), y(t_2), \ldots, y(t_{\lceil\alpha\rceil})$ determine uniquely the solution $y(t_n)$ for $n > \lceil\alpha\rceil$.

The type of initial conditions (3.17) agrees with that standardly used for classical difference equations and correspondence between the left-hand side of (3.17) and the values $y(t_1), y(t_2), \ldots, y(t_{\lceil\alpha\rceil})$ is well-known.

On the other hand, the initial conditions (3.15) are not quite usual and their meaning needs to be clarified. The following statement describes the correspondence between the left-hand side of (3.15) and the values $y(t_1), y(t_2), \ldots, y(t_{\lceil\alpha\rceil})$.

Proposition 3.1. *Let $\alpha \in \mathbb{R}^+ \setminus \mathbb{Z}^+$ and $y(t_n)$ be a real sequence. Then (3.15) represents a one-to-one mapping between the vectors $(y(t_1), y(t_2), \ldots, y(t_{\lceil\alpha\rceil}))$ and $(y_{\alpha-1}, y_{\alpha-2}, \ldots, y_{\alpha-\lceil\alpha\rceil})$.*

Proof. We wish to show that the values of $y(t_1), y(t_2), \ldots, y(t_{\lceil\alpha\rceil})$ determine uniquely the values of

$$\nabla_h^{\alpha-1} y(t_n)\big|_{n=\lceil\alpha\rceil},\ \nabla_h^{\alpha-2} y(t_n)\big|_{n=\lceil\alpha\rceil},\ \ldots,\ \nabla_h^{\alpha-\lceil\alpha\rceil} y(t_n)\big|_{n=\lceil\alpha\rceil}$$

and vice versa. Utilizing the relation

$$\nabla_h^{\alpha-j} y(t_n)\big|_{n=\lceil\alpha\rceil} = \sum_{k=1}^{\lceil\alpha\rceil} \frac{(-1)^{\lceil\alpha\rceil-k}}{h^{\alpha-j}} \binom{\alpha-j}{\lceil\alpha\rceil-k} y(t_k)$$

following from (2.12), we can rewrite (3.15) as the linear mapping

$$\sum_{k=1}^{\lceil\alpha\rceil} r_{jk} y(t_k) = y_{\alpha-j}\,, \qquad j = 1, \ldots, \lceil\alpha\rceil\,, \tag{3.19}$$

where $r_{jk} = \frac{(-1)^{\lceil\alpha\rceil-k}}{h^{\alpha-j}}\binom{\alpha-j}{\lceil\alpha\rceil-k}$ are the elements of the transformation matrix R_α. We show that R_α is regular. Obviously,

$$R_\alpha = \begin{pmatrix} \frac{(-1)^{\lceil\alpha\rceil-1}}{h^{\alpha-1}}\binom{\alpha-1}{\lceil\alpha\rceil-1} & \frac{(-1)^{\lceil\alpha\rceil-2}}{h^{\alpha-1}}\binom{\alpha-1}{\lceil\alpha\rceil-2} & \cdots & \frac{(-1)^{0}}{h^{\alpha-1}}\binom{\alpha-1}{0} \\ \frac{(-1)^{\lceil\alpha\rceil-1}}{h^{\alpha-2}}\binom{\alpha-2}{\lceil\alpha\rceil-1} & \frac{(-1)^{\lceil\alpha\rceil-2}}{h^{\alpha-2}}\binom{\alpha-2}{\lceil\alpha\rceil-2} & \cdots & \frac{(-1)^{0}}{h^{\alpha-2}}\binom{\alpha-2}{0} \\ \vdots & \vdots & \ddots & \vdots \\ \frac{(-1)^{\lceil\alpha\rceil-1}}{h^{\alpha-\lceil\alpha\rceil}}\binom{\alpha-\lceil\alpha\rceil}{\lceil\alpha\rceil-1} & \frac{(-1)^{\lceil\alpha\rceil-2}}{h^{\alpha-\lceil\alpha\rceil}}\binom{\alpha-\lceil\alpha\rceil}{\lceil\alpha\rceil-2} & \cdots & \frac{(-1)^{0}}{h^{\alpha-\lceil\alpha\rceil}}\binom{\alpha-\lceil\alpha\rceil}{0} \end{pmatrix}.$$

To calculate $\det(R_\alpha)$, we employ some elementary operations preserving its value. Using basic properties of binomial numbers, namely

$$\binom{\alpha-i-1}{\lceil\alpha\rceil-\ell-1} - \binom{\alpha-i}{\lceil\alpha\rceil-\ell-1} = -\binom{\alpha-i-1}{\lceil\alpha\rceil-\ell-2}$$

($i = 1, 2, \ldots, \lceil\alpha\rceil-1, \ell = 0, 1, \ldots, \lceil\alpha\rceil-2$), we multiply the i-th row ($i = 1, \ldots, \lceil\alpha\rceil-1$) of R_α by $-h$ and add it to the successive one. We arrive at the form

$$\left(\begin{array}{ccc|c} \frac{(-1)^{\lceil\alpha\rceil-1}}{h^{\alpha-1}}\binom{\alpha-1}{\lceil\alpha\rceil-1} & \frac{(-1)^{\lceil\alpha\rceil-2}}{h^{\alpha-1}}\binom{\alpha-1}{\lceil\alpha\rceil-2} & \cdots & \frac{(-1)^{0}}{h^{\alpha-1}}\binom{\alpha-1}{0} \\ \hline & & & 0 \\ & R_{\alpha-1} & & \vdots \\ & & & 0 \end{array}\right).$$

Then we apply repeatedly this procedure to obtain the triangular matrix

$$\begin{pmatrix} \frac{(-1)^{\lceil\alpha\rceil-1}}{h^{\alpha-1}}\binom{\alpha-1}{\lceil\alpha\rceil-1} & \frac{(-1)^{\lceil\alpha\rceil-2}}{h^{\alpha-1}}\binom{\alpha-1}{\lceil\alpha\rceil-2} & \cdots & \frac{(-1)^{0}}{h^{\alpha-1}}\binom{\alpha-1}{0} \\ \frac{(-1)^{\lceil\alpha\rceil-2}}{h^{\alpha-2}}\binom{\alpha-2}{\lceil\alpha\rceil-2} & \cdots & \frac{(-1)^{0}}{h^{\alpha-2}}\binom{\alpha-2}{0} & 0 \\ \vdots & \iddots & \iddots & \vdots \\ \frac{(-1)^{0}}{h^{\alpha-\lceil\alpha\rceil}}\binom{\alpha-\lceil\alpha\rceil}{0} & 0 & \cdots & 0 \end{pmatrix}.$$

We get

$$\det(R_\alpha) = \frac{(-1)^{\lceil\frac{\alpha+1}{2}\rceil-1}}{h^\alpha}\prod_{k=1}^{\lceil\alpha\rceil} h^k \neq 0\,,$$

i.e., the matrix R_α is regular, hence the corresponding mapping (3.19) is one-to-one. □

Thus we have shown that both the problems (3.14), (3.15) and (3.16), (3.17) are well-posed. In particular, the equations (3.14), (3.16) are the Volterra difference equation (3.18) and the conditions (3.15), (3.17) uniquely determine required starting values $y(t_1)$, $y(t_2)$,..., $y(t_{\lceil\alpha\rceil})$ for (3.18). This immediately implies

Theorem 3.1. *Let $\alpha \in \mathbb{R}^+$ and $1 + p(t_n)h^\alpha \neq 0$ for all $n > \lceil\alpha\rceil$. Then the initial value problems (3.14), (3.15) as well as (3.16), (3.17) have unique solutions defined for all $n > 0$.*

Our next aim is to describe the structure of the solution space of (3.14) and (3.16). We start with the following notion generalizing the Wronskian.

Definition 3.1. *Let $0 \leq \gamma < 1$ be a real scalar. For real sequences $y_j(t_n)$ ($j = 1, 2, \ldots, m$), we define the γ-Wronskian $W_\gamma(y_1, \ldots, y_m)(t_n)$ for all $n > m-1$ as determinant of the matrix*

$$V_\gamma(y_1, \ldots, y_m)(t_n) = \begin{pmatrix} \nabla_h^{-\gamma} y_1(t_n) & \nabla_h^{-\gamma} y_2(t_n) & \cdots & \nabla_h^{-\gamma} y_m(t_n) \\ \nabla_h^{1-\gamma} y_1(t_n) & \nabla_h^{1-\gamma} y_2(t_n) & \cdots & \nabla_h^{1-\gamma} y_m(t_n) \\ \vdots & \vdots & \ddots & \vdots \\ \nabla_h^{m-1-\gamma} y_1(t_n) & \nabla_h^{m-1-\gamma} y_2(t_n) & \cdots & \nabla_h^{m-1-\gamma} y_m(t_n) \end{pmatrix}.$$

Remark 3.1. *Note that $W_\gamma(y_1, \ldots, y_m)(t_n)$ coincides for $\gamma = 0$ with the classical definition of the Wronskian (sometimes called the Casoratian). Moreover, it holds $W_\gamma(y_1, \ldots, y_m)(t_n) = W_0(\nabla_h^{-\gamma} y_1, \ldots, \nabla_h^{-\gamma} y_m)(t_n)$.*

Now we show that the structure of the solution space of (3.14) and (3.16) is the same as in the integer-order case. We restrict only to (3.14), the assertions and proofs regarding (3.16) are analogous.

We first discuss the homogeneous problem, i.e., we assume that $f(t_n) \equiv 0$.

Theorem 3.2. *Let $y_1(t_n), \ldots, y_{\lceil\alpha\rceil}(t_n)$ be solutions of (3.14), where $1 + p(t_n)h^\alpha \neq 0$ and $f(t_n) = 0$ for all $n > \lceil\alpha\rceil$. Further, let $W_{\lceil\alpha\rceil-\alpha}(y_1, \ldots, y_{\lceil\alpha\rceil})(t_{\lceil\alpha\rceil}) \neq 0$. Then any solution $y(t_n)$ of (3.14) can be written in the form*

$$y(t_n) = \sum_{k=1}^{\lceil\alpha\rceil} c_k y_k(t_n)\,, \qquad n > 0\,, \tag{3.20}$$

where $c_1, \ldots, c_{\lceil\alpha\rceil}$ are real constants.

Proof. Let $y(t_n)$ be a solution of (3.14). By Proposition 3.1, there exist real scalars $y_{\alpha-1}, \ldots, y_{\alpha-\lceil\alpha\rceil}$ such that $y(t_n)$ satisfies (3.15). Now we consider the sequence

$$u(t_n) = \sum_{k=1}^{\lceil\alpha\rceil} c_k y_k(t_n)\,,$$

where the $\lceil\alpha\rceil$-tuple $(c_1, \ldots, c_{\lceil\alpha\rceil})$ is the unique solution of

$$V_{\lceil\alpha\rceil-\alpha}(y_1, \ldots, y_{\lceil\alpha\rceil})(t_{\lceil\alpha\rceil}) \cdot \begin{pmatrix} c_1 \\ c_2 \\ \vdots \\ c_{\lceil\alpha\rceil} \end{pmatrix} = \begin{pmatrix} y_{\alpha-\lceil\alpha\rceil} \\ y_{\alpha-\lceil\alpha\rceil+1} \\ \vdots \\ y_{\alpha-1} \end{pmatrix}.$$

The linearity of (3.14) implies that $u(t_n)$ has to be its solution. Moreover, it holds

$$\nabla_h^{\alpha-j} u(t_n)\big|_{n=\lceil\alpha\rceil} = y_{\alpha-j}\,, \qquad j = 1, 2, \ldots, \lceil\alpha\rceil\,,$$

hence $u(t_n)$ is a solution of the initial value problem (3.14), (3.15). By Theorem 3.1, it must be $y(t_n) = u(t_n)$ for all $n > 0$ and (3.20) holds. □

The following assertion describes the form of a general solution of (3.14).

Theorem 3.3. *Let $y_1(t_n), \ldots, y_{\lceil\alpha\rceil}(t_n)$ be the same as in Theorem 3.2. Further, let $\hat{y}(t_n)$ be the solution of (3.14) meeting the homogeneous initial conditions. Then any solution $y(t_n)$ of (3.14) can be written in the form*

$$y(t_n) = \hat{y}(t_n) + \sum_{k=1}^{\lceil\alpha\rceil} c_k y_k(t_n)\,, \qquad n > 0\,,$$

where $c_1, \ldots, c_{\lceil\alpha\rceil}$ are suitable real constants.

Proof. The assertion follows from Theorem 3.2 with respect to linearity of (3.14). □

3.2. Single-Term Fractional Equations and Representation of Their Solutions

The aim of this subsection is to extend our previous considerations and find some representation formulae for the solutions of studied linear single-term fractional difference equations. Following the continuous case, where such formulae (utilizing the Mittag-Leffler type functions) for solutions of linear autonomous fractional differential equations are known, we consider (3.14) and (3.16) in their special forms, namely

$$\nabla_h^\alpha y(t_n) = \lambda y(t_n)\,, \qquad n > \lceil\alpha\rceil, \quad \lambda \in \mathbb{R} \tag{3.21}$$

and

$${}^{\mathrm{C}}\nabla_h^\alpha y(t_n) = \lambda y(t_n)\,, \qquad n > \lceil\alpha\rceil, \quad \lambda \in \mathbb{R}\,. \tag{3.22}$$

We wish to describe the base of their solution spaces by use the of the eigenfunctions of corresponding fractional difference operators. We show that these eigenfunctions can be taken for discrete analogues of the Mittag-Leffler functions. To ensure the correctness of the following procedures, throughout this subsection we assume that the regressivity condition holds, i.e.,

$$\lambda h^{\alpha} \neq 1 .$$

In general, discussions on methods for solving fractional difference equations are just at the beginning. Some techniques how to explicitly solve these equations (at least in very particular cases) are exhibited, e.g., in [5, 6], where a discrete analogue of the Laplace transform turns out to be the most developed method.

We describe here a technique not utilizing the transform method, but directly originating from the role which is played by the Mittag-Leffler function

$$E_{\alpha,\beta}(t) = \sum_{k=0}^{\infty} \frac{t^k}{\Gamma(\alpha k + \beta)} \tag{3.23}$$

in the continuous fractional calculus (see, e.g., [39]). More precisely, we introduce an h-analogue of the modified Mittag-Leffler function $t^{\beta-1}E_{\alpha,\beta}(\lambda t^{\alpha})$, which satisfies, under special choices of its parameters, continuous patterns for (3.21) and (3.22). Our next results generalize and extend those derived in [36] and [10].

Definition 3.2. *Let α, β, $\lambda \in \mathbb{R}$. We introduce the h-Mittag-Leffler function $E^{\lambda}_{\alpha,\beta}(t_n)$ by the series expansion*

$$E^{\lambda}_{\alpha,\beta}(t_n) = \sum_{k=0}^{\infty} \frac{(-1)^{n-1}\lambda^k}{h^{-\alpha k-\beta+1}} \binom{-\alpha k - \beta}{n-1}, \qquad n > 0 .$$

Remark 3.2. *It is easy to check that the series on the right-hand side converges (absolutely) if $|\lambda| h^{\alpha} < 1$. As it might be expected, the particular h-Mittag-Leffler function*

$$E^{\lambda}_{1,1}(t_n) = (1 - \lambda h)^{-n}, \qquad |\lambda| h < 1$$

obeys the trivial difference equation

$$\nabla_h y(t_n) = \lambda y(t_n), \qquad n > 0 .$$

The main properties of the h-Mittag-Leffler function are collected in the following assertion.

Theorem 3.4. *(i) Let $\gamma \in \mathbb{R}^+$ and $n > 0$. Then*

$$\nabla_h^{-\gamma} E^{\lambda}_{\eta,\beta}(t_n) = E^{\lambda}_{\eta,\beta+\gamma}(t_n) .$$

(ii) Let $\alpha \in \mathbb{R}^+$ and $\eta k + \beta - \alpha \notin \{0, -1, \ldots, -\lceil\alpha\rceil + 1\}$ for all $k \in \mathbb{Z}^+$. If $n > \lceil\alpha\rceil$ then

$$\nabla_h^{\alpha} E^{\lambda}_{\eta,\beta}(t_n) = \begin{cases} E^{\lambda}_{\eta,\beta-\alpha}(t_n), & \beta - \alpha \notin \{0, -1, \ldots, -\lceil\alpha\rceil + 1\}, \\ \lambda E^{\lambda}_{\eta,\beta-\alpha+\eta}(t_n), & \beta - \alpha \in \{0, -1, \ldots, -\lceil\alpha\rceil + 1\}. \end{cases} \tag{3.24}$$

(iii) Let $\alpha \in \mathbb{R}^+$ and $\eta k + \beta \notin \{1, \dots, \lceil\alpha\rceil\}$ for all $k \in \mathbb{Z}^+$. If $n > \lceil\alpha\rceil$ then

$$^{\mathrm{C}}\nabla_h^\alpha E_{\eta,\beta}^\lambda(t_n) = \begin{cases} E_{\eta,\beta-\alpha}^\lambda(t_n)\,, & \beta \notin \{1, \dots, \lceil\alpha\rceil\}\,, \\ \lambda E_{\eta,\beta-\alpha+\eta}^\lambda(t_n)\,, & \beta \in \{1, \dots, \lceil\alpha\rceil\}\,. \end{cases}$$

Proof. The part (i) follows immediately from Lemma 2.1. Considering the part (ii) we can write

$$\nabla_h^\alpha E_{\eta,\beta}^\lambda(t_n) = \nabla_h^\alpha \sum_{k=0}^\infty \frac{(-1)^{n-1}\lambda^k}{h^{-\eta k-\beta+1}}\binom{-\eta k-\beta}{n-1} = \sum_{k=0}^\infty \lambda^k \nabla_h^\alpha \frac{(-1)^{n-1}}{h^{-\eta k-\beta+1}}\binom{-\eta k-\beta}{n-1}$$

due to the absolute convergence property.

If $k \in \mathbb{Z}^+$ then Corollary 2.1 (i) implies the relation

$$\nabla_h^\alpha \frac{(-1)^{n-1}}{h^{-\eta k-\beta+1}}\binom{-\eta k-\beta}{n-1} = \frac{(-1)^{n-1}}{h^{-\eta k-\beta+\alpha+1}}\binom{-\eta k-\beta+\alpha}{n-1} \tag{3.25}$$

due to the assumption $\eta k+\beta-\alpha \notin \{0, -1, \dots, -\lceil\alpha\rceil+1\}$. If $k = 0$ then two possibilities may occur. If $\beta - \alpha \notin \{0, -1, \dots, -\lceil\alpha\rceil + 1\}$ we get (3.25) with $k = 0$ which implies the validity of $(3.24)_1$. If $\beta-\alpha \in \{0, -1, \dots, -\lceil\alpha\rceil+1\}$, the fractional derivative of this term is zero and by shifting the index k we obtain $(3.24)_2$.

Considering Corollary 2.1 (ii) instead of Corollary 2.1(i), the proof of the part (iii) corresponds to that of (ii). □

Remark 3.3. *The assumption $\eta k + \beta - \alpha \notin \{0, -1, \dots, -\lceil\alpha\rceil + 1\}$ for all $k \in \mathbb{Z}^+$ in Theorem 3.4 (ii) may seem to be quite restrictive. Note that it is satisfied trivially for $\beta \in \mathbb{R}^+$ and $\eta + \beta > \alpha$ and, as shown in the following assertion, this is the case we are interested in. Similarly, the assumption $\eta k + \beta \notin \{1, \dots, \lceil\alpha\rceil\}$ for all $k \in \mathbb{Z}^+$ in Theorem 3.4 (iii) rules out many combinations of parameters η, β. However, we note that for $\eta = \alpha$ and $\beta \in \mathbb{Z}^+$ the condition is always satisfied.*

Corollary 3.1. *Let $\alpha \in \mathbb{R}^+$ and $\lambda \in \mathbb{R}$ be such that $|\lambda|h^\alpha < 1$. Then*
(i) the eigenfunctions of the operator ∇_h^α are given by

$$E_{\alpha,\beta}^\lambda(t_n), \qquad \beta = \alpha - \lceil\alpha\rceil + 1, \dots, \alpha - 1, \alpha\,. \tag{3.26}$$

(ii) the eigenfunctions of the operator $^{\mathrm{C}}\nabla_h^\alpha$ are given by

$$E_{\alpha,\beta}^\lambda(t_n), \qquad \beta = 1, \dots, \lceil\alpha\rceil\,. \tag{3.27}$$

Proof. The assertion follows from Theorem 3.4 by the use of $\eta = \alpha$. □

Our final aim is to show that any solution of (3.21) and (3.22) can be written as a linear combination of h-Mittag-Leffler functions (3.26) and (3.27), respectively.

Lemma 3.1. *Let $\alpha \in \mathbb{R}^+$ and $\lambda \in \mathbb{R}$ be such that $|\lambda|h^\alpha < 1$. Then*

$$W_{\lceil\alpha\rceil-\alpha}(E_{\alpha,\alpha-\lceil\alpha\rceil+1}^\lambda, E_{\alpha,\alpha-\lceil\alpha\rceil+2}^\lambda, \dots, E_{\alpha,\alpha}^\lambda)(t_{\lceil\alpha\rceil}) = (1-\lambda h^\alpha)^{-\lceil\alpha\rceil} \neq 0.$$

Proof. The case $\lceil\alpha\rceil = 1$ is trivial. For $\lceil\alpha\rceil \geq 2$, we can formally write $\lambda E^{\lambda}_{\alpha,\alpha-\ell}(t_n) = E^{\lambda}_{\alpha,-\ell}(t_n)$ for all $n > \lceil\alpha\rceil - 1$ ($\ell = 0, \ldots, \lceil\alpha\rceil - 2$). Consequently, applying Theorem 3.4, the Wronskian can be expressed as

$$W_{\lceil\alpha\rceil-\alpha}(E^{\lambda}_{\alpha,\alpha-\lceil\alpha\rceil+1}, E^{\lambda}_{\alpha,\alpha-\lceil\alpha\rceil+2}, \ldots, E^{\lambda}_{\alpha,\alpha})(t_{\lceil\alpha\rceil}) = \det(M_{\lceil\alpha\rceil}),$$

where

$$M_{\lceil\alpha\rceil} = \begin{pmatrix} E^{\lambda}_{\alpha,1}(t_{\lceil\alpha\rceil}) & E^{\lambda}_{\alpha,2}(t_{\lceil\alpha\rceil}) & \cdots & E^{\lambda}_{\alpha,\lceil\alpha\rceil}(t_{\lceil\alpha\rceil}) \\ E^{\lambda}_{\alpha,0}(t_{\lceil\alpha\rceil}) & E^{\lambda}_{\alpha,1}(t_{\lceil\alpha\rceil}) & \cdots & E^{\lambda}_{\alpha,\lceil\alpha\rceil-1}(t_{\lceil\alpha\rceil}) \\ \vdots & \vdots & \ddots & \vdots \\ E^{\lambda}_{\alpha,2-\lceil\alpha\rceil}(t_{\lceil\alpha\rceil}) & E^{\lambda}_{\alpha,3-\lceil\alpha\rceil}(t_{\lceil\alpha\rceil}) & \cdots & E^{\lambda}_{\alpha,1}(t_{\lceil\alpha\rceil}) \end{pmatrix}.$$

Using the Pascal rule we obtain the equality

$$E^{\lambda}_{\alpha,i}(t_{\lceil\alpha\rceil}) - hE^{\lambda}_{\alpha,i-1}(t_{\lceil\alpha\rceil}) = E^{\lambda}_{\alpha,i}(t_{\lceil\alpha\rceil-1}), \qquad i \in \mathbb{Z}, \quad i \geq 3 - \lceil\alpha\rceil. \tag{3.28}$$

Starting with the first row, $\binom{\lceil\alpha\rceil}{2}$ elementary row operations of the type (3.28) transform the matrix $M_{\lceil\alpha\rceil}$ into the matrix

$$\widehat{M}_{\lceil\alpha\rceil} = \begin{pmatrix} E^{\lambda}_{\alpha,1}(t_1) & E^{\lambda}_{\alpha,2}(t_1) & \cdots & E^{\lambda}_{\alpha,\lceil\alpha\rceil}(t_1) \\ E^{\lambda}_{\alpha,0}(t_2) & E^{\lambda}_{\alpha,1}(t_2) & \cdots & E^{\lambda}_{\alpha,\lceil\alpha\rceil-1}(t_2) \\ \vdots & \vdots & \ddots & \vdots \\ E^{\lambda}_{\alpha,2-\lceil\alpha\rceil}(t_{\lceil\alpha\rceil}) & E^{\lambda}_{\alpha,3-\lceil\alpha\rceil}(t_{\lceil\alpha\rceil}) & \cdots & E^{\lambda}_{\alpha,1}(t_{\lceil\alpha\rceil}) \end{pmatrix}$$

with the property $\det(\widehat{M}_{\lceil\alpha\rceil}) = \det(M_{\lceil\alpha\rceil})$. By Theorem 3.4 (ii), where the first difference is considered, we have

$$\begin{aligned} &E^{\lambda}_{\alpha,p}(t_i) - hE^{\lambda}_{\alpha,p-1}(t_i) = E^{\lambda}_{\alpha,p}(t_{i-1}), \qquad i > 1, \\ &E^{\lambda}_{\alpha,p}(t_1) - hE^{\lambda}_{\alpha,p-1}(t_1) = 0, \end{aligned} \tag{3.29}$$

where $p \in \mathbb{Z}$, $p \geq 3 - i$. Starting with the last column, using $\lceil\alpha\rceil - 1$ elementary column operations of the type (3.29) we obtain the matrix

$$\left(\begin{array}{c|ccc} E^{\lambda}_{\alpha,1}(t_1) & 0 & \cdots & 0 \\ \hline E^{\lambda}_{\alpha,0}(t_2) & & & \\ \vdots & & \widehat{M}_{\lceil\alpha\rceil-1} & \\ E^{\lambda}_{\alpha,2-\lceil\alpha\rceil}(t_{\lceil\alpha\rceil}) & & & \end{array}\right)$$

preserving the value of $\det(\widehat{M}_{\lceil\alpha\rceil})$. Since

$$E^{\lambda}_{\alpha,1}(t_1) = \sum_{k=0}^{\infty} \lambda^k h^{\alpha k} = \frac{1}{1-\lambda h^{\alpha}},$$

we have the recurrence

$$\det(\widehat{M}_{\lceil\alpha\rceil}) = \frac{1}{1-\lambda h^{\alpha}} \det(\widehat{M}_{\lceil\alpha\rceil-1}),$$

which implies the assertion. □

Now we summarize the results of this subsection to obtain

Theorem 3.5. *(i) Let $y(t_n)$ be any solution of (3.21) and let $|\lambda|h^\alpha < 1$. Then*

$$y(t_n) = \sum_{j=1}^{\lceil\alpha\rceil} c_j E^{\lambda}_{\alpha,\alpha-\lceil\alpha\rceil+j}(t_n)\,,$$

where $c_1, \dots, c_{\lceil\alpha\rceil}$ are real constants.
(ii) Let $y(t_n)$ be any solution of (3.22) and let $|\lambda|h^\alpha < 1$. Then

$$y(t_n) = \sum_{j=1}^{\lceil\alpha\rceil} c_j E^{\lambda}_{\alpha,j}(t_n)\,,$$

where $c_1, \dots, c_{\lceil\alpha\rceil}$ are real constants.

Proof. Utilizing Remark 3.1 and Theorem 3.4 (i) we arrive at

$$W_0(E^{\lambda}_{\alpha,1}, \dots, E^{\lambda}_{\alpha,\lceil\alpha\rceil})(t_{\lceil\alpha\rceil}) = W_{\lceil\alpha\rceil-\alpha}(E^{\lambda}_{\alpha,\alpha-\lceil\alpha\rceil+1}, \dots, E^{\lambda}_{\alpha,\alpha})(t_{\lceil\alpha\rceil})\,.$$

Thus, the assertion follows from Theorem 3.2, Corollary 3.1 and Lemma 3.1. □

We conclude this part by an illustrating example.

Example 3.1. *Let $1 < \alpha \le 2$. Consider initial value problems*

$$\begin{aligned} \nabla_h^\alpha y(t_n) &= \lambda\, y(t_n)\,, \quad n > 2\,, \\ \nabla_h^{\alpha-1} y(t_n)\big|_{n=2} &= y_{\alpha-1}\,, \\ \nabla_h^{\alpha-2} y(t_n)\big|_{n=2} &= y_{\alpha-2}\,, \end{aligned} \qquad \begin{aligned} {}^C\nabla_h^\alpha y(t_n) &= \lambda\, y(t_n)\,, \quad n > 2\,, \\ \nabla_h y(t_n)\big|_{n=2} &= y_1\,, \\ y(t_n)\big|_{n=2} &= y_0\,. \end{aligned} \tag{3.30}$$

By Theorem 3.5 (i), the solution of $(3.30)_1$ can be expressed as the linear combination

$$y(t_n) = c_1\, E^{\lambda}_{\alpha,\alpha-1}(t_n) + c_2\, E^{\lambda}_{\alpha,\alpha}(t_n)\,.$$

The constants c_1, c_2 can be determined from the system

$$V_{2-\alpha}(E^{\lambda}_{\alpha,\alpha-1}, E^{\lambda}_{\alpha,\alpha})(t_2) \cdot \begin{pmatrix} c_1 \\ c_2 \end{pmatrix} = \begin{pmatrix} y_{\alpha-2} \\ y_{\alpha-1} \end{pmatrix}$$

with the matrix elements

$$\begin{aligned} v_{11} = v_{22} &= \frac{1 + (\alpha-1)\lambda\, h^\alpha}{(1-\lambda\, h^\alpha)^2}\,, \\ v_{12} &= \frac{2h + (\alpha-2)\lambda\, h^{\alpha+1}}{(1-\lambda\, h^\alpha)^2}\,, \\ v_{21} &= \frac{\alpha\lambda\, h^{\alpha-1}}{(1-\lambda\, h^\alpha)^2}\,. \end{aligned}$$

By Lemma 3.1, the matrix $V_{2-\alpha}(E^{\lambda}_{\alpha,\alpha-1}, E^{\lambda}_{\alpha,\alpha})(t_2)$ has a nonzero determinant, hence applying the Cramer rule we get

$$c_1 = \frac{y_{\alpha-2}v_{22} - y_{\alpha-1}v_{12}}{W_{2-\alpha}(E^{0\lambda}_{\alpha,\alpha-1}, E^{0\lambda}_{\alpha,\alpha})(t_2)} = y_{\alpha-2}(1 + (\alpha-1)\lambda\, h^{\alpha}) - y_{\alpha-1}h(2 + (\alpha-2)\lambda\, h^{\alpha}),$$

$$c_2 = \frac{y_{\alpha-1}v_{11} - y_{\alpha-2}v_{21}}{W_{2-\alpha}(E^{0\lambda}_{\alpha,\alpha-1}, E^{0\lambda}_{\alpha,\alpha})(t_2)} = y_{\alpha-1}(1 + (\alpha-1)\lambda\, h^{\alpha}) - y_{\alpha-2}\alpha\lambda\, h^{\alpha-1}.$$

Similarly, the solution of $(3.30)_2$ takes the form

$$y(t_n) = c_1\, E^{\lambda}_{\alpha,1}(t_n) + c_2\, E^{\lambda}_{\alpha,2}(t_n),$$

where the constants c_1, c_2 are given by

$$c_1 = y_0(1 + (\alpha-1)\lambda\, h^{\alpha}) - y_1 h(2 + (\alpha-2)\lambda\, h^{\alpha}),$$
$$c_2 = y_1(1 + (\alpha-1)\lambda\, h^{\alpha}) - y_0\alpha\lambda\, h^{\alpha-1}.$$

Figures 4 and 5 depict the both solutions of $(3.30)_1$ and $(3.30)_2$, respectively, for a particular choice of the parameters: $\alpha = 1.8$, $\lambda = -\frac{1}{3}$, $y_{\alpha-1} = y_1 = -1$ and $y_{\alpha-2} = y_0 = 1$ and several values of h. For comparative reasons, the solutions of the corresponding continuous (differential) initial value problems are displayed as well. We observe that with decreasing h the solutions tend to their continuous patterns. Also, we can recognize a shifting effect caused by prescribing the initial conditions at t_2 instead of t_0 (it is especially apparent in the Caputo case for $h = 1$).

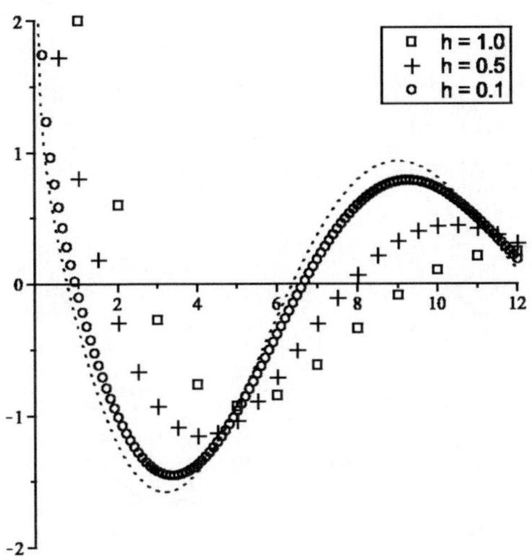

Figure 4. Solutions $y(t_n)$ of $(3.30)_1$ with $\alpha = 1.8$, $\lambda = -\frac{1}{3}$, $y_{\alpha-1} = -1$ and $y_{\alpha-2} = 1$. The dotted line represents the solution of the corresponding continuous problem.

4. Stability of Single-Term Fractional Difference Equations

This section discusses basic qualitative properties of the linear fractional difference equations (3.21) and (3.22). Regarding these equations, we have already derived the representation formulae for their solutions in terms of discrete Mittag-Leffler functions (see Theorem

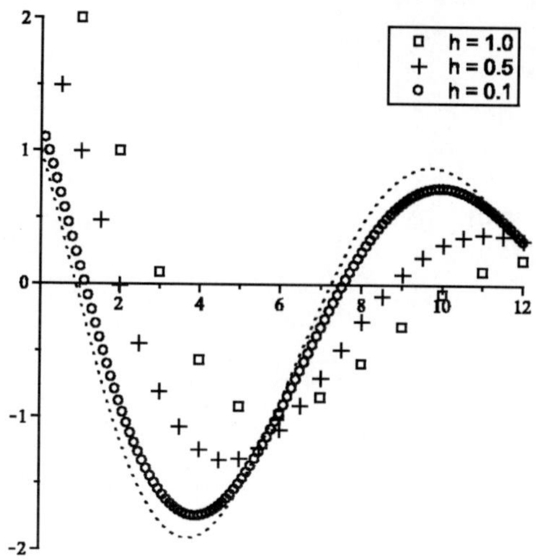

Figure 5. Solutions $y(t_n)$ of $(3.30)_2$ with $\alpha = 1.8$, $\lambda = -\frac{1}{3}$, $y_1 = -1$ and $y_0 = 1$. The dotted line represents the solution of the corresponding continuous problem.

3.5). In the continuous case, this representation provides the key tool for qualitative analysis of corresponding fractional differential equations in the view of several well-known asymptotic expansions of the exact Mittag-Leffler functions (see, e.g., [39]). However, the situation of the discrete case is different. Besides some convergence restrictions, we have no information about the asymptotic behaviour of the discrete Mittag-Leffler functions. Since their asymptotic description by the means of classical tools seems to be very complicated, for analysis of fractional equations (3.21) and (3.22) we prefer the original procedures. These procedures and the corresponding results presented in this section are based on papers [11, 13, 14, 31].

In the sequel, we investigate both equations (3.21) and (3.22) separately, with emphasis on their stability properties. As it is customary, relevant stability notions for these linear equations are meant in the following way.

Definition 4.1. *The fractional difference equation (3.21) is said to be*

(i) stable if any of its solution is bounded;

(ii) asymptotically stable if any of its solution is tending to zero as $n \to \infty$.

The introduction of these notions for the Caputo difference equation (3.22) is analogous.

4.1. Stability and Asymptotics of a Test Equation with the Riemann-Liouville Fractional Difference

Following the differential case, we restrict our considerations to α between 0 and 1. Consequently, we investigate stability and asymptotic properties of the fractional difference equation

$$\nabla_h^\alpha y(t_n) = \lambda y(t_n), \qquad n \geq 2, \quad \lambda, \alpha \in \mathbb{R}, \quad 0 < \alpha < 1\,. \tag{4.31}$$

First we present an alternative form of (4.31), which turns out to be very convenient for the following qualitative investigations.

Proposition 4.1. *Let $\lambda h^\alpha \neq 1$. Then $y(t_n)$ is the solution of (4.31) if and only if $x(n) = y(t_{n+1})$ is the solution of*

$$x(n+1) = \frac{1}{1-\lambda h^\alpha} \sum_{j=0}^{n} (-1)^{n-j} \binom{\alpha}{n-j+1} x(j) \,, \tag{4.32}$$

$n \geq 0$, $\lambda, \alpha \in \mathbb{R}$, $0 < \alpha < 1$.

Proof. Rewritting (4.31) with respect to (3.18), we arrive at

$$y(t_n) = \frac{1}{1-\lambda h^\alpha} \sum_{k=1}^{n-1} (-1)^{n-k-1} \binom{\alpha}{n-k} y(t_k) \,,$$

which after replacing n by $n+2$ and setting $x(n) = y(t_{n+1})$ turns into (4.32). □

Remark 4.1. *The equation (4.32) is the Volterra difference equation of convolution type, generally written as*

$$x(n+1) = \sum_{j=0}^{n} a(n-j) x(j), \qquad n \geq 0 \,, \tag{4.33}$$

where in our case

$$a(n) = \frac{(-1)^n}{1-\lambda h^\alpha} \binom{\alpha}{n+1}, \qquad n \geq 0 \,. \tag{4.34}$$

We recall that if $\lambda h^\alpha \neq 1$ then for any initial condition

$$x(0) = \phi_0 \tag{4.35}$$

the corresponding initial value problem (4.33), (4.35) is well-posed, i.e., there exists its unique solution $x(n)$ depending on the starting value ϕ_0.

As a result of Proposition 4.1, it is enough to analyse stability and asymptotic properties of (4.32) instead of (4.31). Before doing this, we recall corresponding stability definitions for the Volterra equation (4.33) adapted to the linear case (of course, these definitions are identical with those for (3.21) and (3.22)).

Definition 4.2. *The equation (4.33) is said to be*

(i) stable if any of its solution is bounded;

(ii) asymptotically stable if any of its solution is tending to zero as $n \to \infty$.

A very effective method for stability analysis of (4.33) is the $\mathcal{Z}$-transform method. We recall that the $\mathcal{Z}$-transform of a sequence $u(n)$, defined on $\mathbb{Z}_0^+$, is a complex function given by

$$\tilde{u}(z) = \mathcal{Z}(u(n)) = \sum_{k=0}^{\infty} u(k)z^{-k},$$

where z is a complex number for which this series converges absolutely. It is known that $\mathcal{Z}$-transform can be used to solve the linear Volterra convolution equation (4.33) and to find its stability or asymptotic stability conditions by analysing the roots of the associated characteristic equation

$$z - \tilde{a}(z) = 0, \tag{4.36}$$

where $\tilde{a}(z)$ is the $\mathcal{Z}$-transform of $a(n)$ (for more details we refer to [19]). We recall the criterion [20, Theorem 2], which is the most relevant result for our next study (as usually, the symbol ℓ^1 used here denotes the space of sequences whose series is absolutely convergent).

Theorem 4.1. *The following statements are equivalent:*

(i) all the roots of the characteristic equation (4.36) lie inside the unit disk, i.e.

$$z - \tilde{a}(z) \neq 0 \qquad \text{for all } |z| \geq 1;$$

(ii) $x(n) \in \ell^1$ *for any solution* $x(n)$ *of (4.33).*

Our first aim is to formulate the condition (ii) explicitly for the Volterra equation (4.32), i.e., in terms of its parameters α, h and λ.

Proposition 4.2. *Let* $\lambda h^\alpha \neq 1$. *Then* $x(n) \in \ell^1$ *for any solution* $x(n)$ *of (4.32) if and only if*

$$\lambda < 0 \qquad \text{or} \qquad \lambda > (2/h)^\alpha .$$

Proof. By Theorem 4.1, we have to set up the corresponding characteristic equation (4.36) and analyse the location of its roots with respect to the unit disk.

Taking the $\mathcal{Z}$-transform of (4.34) and using the binomial theorem we get

$$\begin{aligned}\tilde{a}(z) &= \frac{1}{1-\lambda h^\alpha}\sum_{k=0}^{\infty}(-1)^k\binom{\alpha}{k+1}z^{-k} = \frac{1}{1-\lambda h^\alpha}\sum_{k=1}^{\infty}(-1)^{k-1}\binom{\alpha}{k}z^{-k+1}\\ &= -\frac{z}{1-\lambda h^\alpha}\sum_{k=1}^{\infty}\binom{\alpha}{k}\left(-\frac{1}{z}\right)^k = -\frac{z}{1-\lambda h^\alpha}\left(\left(1-\frac{1}{z}\right)^\alpha - 1\right)\end{aligned}$$

for all $z \in \mathbb{C}$ with $|z| \geq 1$ (we point out that the resulting power function $\left(1-\frac{1}{z}\right)^\alpha$ is understood in the sense of the principal value). Consequently, the characteristic equation (4.36) becomes

$$z + \frac{z}{1-\lambda h^\alpha}\left(\left(1-\frac{1}{z}\right)^\alpha - 1\right) = 0 \tag{4.37}$$

and, applying Theorem 4.1, $x(n) \in \ell^1$ for any solution $x(n)$ of (4.32) if and only if

$$1 + \frac{1}{1 - \lambda h^\alpha}\left(\left(1 - \frac{1}{z}\right)^\alpha - 1\right) \neq 0 \qquad \text{for all } |z| \geq 1 \,. \tag{4.38}$$

The nonzero roots z_r of the characteristic equation (4.37) satisfy

$$\left(1 - \frac{1}{z_r}\right)^\alpha = \lambda h^\alpha \,. \tag{4.39}$$

We analyse (4.39) with respect to λ. First, let $\lambda < 0$. Then, (4.39) has no root z_r, hence the condition (4.38) is satisfied trivially. Further, let $\lambda \geq 0$. Then

$$z_r = \frac{1}{1 - h\lambda^{1/\alpha}}$$

is the unique nonzero (real) root of the characteristic equation (4.37). To satisfy (4.38), we have to require

$$h\lambda^{1/\alpha} > 2, \qquad \text{i.e.,} \qquad \lambda > (2/h)^\alpha \,.$$

The assertion is proved. □

Proposition 4.2 immediately implies that if $\lambda < 0$ or $\lambda > (2/h)^\alpha$ then (4.32) is asymptotically stable. Our next aim is to explore the remaining choices of λ.

Proposition 4.3. *Let $\lambda = 0$. Then (4.32) is asymptotically stable.*

Proof. Let $\tilde{x}(z)$ be the $\mathcal{Z}$-transform of a solution $x(n)$ of (4.32), i.e.,

$$\tilde{x}(z) = \sum_{k=0}^{\infty} x(k) z^{-k} \,. \tag{4.40}$$

Using the well-known shift and convolution properties (see, e.g., [19]),

$$\mathcal{Z}(x(n+1)) = z\tilde{x}(z) - zx(0) \qquad \text{and} \qquad \mathcal{Z}\left(\sum_{j=0}^{n} a(n-j)x(j)\right) = \tilde{a}(z)\tilde{x}(z)\,,$$

the application of the $\mathcal{Z}$-transform to both sides of (4.32) with $\lambda = 0$ yields

$$\tilde{x}(z) = \frac{x(0)}{(1 - \frac{1}{z})^\alpha} \,.$$

Further, by the binomial theorem,

$$\tilde{x}(z) = x(0)\sum_{k=0}^{\infty}\binom{-\alpha}{k}\left(-\frac{1}{z}\right)^k = x(0)\sum_{k=0}^{\infty}(-1)^k\binom{-\alpha}{k}z^{-k}\,,$$

and comparing this relation with (4.40), we have

$$x(n) = x(0)(-1)^n \binom{-\alpha}{n} . \tag{4.41}$$

To conclude the proof, we recall the binomial asymptotic relation

$$\binom{p}{m} \sim \frac{(-1)^m}{m^{p+1}\Gamma(-p)} \qquad \text{as } m \to \infty, \quad m \in \mathbb{Z}^+ , \, p \in \mathbb{R} \setminus \mathbb{Z}_0^+ \tag{4.42}$$

(see [38, p. 54]), where the symbol $\sim$ means the asymptotic equivalency. This implies the asymptotic property $x(n) \to 0$ as $n \to \infty$ for any initial value $x(0)$. □

Proposition 4.4. *Let $\lambda = (2/h)^\alpha$. Then (4.32) is stable, but not asymptotically stable.*

Proof. Let $\tilde{x}(z)$ be the $\mathcal{Z}$-transform of a solution $x(n)$ of (4.32) given by (4.40). Analogously to the previous proof we get

$$\tilde{x}(z) = \frac{(1 - \lambda h^\alpha)x(0)}{(1 - \frac{1}{z})^\alpha - \lambda h^\alpha} . \tag{4.43}$$

Setting $w = 1/z$ and considering $\lambda = (2/h)^\alpha$ we have

$$\tilde{x}(1/w) = \frac{(1 - 2^\alpha)x(0)}{(1 - w)^\alpha - 2^\alpha} . \tag{4.44}$$

We can see that $\tilde{x}(1/w)$ has a pole $w = -1$ of the first order. Thus, we can write

$$\tilde{x}(1/w) = \frac{\tilde{g}(1/w)}{w + 1} , \tag{4.45}$$

where $\tilde{g}(z) = \mathcal{Z}(g(n))$ is a function analytic on $\mathbb{C}$ and $g(n)$ is a suitable auxiliary sequence.

Using (4.45) we can express the limit of the solution $x(n)$ as

$$\begin{aligned} \lim_{n\to\infty} |x(n)| &= \lim_{n\to\infty} \left| \mathcal{Z}^{-1}\left(\tilde{g}(z) \cdot \left(\frac{1}{z} + 1 \right)^{-1} \right) \right| = \lim_{n\to\infty} \left| \sum_{j=0}^{n} g(j)(-1)^{n-j} \right| \\ &= \lim_{n\to\infty} \left| \sum_{j=0}^{n} g(j)(-1)^{-j} \right| = |\tilde{g}(-1)| . \end{aligned} \tag{4.46}$$

This value can be calculated employing (4.44) and (4.45) as

$$\begin{aligned} |\tilde{g}(-1)| &= \lim_{z\to-1} \left| (1 - 2^\alpha)x(0) \frac{\frac{1}{z} + 1}{(1 - \frac{1}{z})^\alpha - 2^\alpha} \right| \\ &= (2^\alpha - 1)|x(0)| \lim_{z\to-1} \left| \frac{-z^{-2}}{\alpha(1 - \frac{1}{z})^{\alpha-1} z^{-2}} \right| = \frac{(2^\alpha - 1)|x(0)|}{\alpha 2^{\alpha-1}} > 0 . \end{aligned} \tag{4.47}$$

We conclude that $|x(n)|$ tends to a positive real number, hence (4.32) is stable, but not asymptotically stable. □

Proposition 4.5. *Let $0 < \lambda < (2/h)^\alpha$ and $\lambda h^\alpha \neq 1$. Then (4.32) is not stable.*

Proof. Considering (4.40), (4.43) and setting $w = 1/z$ we can write

$$\frac{(1-\lambda h^\alpha)x(0)}{(1-w)^\alpha - \lambda h^\alpha} = \sum_{k=0}^{\infty} x(k) w^k . \tag{4.48}$$

The function on the left-hand side has the (unique) pole

$$w_r = 1 - h\lambda^{1/\alpha} \in (-1, 1) .$$

Consequently, the series on the right-hand side of (4.48) has the radius of convergence $R < 1$. By the Cauchy-Hadamard theorem,

$$\limsup_{n\to\infty} \sqrt[n]{|x(n)|} = \frac{1}{R} > 1 ,$$

hence (4.32) is not stable. □

Now we precise the stability results derived in the previous part. In particular, we consider the asymptotically stable case $\lambda \leq 0$, $\lambda > (2/h)^\alpha$, when the solutions $x(n)$ of (4.32) are tending to zero as $n \to \infty$, and describe the exact rate of their decay. An asymptotic result concerning the non-stable case will be derived as well.

First we note that a preliminary information on the decay rate of the solutions $x(n)$ of (4.32), namely the property

$$x(n) \in \ell^1 , \tag{4.49}$$

follows immediately from Proposition 4.2.

To obtain a precise description of the asymptotics of (4.32), we employ the following general result, which is due to [4]. For any finite $r > 0$, the authors introduced a class $W(r)$ of real-valued weight sequences $\gamma(n)$ by the requirements

$$\gamma(n) > 0, \quad n \geq 1, \qquad \lim_{n\to\infty} \frac{\gamma(n-1)}{\gamma(n)} = \frac{1}{r}, \qquad \sum_{k=0}^{\infty} \gamma(k) r^{-k} < \infty$$

and

$$\lim_{m\to\infty} \left(\limsup_{n\to\infty} \frac{1}{\gamma(n)} \sum_{j=m}^{n-m} \gamma(n-j)\gamma(j) \right) = 0 .$$

We reformulate here the scalar version of [4, Theorem 3.2] (originally proved for vector Volterra difference equations), which describes the asymptotics of a solution $x(n)$ of (4.33) in terms of an appropriate sequence $\gamma(n) \in W(r)$.

Theorem 4.2. *Suppose that, for some sequence $\gamma(n) \in W(r)$, there exists the finite limit*

$$L = \lim_{n\to\infty} \frac{a(n)}{\gamma(n)}$$

and let

$$\sum_{k=0}^{\infty} |a(k)| r^{-k-1} < 1 \,. \tag{4.50}$$

Then the solution $x(n)$ of (4.33) satisfies

$$\lim_{n\to\infty} \frac{x(n)}{\gamma(n)} = \frac{Lx(0)}{r(1-S)^2}\,, \tag{4.51}$$

where $S = \sum_{k=0}^{\infty} a(k) r^{-k-1}$.

This assertion turns out to be very useful in analysis of asymptotic properties of (4.32). It implies

Corollary 4.1. *Let $|1-\lambda h^{\alpha}| > 1$. Then any solution $x(n)$ of (4.32) satisfies*

$$\lim_{n\to\infty} \frac{x(n)}{n^{-(1+\alpha)}} = \frac{\alpha(1-\lambda h^{\alpha})}{\lambda^2 h^{2\alpha}\Gamma(1-\alpha)} x(0)\,. \tag{4.52}$$

Proof. Consider (4.32), i.e., the Volterra convolution equation (4.33) with the coefficients given by (4.34). We put $r = 1$ and introduce the decreasing sequence

$$\gamma(n) = n^{-(1+\alpha)}, \qquad n \ge 1\,.$$

As it was remarked in [4], $n^{-(1+\alpha)} \in W(1)$ provided $0 < \alpha < 1$. We verify the validity of assumptions of Theorem 4.2 and, in particular, specify the values of L and S. The first calculation employs the asymptotic property (4.42). Using this we can write

$$L = \frac{1}{1-\lambda h^{\alpha}} \lim_{n\to\infty} \frac{(-1)^n \binom{\alpha}{n+1}}{n^{-(1+\alpha)}} = -\frac{1}{(1-\lambda h^{\alpha})\Gamma(-\alpha)}\,.$$

Discussing (4.50), we need to sum the infinite series $\sum_{k=0}^{\infty} |a(k)|$. We have

$$\sum_{k=0}^{\infty} |a(k)| = \frac{1}{|1-\lambda h^{\alpha}|} \sum_{k=0}^{\infty} (-1)^k \binom{\alpha}{k+1} = \frac{1}{|1-\lambda h^{\alpha}|}$$

by the use of the binomial theorem. Consequently, (4.50) holds. Analogously we get

$$S = \sum_{k=0}^{\infty} a(k) = \frac{1}{1-\lambda h^{\alpha}}\,.$$

The property (4.52) now follows from (4.51) by the use of

$$\Gamma(1-\alpha) = -\alpha\Gamma(-\alpha)\,.$$

□

Remark 4.2. *The assumptions of Corollary 4.1 do not cover the cases* $(2/h)^\alpha < \lambda \le 2/h^\alpha$ *and* $\lambda = 0$, *when the equation (4.32) is asymptotically stable as well. If* $(2/h)^\alpha < \lambda \le 2/h^\alpha$ *then the asymptotic property (4.49) provides at least a partial information on the behaviour of the solutions of (4.32). However, this property is no longer valid if* $\lambda = 0$ *due to Proposition 4.2. Fortunately, earlier we have derived the exact form of the solutions* $x(n)$ *of (4.32) with* $\lambda = 0$ *via the relation (4.41). Then, using (4.42), we can easily get the following asymptotic result.*

Corollary 4.2. *Let* $0 < \alpha < 1$ *and* $\lambda = 0$. *Then*

$$\lim_{n\to\infty} \frac{x(n)}{n^{-(1-\alpha)}} = \frac{1}{\Gamma(\alpha)} x(0)$$

for any solution $x(n)$ *of (4.32).*

Now we turn our attention to the non-stable case when $0 < \lambda < (2/h)^\alpha$. In a part of this non-stability interval, we derive the lower and upper bounds of the solutions of (4.32).

Proposition 4.6. *Let* $0 < \lambda < 1/h^\alpha$ *and let* $x(n)$ *be a solution of (4.32) with* $x(0) > 0$. *Then*

$$\frac{\lambda^{1/\alpha} h x(0)}{(1-\lambda^{1/\alpha}h)^n} < x(n) < \frac{x(0)}{(1-\lambda^{1/\alpha}h)^n}, \qquad n \ge 1. \tag{4.53}$$

Proof. First we introduce the function

$$v(n) = \frac{x(0)}{(1-\lambda^{1/\alpha}h)^n}, \qquad n \in \mathbb{Z}$$

defining the solution of the Volterra convolution equation with infinite delay

$$v(n+1) = \sum_{j=-\infty}^{n} a(n-j)v(j), \qquad n \ge 0, \tag{4.54}$$

where $a(n)$ is given by (4.34). We rewrite (4.54) as

$$v(n+1) = \sum_{j=0}^{n} a(n-j)v(j) + g(n),$$

where $g(n) = \sum_{j=-\infty}^{-1} a(n-j)v(j) > 0$. Using the variation of constants formula (see, e.g., [19, Theorem 6.27]) we have

$$v(n) = x(n) + \frac{1}{x(0)} \sum_{j=0}^{n-1} x(n-j-1)g(j). \tag{4.55}$$

Since $x(n)$ is positive, (4.55) implies $x(n) < v(n)$ $(n \ge 1)$, which proves the right inequality of (4.53).

Further, it follows from (4.55) that

$$\nabla v(n) = x(n) + \left(\frac{g(0)}{x(0)} - 1\right) x(n-1) + \frac{1}{x(0)} \sum_{j=1}^{n-1} x(n-j-1) \nabla g(j) .$$

Since

$$\nabla g(j) = \sum_{k=-\infty}^{-1} \frac{(-1)^{j-k}}{1-\lambda h^{\alpha}} \binom{\alpha+1}{j-k+1} v(k) < 0 ,$$

we get

$$\nabla v(n) < x(n) + p\,x(n-1) , \qquad \text{where} \quad p = \frac{g(0)}{x(0)} - 1 . \tag{4.56}$$

Using the binomial theorem we can verify that

$$p = -\frac{\lambda^{1+1/\alpha} h^{\alpha+1} - (1+\alpha)\lambda^{1/\alpha} h + \alpha}{(1-\lambda h^{\alpha})(1-\lambda^{1/\alpha} h)} .$$

We show that $p < 0$, i.e.,

$$F_{\alpha}(\lambda) = \lambda^{1+1/\alpha} h^{\alpha+1} - (1+\alpha)\lambda^{1/\alpha} h + \alpha > 0$$

for all $0 < \alpha < 1$ and $0 < \lambda < 1/h^{\alpha}$. Indeed, we have

$$F_{\alpha}(0) = \alpha > 0 , \quad F_{\alpha}(1) = 0 \quad \text{and} \quad F'_{\alpha}(\lambda) = \frac{\alpha+1}{\alpha} \lambda^{1/\alpha} h (1 - \lambda^{-1} h^{-\alpha}) < 0$$

for all such values of α and λ. Consequently, we can neglect the last term in the inequality (4.56) to obtain

$$\nabla v(n) = \frac{\lambda^{1/\alpha} h x(0)}{(1-\lambda^{1/\alpha} h)^n} < x(n) , \qquad n \geq 1 .$$

The left inequality of (4.53) is proved. □

We conclude this part by the summary of the derived stability and asymptotic results and their reformulation for the original fractional difference equation (4.31) with respect to Proposition 4.1.

Theorem 4.3. *Let* $\lambda h^{\alpha} \neq 1$. *Then (4.31) is*
(i) stable if and only if

$$\lambda \leq 0 \qquad \text{or} \qquad \lambda \geq (2/h)^{\alpha} ;$$

(ii) asymptotically stable if and only if

$$\lambda \leq 0 \qquad \text{or} \qquad \lambda > (2/h)^{\alpha} ;$$

(iii) not stable if and only if

$$0 < \lambda < (2/h)^{\alpha} .$$

Contrary to stability analysis, its asymptotic extension is not quite complete. We summarize the known results in the following

Theorem 4.4. *Let $\lambda h^\alpha \neq 1$ and let $y(t_n)$ be a solution of (4.31).*
(i) If $\lambda < 0$ or $\lambda > (2/h)^\alpha$ then $y(t_n) \in \ell^1$. Moreover, if $|1-\lambda h^\alpha| > 1$ then

$$y(t_n) \sim \frac{K}{n^{1+\alpha}} \quad \text{as } n \to \infty, \qquad K = \frac{\alpha(1-\lambda h^\alpha)}{\lambda^2 h^{2\alpha}\Gamma(1-\alpha)}\, y(t_1)\,.$$

(ii) If $\lambda = 0$ then

$$y(t_n) \sim \frac{K}{n^{1-\alpha}} \quad \text{as } n \to \infty, \qquad K = \frac{1}{\Gamma(\alpha)}\, y(t_1)\,.$$

(iii) If $\lambda = (2/h)^\alpha$ then

$$y(t_n) \sim (-1)^{n-1}K \quad \text{as } n \to \infty, \qquad K = \frac{2^\alpha - 1}{\alpha 2^{\alpha-1}} y(t_1)\,.$$

(iv) If $0 < \lambda < 1/h^\alpha$ and $y(t_1) > 0$ then

$$\frac{\lambda^{1/\alpha} h y(t_1)}{(1-\lambda^{1/\alpha}h)^{n-1}} < y(t_n) < \frac{y(t_1)}{(1-\lambda^{1/\alpha}h)^{n-1}} \qquad \text{for } n \geq 2\,.$$

Proof. The parts (i), (ii) and (iv) follow immediately from our preceding considerations summarized in Corollaries 4.1, 4.2 and Propositions 4.2, 4.6. Thus, we focus on (iii). Considering the calculations performed in the proof of Proposition 4.4, namely the formulae (4.46) and (4.47), we have

$$\lim_{n\to\infty} \frac{y(t_n)}{(-1)^{n-1}K} = \lim_{n\to\infty} \frac{x(n-1)}{(-1)^{n-1}K} = \lim_{n\to\infty} \frac{1}{(-1)^{n-1}K} \sum_{j=0}^{n-1} g(j)(-1)^{n-1-j}$$
$$= \frac{1}{K} \lim_{n\to\infty} \sum_{j=0}^{n-1} g(j)(-1)^{-j} = \frac{1}{K}\tilde{g}(-1) = \frac{1}{K}\frac{(2^\alpha-1)x(0)}{\alpha 2^{\alpha-1}} = 1\,,$$

which completes the proof. □

4.2. Stability and Asymptotics of a Test Equation with the Caputo Fractional Difference

In Section 3, we observed that the replacement of the Riemann-Liouville difference operator by the Caputo one has no influence on basic qualitative investigations of corresponding fractional difference equations. This is not the case of stability investigations, when some nontrivial distinctions can be observed.

The reason why such distinctions occur comes from the following analogue of Proposition 4.1.

Proposition 4.7. *Let $\lambda h^\alpha \neq 1$. Then $y(t_n)$ is the solution of*

$$ {}^{\mathrm{C}}\nabla_h^\alpha y(t_n) = \lambda y(t_n), \qquad n \geq 2, \quad \lambda, \alpha \in \mathbb{R}, \quad 0 < \alpha < 1 \tag{4.57}$$

if and only if $x(n) = y(t_{n+1})$ is the solution of

$$x(n+1) = \frac{1}{1-\lambda h^\alpha} \sum_{j=0}^{n} (-1)^{n-j} \binom{\alpha}{n-j+1} x(j) + (-1)^{n+1} \binom{\alpha-1}{n+1} x(0), \tag{4.58}$$

$n \geq 0$, $\lambda, \alpha \in \mathbb{R}$, $0 < \alpha < 1$.

Proof. The proof is analogous to the one of Proposition 4.1. □

Compared to (4.32), the Volterra equation (4.58) involves also the nonhomogeneous term

$$f(n) = (-1)^{n+1} \binom{\alpha-1}{n+1} x(0), \tag{4.59}$$

which may influence the stability and asymptotics of (4.58).

For our next considerations, it is suitable to recall a relation between the solutions of (4.32) and (4.58). Let $\hat{x}(n)$ be the solution of (4.32) satisfying $\hat{x}(0) = 1$ (i.e., the solution of the homogeneous part of (4.58) normalized at the initial point). By the variation of constants formula (see [19, Theorem 6.27]), we have

$$x(n) = x(0)\hat{x}(n) + \sum_{k=0}^{n-1} \hat{x}(n-k-1) f(k) = \sum_{k=0}^{n} \hat{x}(n-k) f(k-1). \tag{4.60}$$

Using this formula we can establish next assertions as counterparts to Propositions 4.2-4.5.

Proposition 4.8. *Let $\lambda < 0$ or $\lambda > (2/h)^\alpha$. Then (4.58) is asymptotically stable.*

Proof. We rewrite the property (4.42) as

$$(-1)^{n+1} \binom{\alpha-1}{n+1} x(0) \sim \frac{1}{(n+1)^\alpha \Gamma(1-\alpha)} x(0) \quad \text{as } n \to \infty, \tag{4.61}$$

which yields the asymptotic expansion for the nonhomogeneous term $f(n)$ given by (4.59). We show that the convolution on the right-hand side of (4.60) tends to zero like $\mathcal{O}(n^{-\alpha})$ as $n \to \infty$. Besides (4.61), we use here the property $\hat{x}(n) \in \ell^1$ following from Proposition 4.2, which particularly implies $|\hat{x}(n)| \leq K_1/n$ for a suitable real $K_1 > 0$.

It holds

$$\begin{aligned} &|\sum_{k=0}^{n} \hat{x}(n-k) f(k-1)| \leq \sum_{k=0}^{n} |\hat{x}(n-k)||f(k-1)| \leq \hat{x}(n)x(0) + K_2 \sum_{k=1}^{n} \frac{1}{k^\alpha} |\hat{x}(n-k)| \\ &\leq \frac{K_1}{n} + K_2 \Big(\sum_{k=1}^{\lfloor n/2 \rfloor} \frac{1}{k^\alpha} |\hat{x}(n-k)| + \sum_{k=\lfloor n/2 \rfloor + 1}^{n} \frac{1}{k^\alpha} |\hat{x}(n-k)| \Big), \end{aligned}$$

where K_2 is a real constant and $\lfloor\cdot\rfloor$ means an integer part. We can estimate the first sum of the preceding relation as

$$\sum_{k=1}^{\lfloor n/2\rfloor}\frac{1}{k^\alpha}|\hat{x}(n-k)| \le \sum_{k=1}^{\lfloor n/2\rfloor}\frac{1}{k^\alpha}\frac{K_1}{n-k} \le \frac{2K_1}{n}\sum_{k=1}^{\lfloor n/2\rfloor}\frac{1}{k^\alpha}$$
$$\le \frac{2K_1}{n}\left(1+\int_1^{\lfloor n/2\rfloor}\frac{\mathrm{d}x}{x^\alpha}\right) \le \frac{2^\alpha K_1}{1-\alpha}\frac{1}{n^\alpha}.$$

Similarly we can dispose with the second sum, namely

$$\sum_{k=\lfloor n/2\rfloor+1}^{n}\frac{1}{k^\alpha}|\hat{x}(n-k)| \le \frac{2^\alpha}{n^\alpha}\sum_{k=\lfloor n/2\rfloor+1}^{n}|\hat{x}(n-k)| \le \frac{2^\alpha K_3}{n^\alpha},$$

where $K_3 = \sum_{k=1}^{\infty}|\hat{x}(k)|$. Substituting these estimates into (4.60) we get the assertion. □

Remark 4.3. *Notice that we have proved something more than the required stability result, namely also the decay rate $\mathcal{O}(n^{-\alpha})$ of the solutions $x(n)$ of (4.58) as $n\to\infty$.*

Proposition 4.9. *Let $\lambda = 0$. Then (4.58) is stable, but not asymptotically stable.*

Proof. Rearranging (4.58) with $\lambda = 0$ we obtain

$$\sum_{j=0}^{n+1}(-1)^{n-j+1}\binom{\alpha}{n-j+1}x(j) = (-1)^{n+1}\binom{\alpha-1}{n+1}x(0).$$

Compared to the well-known relation $\sum_{j=0}^{N}(-1)^j\binom{\alpha}{j} = (-1)^N\binom{\alpha-1}{N}$ we immediately get the constant solutions of (4.58), namely $x(n) = x(0)$ for all $n \ge 0$. Thus, (4.58) is stable, but not asymptotically stable. □

Proposition 4.10. *Let $\lambda = (2/h)^\alpha$. Then (4.58) is stable, but not asymptotically stable.*

Proof. We use again the representation of the solutions $x(n)$ of (4.58) via (4.60), where $\hat{x}(n)$ is the solution of (4.32) with $\hat{x}(0) = 1$. In the proof of Proposition 4.4, we have shown that for $\lambda = (2/h)^\alpha$ it holds

$$\hat{x}(n) = (-1)^n\sum_{j=0}^{n} g(j)(-1)^j,$$

where $g(n)$ is an unknown sequence such that the function

$$\tilde{g}(1/w) = \sum_{j=0}^{\infty} g(j)w^j$$

is analytic in $\mathbb{C}$, i.e., the series on the right-hand side converges for all $w \in \mathbb{C}$. It implies that $|g(n)| = \mathcal{O}(\delta^n)$ as $n \to \infty$ for arbitrary small $\delta > 0$. Moreover, a precision of the asymptotic property given in Theorem 4.4 (iii) can be derived, namely

$$\left|(-1)^n \frac{2^\alpha - 1}{\alpha 2^{\alpha-1}} - \hat{x}(n)\right| = \left|\sum_{j=0}^{\infty} g(j)(-1)^j - \sum_{j=0}^{n} g(j)(-1)^j\right|$$
$$\leq \sum_{j=n+1}^{\infty} |g(j)| \leq \sum_{j=n+1}^{\infty} C\delta^j = \frac{C\,\delta^{n+1}}{1-\delta},$$

where $C = C(\delta) > 0$ is a suitable constant.

Using this we can discuss the asymptotic behaviour of the solution $x(n)$ of (4.58) given by (4.60). It holds

$$x(n) = \sum_{j=0}^{n} \hat{x}(n-j) f(j-1) = \sum_{j=0}^{n} \hat{x}(n-j)(-1)^j \binom{\alpha-1}{j} x(0)$$
$$\leq x(0) \sum_{j=0}^{n} \left((-1)^{n-j} \frac{2^\alpha - 1}{\alpha 2^{\alpha-1}} + \frac{C\delta^{n-j+1}}{1-\delta}\right)(-1)^j \binom{\alpha-1}{j}$$
$$= (-1)^n \frac{2^\alpha - 1}{\alpha 2^{\alpha-1}} x(0) \sum_{j=0}^{n} \binom{\alpha-1}{j} + \frac{C\delta}{1-\delta} x(0) \sum_{j=0}^{n} \delta^{n-j}(-1)^j \binom{\alpha-1}{j}.$$

Note that the second sum on the last row is a convolution of a convergent geometric sequence and the series with asymptotics given by (4.61). Thus, following the steps of the proof of Proposition 4.8, we can see that this term behaves like $\mathcal{O}(n^{-\alpha})$ as $n \to \infty$. Analogously, we can dispose with the lower estimate of $x(n)$. Thus, we have

$$(-1)^n \frac{2^\alpha - 1}{\alpha 2^{\alpha-1}} \sum_{j=0}^{n} \binom{\alpha-1}{j} - \frac{K}{n^\alpha} \leq \frac{x(n)}{x(0)} \leq (-1)^n \frac{2^\alpha - 1}{\alpha 2^{\alpha-1}} \sum_{j=0}^{n} \binom{\alpha-1}{j} + \frac{K}{n^\alpha},$$

where $K > 0$ is a suitable constant. We can see that the solution is oscillating and

$$\lim_{n\to\infty} |x(n)| = \lim_{n\to\infty} \left| x(0)(-1)^n \frac{2^\alpha - 1}{\alpha 2^{\alpha-1}} \sum_{j=0}^{n} \binom{\alpha-1}{j} \right| = \frac{(2^\alpha - 1)|x(0)|}{\alpha} > 0,$$

which proves the assertion. □

Proposition 4.11. *Let* $0 < \lambda < (2/h)^\alpha$ *and* $\lambda h^\alpha \neq 1$. *Then (4.58) is not stable.*

Proof. We perform the $\mathcal{Z}$-transform of (4.60) and obtain

$$\mathcal{Z}(x(n)) = \mathcal{Z}(\hat{x}(n)) \cdot \mathcal{Z}(f(n-1)).$$

Since $\mathcal{Z}(f(n-1))$ is the binomial series with the radius of convergence $R_f = 1$ and $\mathcal{Z}(\hat{x}(n))$ has the radius of convergence $R_{\hat{x}} < 1$ (see Proposition 4.5), we get that the series $\mathcal{Z}(x(n))$ has the radius of convergence $R_x = \min(R_f, R_{\hat{x}}) < 1$. Using the similar line of arguments as given in the proof of Proposition 4.5 we can conclude that (4.58) is not stable. □

The following assertion, which summarizes the previous stability results with respect to Proposition 4.7, is an analogue of Theorem 4.3 for (4.57) with the Caputo difference operator.

Theorem 4.5. *Let $\lambda h^\alpha \neq 1$. Then (4.57) is*
(i) stable if and only if

$$\lambda \leq 0 \qquad \textit{or} \qquad \lambda \geq (2/h)^\alpha ;$$

(ii) asymptotically stable if and only if

$$\lambda < 0 \qquad \textit{or} \qquad \lambda > (2/h)^\alpha ;$$

(iii) not stable if and only if

$$0 < \lambda < (2/h)^\alpha .$$

One can observe a distinction between stability properties of the both fractional equations (4.31) and (4.57). While (4.31) is asymptotically stable for $\lambda = 0$, (4.57) remains stable, but not asymptotically stable in this case (more precisely, the solution is constant). More significant differences occur when investigating the decay rate of the solutions. While Theorem 4.4 implies that solutions of (4.31) for $\lambda < 0$ or $\lambda > (2/h)^\alpha$ belong to ℓ^1 space and, in particular, for λ such that $|1 - \lambda h^\alpha| > 1$ behave like $n^{-(\alpha+1)}$, asymptotics for the solutions of (4.57) is different and given by the following

Theorem 4.6. *Let $\lambda h^\alpha \neq 1$ and let $y(t_n)$ be a solution of (4.57).*
(i) If $\lambda < 0$ or $\lambda > (2/h)^\alpha$ then

$$y(t_n) \notin \ell^1 \qquad \textit{and} \qquad |y(t_n)| = \mathcal{O}(n^{-\alpha}) \quad \textit{as } n \to \infty .$$

(ii) If $\lambda = 0$ then

$$y(t_n) = K, \quad n \geq 1, \qquad K = y(t_1) .$$

(iii) If $\lambda = (2/h)^\alpha$ then

$$y(t_n) \sim (-1)^{n-1} K \quad \textit{as } n \to \infty , \qquad K = \frac{2^\alpha - 1}{\alpha} y(t_1) .$$

(iv) If $0 < \lambda < 1/h^\alpha$ and $y(t_1) > 0$ then

$$\frac{\lambda^{1/\alpha} h y(t_1)}{(1 - \lambda^{1/\alpha} h)^{n-1}} < y(t_n) < \frac{(\lambda^{1/\alpha} h)^{\alpha-1} y(t_1)}{(1 - \lambda^{1/\alpha} h)^{n-1}}, \qquad n \geq 2 .$$

Proof. (i) Considering $x(n) = y(t_{n+1})$, the property $|y(t_n)| = \mathcal{O}(n^{-\alpha})$ as $n \to \infty$ was already derived in the proof of Proposition 4.8. Next, we consider $x(n)$ in the form (4.60) and using a similar technique as in the proof of Proposition 4.2 express its $\mathcal{Z}$-transform

$$\tilde{x}(z) = \frac{\left(1 - \frac{1}{z}\right)^\alpha - (1 - x(0))\left(1 - \frac{1}{z}\right)}{\left(1 + \frac{1}{1 - \lambda h^\alpha}\left(\left(1 - \frac{1}{z}\right)^\alpha - 1\right)\right)\left(1 - \frac{1}{z}\right)} .$$

Obviously, $\tilde{x}(z)$ has a pole in $z = 1$, hence it holds $\lim_{z\to 1}|\tilde{x}(z)| = \infty$ which together with

$$|\tilde{x}(z)| = |\sum_{j=0}^{\infty} x(j)z^{-j}| \le \sum_{j=0}^{\infty} |x(j)||z|^{-j}$$

implies that $x(n) \notin \ell^1$. The assertion is proved.

The parts (ii) and (iii) follow from the proofs of Propositions 4.9 and 4.10.

(iv) Employing (4.60) and (4.53) we can write

$$y(t_n) = x(n-1) = \sum_{j=0}^{n-1} \hat{x}(n-1-j)f(j-1)$$

$$< \frac{x(0)}{(1-\lambda^{1/\alpha})^{n-1}} \sum_{j=0}^{n-1} \binom{\alpha-1}{j} (\lambda^{1/\alpha}h - 1)^j \le \frac{(\lambda^{1/\alpha}h)^{\alpha-1}y(t_1)}{(1-\lambda^{1/\alpha})^{n-1}}\,.$$

The lower estimate can be obtained analogously. □

4.3. Stability and Asymptotics of Vector Test Equations

Stability investigations of (4.31) and (4.57) can also be extended to their vector analogues, namely to linear fractional difference systems

$$\nabla_h^{\alpha} y(t_n) = A\,y(t_n), \qquad n \ge 2, \quad 0 < \alpha < 1 \tag{4.62}$$

and

$$^{C}\nabla_h^{\alpha} y(t_n) = A\,y(t_n), \qquad n \ge 2, \quad 0 < \alpha < 1\,, \tag{4.63}$$

where A is a $d \times d$ matrix with real constant entries. We describe asymptotic stability conditions for both systems, i.e., conditions ensuring that all their solutions are tending to zero (in an appropriate norm) as $n \to \infty$.

All the results of this subsection are given without proofs, which can be found in [11, 14]. These proofs are either technical extensions of the corresponding procedures utilized in the scalar case (Propositions 4.12–4.13), or more complicated matters extending the range of this chapter (Theorems 4.9–4.10). Main stability results for (4.62) and (4.63) turn out to be very useful in stability analysis of multi-term (scalar) fractional difference equations as it will be illustrated in Section 5.

First we mention the following analogues of Propositions 4.1 and 4.7 to show that (4.62) and (4.63) are in fact specific Volterra difference systems of convolution type.

Proposition 4.12. *Let A be a real $d \times d$ matrix such that $I - h^{\alpha}A$ is invertible. Then $y(t_n)$ is the solution of (4.62) if and only if $x(n) = y(t_{n+1})$ is the solution of*

$$x(n+1) = \sum_{j=0}^{n} M(n-j)x(j), \qquad n \ge 0\,,$$

where

$$M(n) = (-1)^n \binom{\alpha}{n+1} (I - h^{\alpha}A)^{-1}\,.$$

Proposition 4.13. *Let A be a real $d \times d$ matrix such that $I - h^\alpha A$ is invertible. Then $y(t_n)$ is the solution of (4.63) if and only if $x(n) = y(t_{n+1})$ is the solution of*

$$x(n+1) = \sum_{j=0}^{n} M(n-j)x(j) + (-1)^{n+1}\binom{\alpha-1}{n+1}x(0), \qquad n \geq 0,$$

where $M(n)$ has the same meaning as in Proposition 4.12.

Before we formulate the main stability results for (4.62) and (4.63), we introduce the sets

$$\mathcal{S}_\alpha^* = \left\{z \in \mathbb{C} : |\arg(z)| > \frac{\alpha\pi}{2}\right\}$$

and

$$\mathcal{S}_\alpha(h) = \left\{z \in \mathbb{C} : |\arg(z)| > \frac{\alpha\pi}{2} \text{ or } |z| > \frac{2^\alpha}{h^\alpha}\cos^\alpha\left(\frac{\arg(z)}{\alpha}\right)\right\},$$

where $\arg(z)$ denotes the principal argument of a complex number z. The set $\mathcal{S}_\alpha^*$ is well-known from stability theory of fractional differential equations. It is the stability region for the linear systems

$$\mathrm{D}^\alpha y(t) = Ay(t), \qquad t > 0, \quad 0 < \alpha < 1 \tag{4.64}$$

and

$$^{\mathrm{C}}\mathrm{D}^\alpha y(t) = Ay(t), \qquad t > 0, \quad 0 < \alpha < 1, \tag{4.65}$$

where A is a $d \times d$ matrix with real constant entries. More precisely, it holds two stability criteria which are due to [40, 33], respectively.

Theorem 4.7. *Let A be a real $d \times d$ matrix and $y(t)$ be a solution of (4.64). Then (4.64) is asymptotically stable if and only if all the eigenvalues of A satisfy*

$$\lambda(A) \in \mathcal{S}_\alpha^* \cup \{0\}$$

and the maximal size of Jordan blocks corresponding to the zero eigenvalue is less than α^{-1}. In particular, if all these eigenvalues satisfy $\lambda(A) \in \mathcal{S}_\alpha^*$, then the components of the state $y(t)$ decay towards zero like $t^{-(1+\alpha)}$.

Theorem 4.8. *Let A be a real $d \times d$ matrix and $y(t)$ be a solution of (4.65). Then (4.65) is asymptotically stable if and only if all the eigenvalues of A satisfy*

$$\lambda(A) \in \mathcal{S}_\alpha^*.$$

In such a case, the components of the state $y(t)$ decay towards zero like $t^{-\alpha}$.

The set $\mathcal{S}_\alpha(h)$ plays essentially the same role for the discrete systems (4.62) and (4.63) as $\mathcal{S}_\alpha^*$ for (4.64) and (4.65). Indeed, it holds the following stability criteria (by the symbol $|C|$ we understand here the matrix given by $|C| = (|c_{ij}|)$).

Theorem 4.9. *Let A be a real $d \times d$ matrix such that $I - h^\alpha A$ is invertible and let $y(t_n)$ be a solution of (4.62). Then (4.62) is asymptotically stable if all the eigenvalues of A satisfy*

$$\lambda(A) \in \mathcal{S}_\alpha(h) \cup \{0\}$$

and the maximal size of Jordan blocks corresponding to the zero eigenvalue is less than α^{-1}. Furthermore, if all these eigenvalues satisfy $\lambda(A) \in \mathcal{S}_\alpha(h)$, then $y(t_n) \in \ell^1$, and if all eigenvalues of $|(I - h^\alpha A)^{-1}|$ lie inside the open unit disk, then $|y(t_n)| = \mathcal{O}(n^{-(1+\alpha)})$ as $n \to \infty$.

Theorem 4.10. *Let A be a real $d \times d$ matrix such that $I - h^\alpha A$ is invertible and let $y(t_n)$ be a solution of (4.63). Then (4.63) is asymptotically stable if all the eigenvalues of A satisfy*

$$\lambda(A) \in \mathcal{S}_\alpha(h)\,.$$

In such a case, $y(t_n) \notin \ell^1$ and $|y(t_n)| = \mathcal{O}(n^{-\alpha})$ as $n \to \infty$.

Remark 4.4. *It might be interesting to observe a relationship between the stability regions $\mathcal{S}^*_\alpha$ and $\mathcal{S}_\alpha(h)$. Of course, the structure of $\mathcal{S}_\alpha(h)$ is more complicated. However, as we can see in Figure 6, letting $h \to 0$ the discrete stability region $\mathcal{S}_\alpha(h)$ approaches $\mathcal{S}^*_\alpha$ for any $0 < \alpha < 1$. More comments on this topic will be stated in the last section.*

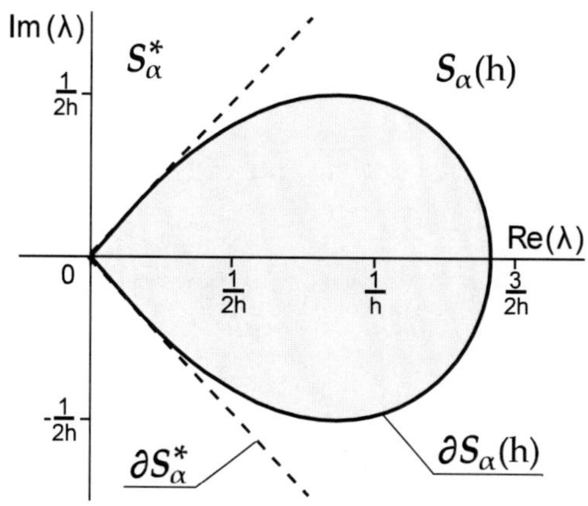

Figure 6. Stability domains $\mathcal{S}^*_\alpha$ and $\mathcal{S}_\alpha(h)$ with their boundary curves for $\alpha = 0.6$.

Remark 4.5. *The stability regions $\mathcal{S}^*_\alpha$ and $\mathcal{S}_\alpha(h)$ are precise enough. It can be proved, that if there exists an eigenvalue $\lambda(A)$ of A such that*

$$\lambda(A) \notin \mathrm{cl}(\mathcal{S}^*_\alpha) \qquad \text{or} \qquad \lambda(A) \notin \mathrm{cl}(\mathcal{S}_\alpha(h))\,,$$

then the corresponding continuous or discrete fractional system is not stable regardless of a type of the fractional operator.

5. Stability of Multi-Term Fractional Difference Equations

In the previous section, we showed that the stability analysis of linear fractional difference equations is well-established for single-term equations, i.e., equations involving only one difference operator. Fortunately, there exists a correspondence between multi-term fractional difference equations and vector single-term fractional difference equations, which enables many important results and methods concerning vector single-term fractional difference equations to be extended to appropriate multi-term fractional difference equations. In this section, we first precise this correspondence and then apply it to a special two-term fractional difference equation. The results presented here are essentially new (some related ideas and calculations can be found in [11]).

5.1. Multi-Term Caputo Fractional Difference Equations

We consider the linear Caputo multi-term fractional difference equation

$$\sum_{j=0}^{m} p_j \, {}^{\mathrm{C}}\nabla_h^{\alpha_j} y(t_n) = f(t_n)\,, \tag{5.66}$$

where $\alpha_0 > \alpha_1 > \cdots > \alpha_m = 0$ and p_j $(j = 0, \ldots, m)$ are real scalars with $p_0 = 1$ and $f(t_n)$ is a real sequence. We show that (5.66) is equivalent to a single-term vector fractional difference equation if $\alpha_j = (m-j)/q\,(j = 0, \ldots, m)$, where q is a suitable positive integer ((5.66) is then usually called commensurate).

The idea of the method goes back to [17], where the authors introduced a simple technique based on a conversion of a given scalar multi-term fractional differential equation into an appropriate vector single-term equation. The discrete analogue of this procedure is given by the following assertion.

Theorem 5.1. *Consider the commensurate equation (5.66), put $\kappa = 1/q$ and introduce the system*

$$\begin{aligned} {}^{\mathrm{C}}\nabla_h^{\kappa} y_1(t_n) &= y_2(t_n)\,, \\ &\ \vdots \\ {}^{\mathrm{C}}\nabla_h^{\kappa} y_{m-1}(t_n) &= y_m(t_n)\,, \\ {}^{\mathrm{C}}\nabla_h^{\kappa} y_m(t_n) &= -\sum_{j=1}^{m} p_j y_{m-j+1}(t_n) + f(t_n)\,, \end{aligned} \tag{5.67}$$

supplied with the initial conditions

$$y_j(0) = \begin{cases} y_{(j-1)\kappa}\,, & \text{for } (j-1)\kappa \in \mathbb{Z}_0^+\,, \\ 0\,, & \text{otherwise} \end{cases} \qquad j = 1, \ldots, m\,. \tag{5.68}$$

Then the vector $(y_1(t_n), \ldots, y_m(t_n))$ solves (5.67), (5.68) if and only if $y(t_n) = y_1(t_n)$ solves (5.66) supplied with the initial conditions $\nabla_h^k y(t_n)|_{n=0} = y_k$ $(k = 0, \ldots, \lceil \alpha_0 \rceil - 1)$.

Proof. The proof is a direct adaptation of its continuous paradigm (see [17, 18]), hence we outline only the key step. The method itself relies on the fact that the utilized Caputo fractional differences commute. This is not generally true, but it is true for our particular case when the first term in a composition process is always ${}^{\mathrm{C}}\nabla_h^{\kappa}$, where $\kappa = 1/q < 1$. Thus, it suffices to show that

$$ {}^{\mathrm{C}}\nabla_h^{\kappa}\,{}^{\mathrm{C}}\nabla_h^{(j-1)\kappa} y(t_n) = {}^{\mathrm{C}}\nabla_h^{j\kappa} y(t_n)\,, \quad j = 2, \ldots, m\,. $$

If $(j-1)\kappa \in \mathbb{Z}^+$, then the relation trivially follows from Definition 2.3. The case $(j-1)\kappa \notin \mathbb{Z}^+$ is technically more complicated, hence we outline the main idea only for $j = 2$. Using (2.11) with respect to (2.6) we obtain

$$ {}^{\mathrm{C}}\nabla_h^{\kappa}\,{}^{\mathrm{C}}\nabla_h^{\kappa} y(t_n) = \nabla_h^{\kappa}\,{}^{\mathrm{C}}\nabla_h^{\kappa} y(t_n) - \frac{(-1)^{n-1}}{h^{\kappa}}\binom{\kappa-1}{n-1}\,{}^{\mathrm{C}}\nabla_h^{\kappa} y(t_k)|_{k=0}\,. $$

The right-hand side contains the term ${}^{\mathrm{C}}\nabla_h^{\kappa} y(t_k)|_{k=0}$ involving the initial value of the Caputo difference of non-integer order. Definitions 2.1 and 2.3 (extended for $t_n = a$) imply ${}^{\mathrm{C}}\nabla_h^{\kappa} y(t_k)_{k=0} = 0$ regardless of the values of $y(t_n)$ and its first difference. Thus, we get

$$ \begin{aligned} {}^{\mathrm{C}}\nabla_h^{\kappa}\,{}^{\mathrm{C}}\nabla_h^{\kappa} y(t_n) &= \nabla_h \nabla_h^{-(1-\kappa)} \nabla_h^{-(1-\kappa)} \nabla_h y(t_n) = \nabla_h \nabla_h^{-(2-2\kappa)} \nabla_h y(t_n) \\ &= \nabla_h \nabla_h^{-1} \nabla_h^{-(1-2\kappa)} \nabla_h y(t_n) = \nabla_h^{-(1-2\kappa)} \nabla_h y(t_n) = {}^{\mathrm{C}}\nabla_h^{2\kappa} y(t_n)\,, \end{aligned} $$

where we have utilized Definitions 2.2 and 2.3, Lemma 2.2 and the fact that $2\kappa \le 1$.

The relationship between the initial conditions is an immediate consequence of the formula $y_j(t_n) = {}^{\mathrm{C}}\nabla_h^{(j-1)\kappa} y(t_n)$ ($j = 1, \ldots, m$). □

Remark 5.1. *(i) The initial conditions $\nabla_h^k y(t_n)|_{n=0} = y_k$ ($k = 0, \ldots, \lceil \alpha_0 \rceil - 1$) slightly differ from the conditions (3.17) utilized in Section 3. However, it can be shown that there is a one-to-one mapping between these two sets of conditions (for more details we refer, e.g., to [28]).*
(ii) This transformation method utilizes composition rule for the Caputo difference. Therefore, its application to Riemann-Liouville equations is not straightforward due to a different form of this rule.

Theorem 5.1 converts the problem of asymptotic stability of the multi-term difference equation (5.66) to the problem of the asymptotic stability of the corresponding single-term difference system (5.67). More precisely, if this system is asymptotically stable, i.e., all components of any solution are tending to zero, then the solution $y(t_n)$ of (5.66) along with ${}^{\mathrm{C}}\nabla_h^{j\kappa} y(t_n)$ ($j = 1, \ldots, m-1$) are also tending to zero. Consequently, applying Theorem 4.10, if all the eigenvalues of the companion matrix A of (5.67) satisfy $\lambda(A) \in \mathcal{S}_\kappa(h)$, then (5.66) is asymptotically stable. Contrary, by Remark 4.5, if there exists an eigenvalue $\lambda(A)$ of the companion matrix A such that $\lambda(A) \notin \mathrm{cl}\,(\mathcal{S}_\kappa(h))$, then (5.67) is not stable, i.e., $y_j(t_n)$ is unbounded for some $j = 1, \ldots, m$. We show that it implies that $y_1(t_n) = y(t_n)$ must be unbounded as well, i.e., (5.66) is not stable. Indeed, if $y_1(t_n), \ldots, y_j(t_n)$ ($j = 1, \ldots, m-1$)

are bounded, then

$$|y_{j+1}(t_n)| = |{}^{\mathrm{C}}\nabla_h^{\kappa} y_j(t_n)| = \left|\sum_{k=1}^{n} \frac{(-1)^{n-k}}{h^{\kappa}} \binom{\kappa}{n-k} y_j(t_k) - \frac{(-1)^{n-1}}{h^{\kappa}} \binom{\kappa-1}{n-1} y_j(t_0)\right|$$

$$\leq \sum_{k=1}^{n} \frac{(-1)^{n-k}}{h^{\kappa}} \binom{\kappa}{n-k} |y_j(t_k)| + \frac{(-1)^{n-1}}{h^{\kappa}} \binom{\kappa-1}{n-1} |y_j(t_0)|$$

$$\leq \frac{K}{h^{\kappa}} \sum_{k=1}^{n} (-1)^{n-k} \binom{\kappa}{n-k} \leq \frac{K}{h^{\kappa}}$$

for a suitable real constant $K > 0$, therefore $y_{j+1}(t_n)$ is also bounded. Equivalently, unboundedness of $y_j(t_n)$ for some $j = 2, \ldots, m$ implies also unboundedness of $y_1(t_n) = y(t_n)$.

Summarizing previous considerations, we are able to express the stability conditions for the multi-term fractional equation (5.66) in terms of a location of eigenvalues of the corresponding companion matrix A of (5.67) with respect to the stability area $\mathcal{S}_{\kappa}(h)$. Our next aim is to propose a technique, which enables to express the stability conditions for (5.66) directly in terms of its coefficients. As a test equation, we consider a special two-term fractional difference equation.

5.2. Stability of the Discrete Basset Equation

We investigate the two-term linear fractional difference equation

$$\nabla_h y(t_n) + a\,{}^{\mathrm{C}}\nabla_h^{1/2} y(t_n) + b\,y(t_n) = 0 \tag{5.69}$$

with real parameters a, b. This equation arises from the Grünwald-Letnikov discretization of the well-known Basset equation

$$y'(t) + a\,{}^{\mathrm{C}}\mathrm{D}^{1/2} y(t) + b\,y(t) = 0\,, \tag{5.70}$$

describing the forces that occur when a spherical object sinks in an incompressible viscous fluid (see [18]). Because of its applicability and formally simple form, (5.70) serves as a useful paradigm for various analytical and numerical investigations of multi-term fractional differential equations. For the same reasons, we choose its discrete analogue (5.69) as the (probably simplest) test equation for the stability analysis of multi-term fractional difference equations. Our goal is to describe the asymptotic stability region $\Sigma(h)$ for (5.69), i.e., the set of real couples (a, b) such that any solution $y(t_n)$ of (5.69) is tending to zero as $n \to \infty$.

Our procedure how to describe $\Sigma(h)$ consists of several steps. After rewriting (5.69) into appropriate fractional difference system, we analyse the stability and asymptotic properties of this difference system. Then, using the boundary locus technique, we shall describe $\Sigma(h)$ explicitly in the (a, b)-plane, including the decay rate of the solutions in the asymptotically stable case.

First we follow Theorem 5.1. We introduce the vector $(y_1(t_n), y_2(t_n))$, whose j-th component is given by $y_j(t_n) = {}^{\mathrm{C}}\nabla_h^{(j-1)/2} y(t_n)$ $(j = 1, 2)$. Then (5.69) can be rewritten as (5.67) with $m = 2$, $\kappa = 1/2$ and the companion matrix

$$A = \begin{pmatrix} 0 & 1 \\ -b & -a \end{pmatrix}$$

having the characteristic equation

$$\Lambda(\lambda) \equiv \lambda^2 + a\lambda + b = 0\,. \tag{5.71}$$

By Theorem 4.10 and Remark 4.5, the stability boundary for such a system is the curve formed by the set of all complex z such that

$$|z| = \left(\frac{2}{h}\cos(2\arg(z))\right)^{\frac{1}{2}} \qquad \text{and} \qquad |\arg z| \le \frac{\pi}{4}\,, \tag{5.72}$$

i.e., in the polar form

$$z = r(\varphi)\mathrm{e}^{\mathrm{i}\varphi}, \qquad r(\varphi) = \left(\frac{2}{h}\cos(2\varphi)\right)^{\frac{1}{2}}, \qquad -\pi/4 \le \varphi \le \pi/4$$

(we suppress the dependance on h in the notation $r(\varphi)$ because it is not important in the studied connection). Using this representation we can define the so-called discrete boundary locus as

$$BL(h) = \{(a,b) \in \mathbb{R}^2 : \Lambda(r(\varphi)\mathrm{e}^{\mathrm{i}\varphi}) = 0 \quad \text{for some } -\pi/4 \le \varphi \le \pi/4\}\,.$$

In other words, $BL(h)$ involves the set of all couples (a,b) such that there exists a root of (5.71) lying on the stability curve (5.72).

Now we put $\lambda = r(\varphi)\mathrm{e}^{\mathrm{i}\varphi}$ into (5.71) and obtain

$$r^2(\varphi)\mathrm{e}^{\mathrm{i}2\varphi} + ar(\varphi)\mathrm{e}^{\mathrm{i}\varphi} + b = 0\,,$$

which, after equating the real and imaginary parts, becomes

$$\begin{aligned} r^2(\varphi)\cos(2\varphi) + ar(\varphi)\cos(\varphi) + b &= 0\,, \\ r^2(\varphi)\sin(2\varphi) + ar(\varphi)\sin(\varphi) &= 0\,. \end{aligned}$$

This system is satisfied for some a, b and φ with $|\varphi| \le \pi/4$ if and only if either

$$r(\varphi) = 0, \qquad b = 0\,,$$

or

$$\varphi = 0, \qquad r^2(0) + ar(0) + b = 0\,,$$

or

$$\begin{aligned} &\varphi r(\varphi) \neq 0\,, \\ &a = -r(\varphi)\frac{\sin(2\varphi)}{\sin(\varphi)} = -2r(\varphi)\cos(\varphi)\,, \\ &b = -r^2(\varphi)\cos(2\varphi) + 2r^2(\varphi)\cos^2(\varphi) = r^2(\varphi)\,. \end{aligned}$$

These relations can be equivalently written as

$$\varphi = \pm\frac{\pi}{4}, \qquad b = 0\,,$$

$$\varphi = 0, \qquad bh + a(2h)^{1/2} + 2 = 0$$

and

$$0 < |\varphi| < \frac{\pi}{4}, \quad a = -\left(\frac{8}{h}\cos(2\varphi)\right)^{1/2}\cos(\varphi), \quad b = \frac{2}{h}\cos(2\varphi)\,,$$

respectively.

The dependent variables a, b are even with respect to φ and define the parametrical curve connecting the points $(0,0)$ and $(-(8/h)^{1/2}, 2/h)$, which can be given in the explicit form. To summarize this, the set $BL(h)$ is formed by two lines: $b = 0$,

$$l(h):\ bh + a(2h)^{1/2} + 2 = 0$$

and by the curve

$$\gamma(h):\ a = -(2b + b^2h)^{1/2}, \qquad 0 \le b \le 2/h\,.$$

Thus $BL(h)$ divides the (a,b)-plane into five regions with respect to a changing number of roots of $\Lambda(\lambda)$ located outside the stability set $\mathcal{S}_{1/2}(h)$. Figure 7 depicts these regions (in each of the regions we state also the number of roots of $\Lambda(\lambda)$ located outside $\mathcal{S}_{1/2}(h)$).

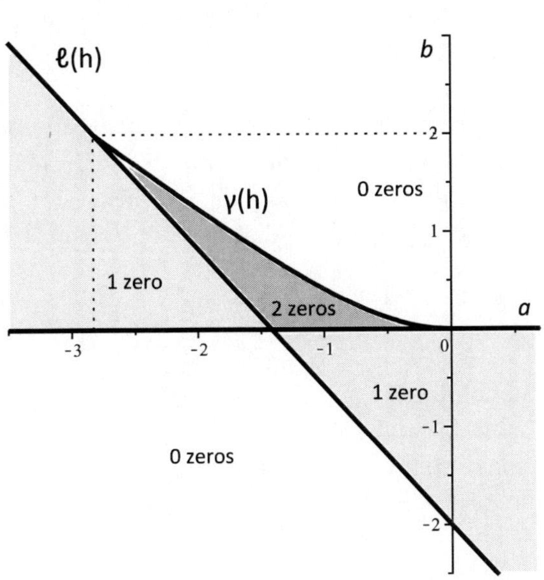

Figure 7. $BL(h)$ in the (a,b)-plane for $h = 1$.

This immediately implies

Theorem 5.2. *The discrete Basset equation (5.69) is asymptotically stable if*

$$a > -(2b + b^2 h)^{1/2}, \qquad 0 \le b \le 2/h\,, \tag{5.73}$$

or

$$a > -b(h/2)^{1/2} - (2/h)^{1/2}, \qquad b \ge 2/h\,, \tag{5.74}$$

or

$$a < -b(h/2)^{1/2} - (2/h)^{1/2}, \qquad b < 0\,. \tag{5.75}$$

Remark 5.2. *By Theorem 5.2, the conditions (5.73)–(5.75) define the asymptotic stability region $\Sigma(h)$ for (5.69). Moreover, Theorem 4.10 implies that if $(a,b) \in \Sigma(h)$ then*

$$y(t_n) = \mathcal{O}(n^{-1/2}) \qquad \text{as } n \to \infty$$

for any solution $y(t_n)$ of (5.69).

6. Conclusion

We have discussed some basic notions and properties of discrete fractional calculus, with special emphasis on linear fractional difference equations and their qualitative analysis. These issues turn out to be more difficult than the ones in the corresponding continuous fractional case. In particular, several classical results supporting qualitative analysis of fractional differential equations (such as asymptotic expansions of the Mittag-Leffler type functions) have no discrete analogue and the relevant results have to be proved by the use of original procedures.

Despite different approaches, several resemblances between continuous and discrete fractional calculus can be observed, especially with respect to the qualitative behaviour of the above investigated fractional equations. Firstly, there is an interesting connection between the stability regions $\mathcal{S}_\alpha(h)$ and $\mathcal{S}_\alpha^*$ discussed in Subsection 4.3. In particular, it holds

$$\mathcal{S}_\alpha^* \subset \mathcal{S}_\alpha(h) \qquad \text{for any } 0 < \alpha < 1 \text{ and any } h > 0$$

and, moreover, $\mathcal{S}_\alpha(h)$ monotonically approaches $\mathcal{S}_\alpha^*$ as $h \to 0$. Secondly, the decay rate of the solutions is the same in both the exact and the discrete case. In this connection, we emphasize that the above studied fractional difference schemes (4.31), (4.57), (4.62), (4.63) and (5.69) are the Grünwald-Letnikov discretizations of the underlying fractional differential equations. Thus, we have revealed the capability of theses discretizations to preserve basic stability and asymptotic properties of their continuous patterns, which are the standard requirements on numerical discretizations. We believe that the techniques utilized in this chapter are also applicable to more general types of fractional difference equations, which might contribute to some further developments of numerical analysis of fractional differential equations.

Throughout this chapter, we have restricted our investigations only to discrete settings with the uniform mesh of points. However, as mentioned in Section 1, basic notions of discrete fractional calculus are not limited only to such a case. Besides q-calculus and (q, h)-calculus, where the basic theory is already well-established, several papers (see, e.g., [3, 8]) discussed the possibility of incorporating fractional calculus to the time scales theory (using its terminology, this chapter has studied fractional calculus on the particular time scale $\mathbb{T} = h\mathbb{Z}$, see [9]). The main problem, preventing us from generalizing the basic fractional operators to this general framework, arises from the notion of a power function, whose previous introductions on general settings were questionable. Only recently, the works [41, 42] and [29] suggested (independently) an axiomatic definition of power functions on an arbitrary time scale, including a comprehensive critical study of previous attempts. This idea was further developed in [30], where feasibility of such a definition was discussed and some unusual properties of proposed introductions of power functions were outlined. Despite these results, the theory of fractional calculus on time scales is just at its beginning and there are almost no results regarding corresponding dynamic equations, which provides a wide area for future research.

Acknowledgment

The research was supported by the grant P201/11/0768 of the Czech Science Foundation and by the project CZ.1.07/2.3.00/30.0039 of Brno University of Technology.

References

[1] Agarwal, R. P., *Certain fractional q-integrals and q-derivatives.* Mathematical Proceedings of the Cambridge Philosophical Society, 1969, 66, 365-370.

[2] Al-Salam, W. A., *Some fractional q-integrals and q-derivatives.* Proceedings of the Edinburgh Mathematical Society, 1966, 15, 133-142.

[3] Anastassiou, G. A., *Foundations of nabla fractional calculus on time scales and inequalities.* Computers and Mathematics with Applications, 2010, 59, 3750-3762.

[4] Appleby, J. A. D., Györi, I., Reynolds, D. W., *On exact convergence rates for solutions of linear systems of Volterra difference equations.* Journal of Difference Equations and Applications, 2006, 12, 1257-1275.

[5] Atici, F. M., Eloe, P. W., *Initial value problems in discrete fractional calculus.* Proceedings of the American Mathematical Society, 2009, 137, 981-989.

[6] Atici, F. M., Eloe, P. W., *Discrete fractional calculus with the nabla operator.* Electronic Journal of Qualitative Theory of Differential Equations, 2009, spec. ed. 1, 1-12.

[7] Atici, F. M., Eloe, P. W., *Linear systems of fractional nabla difference equations.* Rocky Mountain Journal of Mathematics, 2011, 41, 353-370.

[8] Bastos, N. R. O., *Fractional Calculus on Time Scales.* PhD thesis, University of Aveiro, 2012, 119 pp., supervisor: D.F.M. Torres.

[9] Bohner, M., Peterson, A., *Dynamic Equations on Time Scales: An Introduction with Applications.* 1st ed. Boston: Birkhäuser, 2001.

[10] Čermák, J., Kisela, T., *Note on a discretization of a linear fractional differential equation.* Mathematica Bohemica, 2010, 135, 179-188.

[11] Čermák, J., Kisela, T., *Exact and discretized stability of the Bagley-Torvik equation.* Journal of Computational and Applied Mathematics, 2014, 269, 53-67.

[12] Čermák, J., Kisela, T., Nechvátal, L., *Discrete Mittag-Leffler functions in linear fractional difference equations.* Abstract and Applied Analysis, 2011, 2011, 21 pp.

[13] Čermák, J., Kisela, T., Nechvátal, L., *Stability and asymptotic properties of a linear fractional difference equation.* Advances in Difference Equations, 2012, 122, 16 pp.

[14] Čermák, J., Kisela, T., Nechvátal, L., *Stability regions for linear fractional differential systems and their discretizations.* Applied Mathematics and Computation, 2013, **219**, 7012-7022.

[15] Čermák, J., Nechvátal, L., *On (q, h)-analogue of fractional calculus.* Journal of Nonlinear Mathematical Physics, 2010, 17, 1-18.

[16] Díaz, J. B., Osler, T. S., *Differences of fractional order.* Mathematics of Computation, 1974, 28, 185-202.

[17] Diethelm, K., Ford, N. J., *Numerical solution of the Bagley-Torvik equation.* BIT, 2002, 42, 490-507.

[18] Diethelm, K., *The Analysis of Fractional Differential Equations.* Lecture Notes in Mathematics, Springer, Berlin, 2010.

[19] Elaydi, S., *An Introduction to Difference Equations.* 3rd ed. New York: Springer Science+Business Media, Inc., 2005.

[20] Elaydi, S., Murakami, S., *Asymptotic stability versus exponential stability in linear Volterra difference equations of convolution type.* Journal of Difference Equations and Applications, 1996, 2, 401-410.

[21] Galeone, L., Garrappa, R., *On multistep methods for differential equations of fractional order.* Mediterranean Journal of Mathematics, 2006, 3, 565-580.

[22] Galeone, L., Garrappa, R., *Explicit methods for fractional differential equations and their stability properties.* Journal of Computational and Applied Mathematics, 2009, 228, 548-560.

[23] Granger, C. W. J., Joyeux, R., *An introduction to long-memory time series models and fractional differencing.* Journal of Time Series Analysis, 1980, 1, 15-29.

[24] Gray, H. L., Zhang, N. F., *On a new definition of the fractional difference.* Mathematics of Computation, 1988, 50, 513-529.

[25] Holm, M. T., *The Theory of Discrete Fractional Calculus: Development and Application.* PhD thesis, University of Nebraska-Lincoln, 2011, 116 pp., supervisors: L. Erbe, A. Peterson.

[26] Hosking, J. R. M., *Fractional differencing.* Biometrika, 1981, 68, 165-176.

[27] Kilbas, A. A., Srivastava, H. M., Trujillo, J. J., *Theory and Applications of Fractional Differential Equations.* 1st ed. Netherlands: Elsevier B. V., 2006.

[28] Kisela, T., *On a discretization of the initial value problem for a linear nonhomogeneous fractional differential equation.* In Proceedings of the 4th IFAC Workshop on Fractional Differentiation and Its Applications. Badajoz, Spain, 2010, 264-269.

[29] Kisela, T., *Basics of qualitative theory of linear fractional difference equations.* PhD thesis, Brno University of Technology, 2012, 69 pp., supervisor: J. Čermák.

[30] Kisela, T., *Power functions and essentials of fractional calculus on isolated time scales.* Advances in Difference Equations, 2013, 259, 18 pp.

[31] Kisela, T., *An analysis of the stability boundary for a linear fractional difference system.* Mathematica Bohemica, 2014, *to appear.*

[32] Lubich, Ch., *Discretized fractional calculus.* SIAM Journal on Mathematical Analysis, 1986, 17, 704-719.

[33] Matignon, D., *Stability results for fractional differential equations with applications to control processing.* In Computational Engineering in Systems Applications, Lille - France, 1996, 963-968.

[34] Miller, K. S., Ross, B., *An Introduction to the Fractional Calculus and Fractional Differential Equations.* 1st ed. New York: John Wiley & Sons, Inc., 1993.

[35] Mozyrska, D., Girejko, E., *Overview of fractional h-difference operators.* Operator Theory: Advances and Applications, 2013, 229, 253-268.

[36] Nagai, A., *On a certain fractional q-difference and its eigen function.* Journal of Nonlinear Mathematical Physics, 2003, 10, 133-142.

[37] Oldham, K. B., Spanier, J., *The Fractional Calculus.* 1st ed. USA: Academic Press, Inc., 1974.

[38] Oldham, K. B., Myland, J., Spanier, J., *An Atlas of Functions.* 2nd ed. USA: Springer Science+Business Media, 2009.

[39] Podlubný, I., *Fractional Differential Equations.* 1st ed. USA: Academic Press, 1999.

[40] Qian, D., Li, Ch., Agarwal, R. P., Wong, P. J. Y., *Stability analysis of fractional differential system with Riemann-Liouville derivative.* Mathematical and Computer Modelling, 2010, 52, 862-874.

[41] Williams, P. A., *Unifying fractional calculus with time scales.* PhD thesis, The University of Melbourne, 2012, 119 pp., supervisors: J.J. Koliha, C.D. Hodgson, J.H. Rubinstein.

[42] Williams, P. A., *Fractional calculus on time scales with Taylor's theorem.* Fractional calculus and Applied Analysis, 2012, 15, 616-638.

In: Fractional Calculus: Theory
Editors: R. Abi Zeid Daou and X. Moreau

ISBN: 978-1-63463-002-3

Chapter 7

USING THE HANKEL OPERATOR TO INITIALIZE FRACTIONAL-ORDER SYSTEMS

Jay L. Adams*[*], *Robert J. Veillette and Tom T. Hartley
The University of Akron, Akron, Ohio, US

Abstract

This chapter presents a comprehensive discussion of using the Hankel operator to characterize the initial-condition response of a fractional-order system. Using the Hankel operator when either the input to the system for $t < 0$ or the output of the system for $t < 0$ is known is discussed. Sufficient conditions under which output-based initialization can be done via the Hankel operator are developed for a class of fractional-order systems. The error bounds when using the Hankel-operator of a reduced-order approximation to the fractional-order system to approximate the initial-condition response are discussed. The Hankel-operator technique is applied to examples, illustrating the results.

Keywords: Fractional-order systems, Hankel operator, initial-condition response

1. Introduction

The fractional-order calculus is a generalization of integrals and derivatives of integer order to derivatives of arbitrary order. Several in-depth studies of the fractional-order calculus have been written ([9, 8, 14, 5]). Systems whose models include fractional-order elements are infinite dimensional ([1]). Consequently, initial-condition responses of such systems cannot be uniquely parameterized using a finite number of scalars. Some methods of determining and using initial-condition responses are given by [6], [7], [10], and [16].

This chapter discusses a Hankel-operator approach to the initialization of fractional-order systems. This approach uses the Hankel operator $\Gamma_{\mathcal{G}}$ to account for the past behavior of the system on the future behavior of the system.

[*]E-mail address: jla36@uakron.edu

Because it is useful to know for what times a signal is given, we will use the following notation. For a given function $x(t)$, we define $x_-(t)$ by

$$x_-(t) = \begin{cases} x(t), & t < 0 \\ 0, & t \geq 0 \end{cases} \tag{1}$$

and $x_+(t)$ by

$$x_+(t) = \begin{cases} 0, & t < 0 \\ x(t), & t \geq 0 \end{cases}. \tag{2}$$

Since we assume that all signals have finite energy and are thus in $L_2(\mathbb{R})$, we may say that $x_-(t)$ is the projection of $x(t)$ onto $L_2(\mathbb{R}_-)$ and $x_+(t)$ is the projection of $x(t)$ onto $L_2(\mathbb{R}_+)$.

Consider the fractional-order system with transfer function $G(s)$ given by

$$\frac{X(s)}{U(s)} = G(s) = \frac{\sum_{k=1}^{M} b_k s^{\nu_k}}{\sum_{k=1}^{N} a_k s^{q_k}}. \tag{3}$$

Assuming that the system is causal, the convolution operator associated with $G(s)$ is $\mathcal{G} : L_2(\mathbb{R}) \rightarrow L_2(\mathbb{R})$, which is given by

$$(\mathcal{G}u)(t) = \int_{-\infty}^{t} \tilde{g}(t-\tau)u(\tau)d\tau, \tag{4}$$

where $\tilde{g}(\xi)$ is the impulse response of the system. Considering $x(t)$ as the output of the system, we write $x(t) = (\mathcal{G}u)(t)$. For $t > 0$ this operator may be written as

$$\begin{aligned} x_+(t) &= (\mathcal{G}u)(t) = \int_{-\infty}^{t} \tilde{g}(t-\tau)u(\tau)d\tau \\ &= \int_{-\infty}^{0} \tilde{g}(t-\tau)u(\tau)d\tau + \int_{0}^{t} \tilde{g}(t-\tau)u(\tau)d\tau \\ &= (\Gamma_{\mathcal{G}}u_-)(t) + (T_{\mathcal{G}}^{+}u_+)(t), \end{aligned} \tag{5}$$

where $\Gamma_{\mathcal{G}} : L_2(\mathbb{R}_-) \rightarrow L_2(\mathbb{R}_+)$ is the Hankel operator defined by

$$(\Gamma_{\mathcal{G}}u_-)(t) = \int_{-\infty}^{0} \tilde{g}(t-\tau)u_-(\tau)d\tau, \tag{6}$$

and where $T_{\mathcal{G}}^{+} : L_2(\mathbb{R}_+) \rightarrow L_2(\mathbb{R}_+)$ is the Toeplitz operator defined by

$$(T_{\mathcal{G}}^{+}u_+)(t) = \int_{0}^{t} \tilde{g}(t-\tau)u(\tau)d\tau. \tag{7}$$

Here the use of $x_+(t)$ highlights the fact that the output is considered only for time $t > 0$. The Hankel operator $\Gamma_{\mathcal{G}}$ maps inputs $u(t)$ for $t < 0$ to outputs $x(t)$ for $t > 0$. The Toeplitz operator $T_{\mathcal{G}}^{+}$ maps inputs $u(t)$ for $t > 0$ to outputs $x(t)$ for $t > 0$. In other words $\Gamma_{\mathcal{G}}$ gives the natural response of the system $G(s)$ to a past input $u_-(t)$ while $T_{\mathcal{G}}^{+}$ gives the response of the system $G(s)$ to a present and future forcing function $u_+(t)$ without initial conditions ([12, 13]).

For $t < 0$ $x(t) = (\mathcal{G}u)(t)$ may be written as

$$\begin{aligned} x_-(t) &= (\mathcal{G}u)(t) = \int_{-\infty}^{t} \tilde{g}(t-\tau)u(\tau)d\tau \\ &= \left(T_{\mathcal{G}}^{-} u_-\right)(t), \end{aligned} \tag{8}$$

where the Toeplitz operator $T_{\mathcal{G}}^{-} : L_2(\mathbb{R}_-) \to L_2(\mathbb{R}_-)$ is defined by

$$\left(T_{\mathcal{G}}^{-} u_-\right)(t) = \int_{-\infty}^{t} \tilde{g}(t-\tau)u_-(\tau)d\tau. \tag{9}$$

The use of $x_-(t)$ highlights the fact that the output is considered only for time $t < 0$ and the Toeplitz operator $T_{\mathcal{G}}^{-}$ maps the past input $u_-(t)$ into the past output $x_-(t)$.

Because we are using the bilateral Laplace transform, when posed with a transfer function $G(s)$ we must also consider the region of convergence $\mathcal{R}$ of $G(s)$. We assume that $\mathcal{R}$ includes the $i\omega$-axis so that the convolution integral of $\mathcal{G}$ converges. Every such $G(s)$ can be decomposed as the sum of a term that is analytic in the right half-plane $\mathbb{C}^+$ and a term that is analytic in the left half-plane $\mathbb{C}^-$ ([12]). The term that is analytic in the right half-plane corresponds to the causal portion of $G(s)$. Similarly, the term that is analytic in left half-plane corresponds to the anticausal portion of $G(s)$. Because we are interested in the Hankel operator, the effects of initialization, and application to realizable systems, we restrict ourselves to considering only causal systems, that is only transfer functions $G(s)$ that are analytic in the right half-plane.

The Hankel and Toeplitz operators may be considered in three different isometrically equivalent forms. The first is the time-domain form where $\Gamma_{\mathcal{G}} : L_2(\mathbb{R}_-) \to L_2(\mathbb{R}_+)$ is defined by Equation 6 and $T_{\mathcal{G}}^{-} : L_2(\mathbb{R}_-) \to L_2(\mathbb{R}_-)$ is defined by Equation 7. These time-domain operators are integral operators with kernels closely related to $\tilde{g}(t)$.

The second is the Laplace-domain form where the Hankel and Toeplitz operators are denoted as $\Gamma_G : H_2(\mathbb{C}^-) \to H_2(\mathbb{C}^+)$ and $T_G^{-} : H_2(\mathbb{C}^-) \to H_2(\mathbb{C}^-)$, which are defined as the projections of multiplication operators in $L_2(i\mathbb{R})$ with the multiplicand $G(s) = \mathcal{L}\{\tilde{g}(t)\}$. The Laplace-domain operators in $L_2(i\mathbb{R})$ are isometrically equivalent to the time-domain operators in $L_2(\mathbb{R})$, that is, $\Gamma_{\mathcal{G}}$ and Γ_G are isometrically equivalent, and $T_{\mathcal{G}}^{-}$ and T_G^{-} are isometrically equivalent ([12]).

The third form arises by a transformation of the Laplace-domain multiplicand $G(s)$ defined by

$$g(z) = \Pi_+\left(\frac{1}{z}G\left(\frac{1-z}{1+z}\right)\right), \tag{10}$$

where Π_+ is the projection operator onto $H_\infty(\mathbb{D})$. Often $\Pi_+(f(z))$ is calculated by writing $f(z)$ as a sum with a term in $H_\infty(\mathbb{D})$ and a term in $H_\infty(\mathbb{D})^\perp$, and discarding the term in $H_\infty(\mathbb{D})^\perp$. If $g(z)$ is given, $G(s)$ may be obtained by the inverse transformation

$$G(s) = \tilde{\Pi}_+\left(\frac{1-s}{1+s}g\left(\frac{1-s}{1+s}\right)\right), \tag{11}$$

where $\tilde{\Pi}_+$ is the projection operator onto $H_\infty(\mathbb{C}^+)$. Often, $\tilde{\Pi}_+(F(s))$ is calculated by writing $F(s)$ as a sum with a term in $H_\infty(\mathbb{C}^+)$ and a term in $H_\infty(\mathbb{C}^-)$ and discarding the $H_\infty(\mathbb{C}^-)$ term.

The resulting Hankel and Toeplitz operators are denoted $\Gamma_g : H_2(\mathbb{D})^\perp \to H_2(\mathbb{D})$ and $T_g^- : H_2(\mathbb{D})^\perp \to H_2(\mathbb{D})^\perp$, which are defined as the appropriate projections of the multiplication operators in $L_2(\mathbb{T})$ with multiplicand $g(z)$. The operators $\Gamma_{\mathcal{G}}$, Γ_G, and Γ_g are isometrically equivalent and $T_{\mathcal{G}}^-$, T_G^-, and T_g^- are isometrically equivalent ([12]). The forms Γ_g and T_g^- allow the use of the wealth of results that exist concerning Hankel and Toeplitz operators on the unit circle.

From Equation 6, it can be seen that the Hankel operator produces the initial-condition response corresponding to any particular input applied to the system for $t < 0$. There are three major reasons why the Hankel operator does not see more widespread use in the area of fractional-order systems. The first reason is that to use the integral operator approach as specified by Equation 6 requires that the impulse response of the system be known. A problem arises in that the impulse response of a fractional-order system leads to a convolution kernel that is not separable, such as the R-function or Mittag-Leffler function. By contrast the impulse response of a finite integer-order system consists of exponential functions and has convolution kernels which are separable. The second reason is that the Hankel-operator approach requires the calculation of a difficult improper integral. By contrast, the exponential functions that are characteristic of a finite integer-order system are easy to integrate by analytical or numerical means. The third reason is that the Hankel-operator approach requires the knowledge of the input for all of time $t < 0$. By contrast, for finite integer-order systems, the effects of the input for all $t < 0$ can be uniquely determined by a finite number of initial conditions. The next three sections will address each of these difficulties.

2. Impulse Response Approximation

For a broad class of fractional order systems, the Hankel operators have been shown to be compact [1]. A compact operator is one that may be approximated with any degree of accuracy by a finite-rank operator. Consequently, such a system may be approximated with an arbitrary degree of accuracy using an integer-order system.

We approximate the symbol $G(s)$ for the Hankel operator $\Gamma_G : H_2(\mathbb{C}^+) \to H_2(\mathbb{C}^+)$ with a rational function $G_N(s)$ of degree N, using any approximation scheme we prefer. Then $\tilde{g}_N(t) = \mathcal{L}^{-1}\{G_N(s)\}$ is used to form the kernel of the approximate Hankel operator $\Gamma_{\mathcal{G}_N}$. Note that $G_N(s)$ admits a partial-fraction expansion, so the kernel of $\Gamma_{\mathcal{G}_N}$ may be written as

$$\tilde{g}_N(t-\tau) = \sum_{k=1}^{N} r_k e^{\lambda_k t} e^{-\lambda_k \tau}, \tag{12}$$

where r_k is the residue of $G_N(s)$ at the pole λ_k, assuming all the poles are real and non-repeated. Then, $\Gamma_{\mathcal{G}_N}$ becomes

$$\begin{aligned}(\Gamma_{\mathcal{G}_N} u_-)(t) &= \int_{-\infty}^{0} \sum_{k=1}^{N} r_k e^{\lambda_k t} e^{-\lambda_k \tau} u_-(\tau) d\tau \\ &= \sum_{k=1}^{N} a_k e^{\lambda_k t},\end{aligned} \tag{13}$$

where

$$a_k = r_k \int_{-\infty}^{0} e^{-\lambda_k \tau} u_-(\tau) d\tau. \tag{14}$$

The coefficients $\{a_k\}_{k=1}^N$ are independent of t, and constitute weightings on the associated basis functions $e^{\lambda_k t}$. The response of the Hankel operator $\Gamma_{\mathcal{G}_N}$ to the input $u_-(t)$ is fully determined by the N coefficients $\{a_k\}_{k=1}^N$, and is used to approximate the response of $\Gamma_{\mathcal{G}}$. A similar development can be made for approximations with complex or repeated poles. If $G_N(s)$ is chosen so that $\|\Gamma_{\mathcal{G}} - \Gamma_{\mathcal{G}_N}\|$ is less than a known bound η, then the error in the approximate initial condition response satisfies

$$\|\Gamma_{\mathcal{G}} u_- - \Gamma_{\mathcal{G}_N} u_-\| < \eta \|u_-\|. \tag{15}$$

The remainder of this section presents an approximation to the system with transfer function

$$G(s) = \frac{1}{s^{\frac{1}{2}} + 1} \tag{16}$$

and the response of the associated approximate Hankel operator to three inputs $u_-(t)$. This example system was selected because its impulse response can be calculated analytically as

$$\tilde{g}(t) = \mathcal{L}^{-1}\{G_1(s)\} = \frac{1}{\sqrt{\pi(t)}} - e^t \operatorname{Erfc}\left(\sqrt{t}\right), \tag{17}$$

where Erfc $(\cdot)$ is the complementary error function. The Hankel operator $\Gamma_{\mathcal{G}}$ is then

$$(\Gamma_{\mathcal{G}} u_-)(t) = \int_{-\infty}^{0} \left(\frac{1}{\sqrt{\pi(t-\tau)}} - e^{t-\tau} \operatorname{Erfc}\left(\sqrt{t-\tau}\right) \right) u_-(\tau) d\tau \tag{18}$$

An approximation of the transfer function $G(s)$, obtained using the method of [11], is

$$G_N(s) = \frac{15.93s^4 + 868.7s^3 + 3856s^2 + 1700s + 56}{s^5 + 206.5s^4 + 3495s^3 + 7519s^2 + 2195s + 59.78}. \tag{19}$$

The inverse Laplace transform of $G_N(s)$ is the approximate impulse response

$$\begin{aligned}\tilde{g}_N(t) &= 12.64e^{-188.14t} + 2.599e^{-15.90t} \\ &\quad + 0.6005e^{-2.122t} + 0.0849e^{-0.3102t} \\ &\quad + 0.004542e^{-0.03034t},\end{aligned} \tag{20}$$

and the approximate Hankel operator $\Gamma_{\mathcal{G}_N}$ is given by

$$\left(\Gamma_{\mathcal{G}_N} u_-\right)(t) = \int_{-\infty}^{0} \tilde{g}_N(t-\tau) u_-(\tau) d\tau. \tag{21}$$

The method of [2] was used to calculate the bounds

$$0.006615 < \|\Gamma_{\mathcal{G}} - \Gamma_{\mathcal{G}_N}\| < 0.008344 \tag{22}$$

on the Hankel-norm error between the actual system and the approximation.

The following three subsections present the responses of the approximate Hankel operator $\Gamma_{\mathcal{G}_N}$ to three different past inputs $u_-(t)$. For the first two inputs, both $\Gamma_{\mathcal{G}_N}$ and the response of the actual Hankel operator $\Gamma_{\mathcal{G}}$ can be calculated analytically, and the errors between $\Gamma_{\mathcal{G}_N} u_-$ and $\Gamma_{\mathcal{G}} u_-$ are shown. For the third input, $\Gamma_{\mathcal{G}_N} u_-$ is still easy to calculate analytically, but analytic calculation of $\Gamma_{\mathcal{G}} u_-$ is too difficult. Consequently, $\Gamma_{\mathcal{G}_N} u_-$ and a bound on the error between the approximate response and actual response are calculated.

2.1. Past Input 1

The first input for this example that we will consider is $u_{1-}(t)$ given by

$$u_{1-}(t) = \mathcal{U}_s(-t) - \mathcal{U}_s(-(t-2)). \tag{23}$$

This input has a value of unity for $t \in [-2, 0)$ and a value of zero other wise. By computing the integrals analytically, $\left(\Gamma_{\mathcal{G}} u_{1-}\right)(t)$ is given by

$$\left(\Gamma_{\mathcal{G}} u_{1-}\right)(t) = e^t \left(\operatorname{Erfc}\left(\sqrt{t}\right) - e^2 \operatorname{Erfc}\left(\sqrt{t+2}\right)\right), \tag{24}$$

and $\left(\Gamma_{\mathcal{G}_N} u_{1-}\right)(t)$ is given by

$$\begin{aligned} \left(\Gamma_{\mathcal{G}_N} u_{1-}\right)(t) &= 0.0612e^{-188.14t} + 0.1635e^{-15.90t} + 0.2788e^{-2.122t} \\ &\quad + 0.1265e^{-0.0849t} + 0.008814e^{-0.03034t}. \end{aligned} \tag{25}$$

Figure 1 shows $\left(\Gamma_{\mathcal{G}} u_{1-}\right)(t)$ and $\left(\Gamma_{\mathcal{G}_N} u_{1-}\right)(t)$. Figure 2 shows the error between $\left(\Gamma_{\mathcal{G}} u_{1-}\right)(t)$ and $\left(\Gamma_{\mathcal{G}_N} u_{1-}\right)(t)$. Because $\tilde{g}_N(t-\tau)$ is a separable kernel, $\left(\Gamma_{\mathcal{G}_N} u_-\right)(t)$ for any $u_-(t)$ is a weighted sum of five exponentials [15].

2.2. Past Input 2

Next we consider the response for the input

$$u_{2-}(t) = \begin{cases} e^{2t}, & t \in (-3, 0) \\ 0, & t \notin (-3, 0) \end{cases}. \tag{26}$$

Again the integrals were calculated analytically to yield $\left(\Gamma_{\mathcal{G}} u_{2-}\right)(t)$ given by

$$\begin{aligned} \left(\Gamma_{\mathcal{G}} u_{2-}\right)(t) &= \sqrt{2} e^{2t} \left(\operatorname{Erfc}\left(\sqrt{2(t+3)}\right) - \operatorname{Erfc}\left(\sqrt{2t}\right)\right) + e^{t-3} \operatorname{Erfc}\left(\sqrt{t+3}\right) \\ &\quad - e^t \operatorname{Erfc}\left(\sqrt{t}\right), \end{aligned} \tag{27}$$

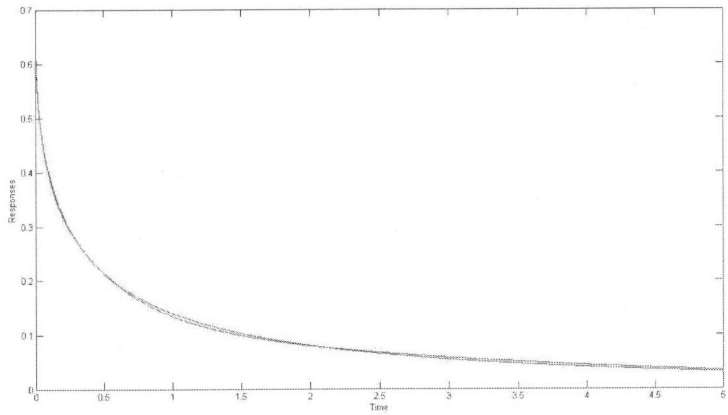

Figure 1. $\left(\Gamma_{\mathcal{G}}u_{1-}\right)(t)$ and $\left(\Gamma_{\mathcal{G}_N}u_{1-}\right)(t)$ for u_{1-} a pulse.

and $\left(\Gamma_{\mathcal{G}_N}u_{2-}\right)(t)$ given by

$$\begin{aligned}\left(\Gamma_{\mathcal{G}_N}u_{2-}\right)(t) &= 0.06648e^{-188.14t}+0.1452e^{-15.90t}+0.1457e^{-2.122t} \\ &\quad +0.03671e^{-0.3102t}+0.002232e^{-0.03034t}.\end{aligned} \tag{28}$$

Figure 3 shows $\left(\Gamma_{\mathcal{G}}u_{2-}\right)(t)$ and $\left(\Gamma_{\mathcal{G}_N}u_{2-}\right)(t)$. Figure 4 shows the error between $\left(\Gamma_{\mathcal{G}}u_{2-}\right)(t)$ and $\left(\Gamma_{\mathcal{G}_N}u_{2-}\right)(t)$.

2.3. Past Input 3

We next consider the response for the input

$$u_{3-}(t)=\begin{cases}\cos(t)e^{t}, & t\in\left(-\frac{\pi}{2},0\right)\\ 0, & t\notin\left(-\frac{\pi}{2},0\right)\end{cases}. \tag{29}$$

While it is possible to calculate $\Gamma_{\mathcal{G}}u_{3-}$ analytically, the convolution integral calculation is difficult and the response is written in terms of error functions. Instead, we note that $\|u_{3-}\|_2 = 0.3696$. Then because the Hankel norm error of the approximation is less than $\eta = 0.008344?$, the possible error for $\Gamma_{\mathcal{G}_N}$ is 0.3696η or 0.003084.

The response $\left(\Gamma_{\mathcal{G}_N}u_{2-}\right)(t)$ was calculated to be

$$\begin{aligned}\left(\Gamma_{\mathcal{G}_N}u_{2-}\right)(t) &= 0.066827e^{-188.14t}+0.15325e^{-15.90t}+0.17486e^{-2.122t} \\ &\quad +0.04494e^{-0.3102t}+0.0027066e^{-0.03034t}.\end{aligned} \tag{30}$$

3. Output-Based Initialization

When only the output of a system for time $t<0$ is known, it may still be possible to use the Hankel operator to account for the effects of initialization. If the output $x_{-}(t)$ is known for time $t<0$, in order to use the Hankel operator an input $u_{-}(t)$ must be found which, when applied to $T_{\mathcal{G}}^{-}$, produces $x_{-}(t)$ as the output.

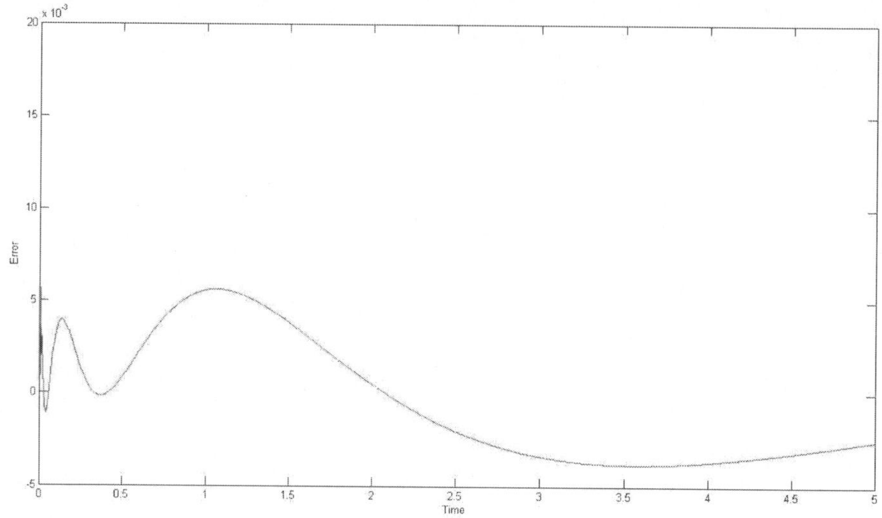

Figure 2. Error between $(\Gamma_{\mathcal{G}} u_{1-})(t)$ and $(\Gamma_{\mathcal{G}_N} u_{1-})(t)$ for u_{1-} a pulse.

The method of finding using the Hankel operator to calculate $x_+(t)$ when only $x_-(t)$ is known can only be used if $x_-(t)$ can be generated only by a single $u_-(t)$. In other words the Hankel operator can only be used to account for inputs in the past if $T_{\mathcal{G}}^-$ is injective. An operator $\mathcal{A}$ is said to be injective if $\mathcal{A}u_1 = \mathcal{A}u_2$ implies that $u_1 = u_2$. Additionally, for a linear operator, $\mathcal{A}$ is injective if and only if $\ker(\mathcal{A}) = \{0\}$ ([4]).

It is not always the case that the past input to a system may be recovered from its past output. For example, consider the system $x(t) = (Ku)(t) = u(t-1)$. For a given $x_-(t)$, $u_-(t)$ is unique for $t \in (-\infty, -1)$. However, $x_-(t)$ does not give any information about the behavior of $u_-(t)$ for $t \in (-1, 0)$. For any given output $x_-(t)$ in $L_2(\mathbb{R}_-)$, there is an infinite number of inputs $u_-(t)$ in $L_2(\mathbb{R}_-)$ that can generate $x_-(t)$. Recovering the input $u_-(t)$ that generated $x_-(t)$ is not possible for this system because $K : L_2(\mathbb{R}_-) \to L_2(\mathbb{R}_-)$ is not injective; the kernel of K is not $\{0\}$, in fact $\ker(K)$ contains any function that is zero for $t \in (-\infty, -1)$.

If an operator $\mathcal{A} : \mathcal{X} \to \mathcal{Y}$ is injective, it has a left-inverse. That is, there exists an operator $\mathcal{B} : \mathcal{Y} \to \mathcal{X}$ such that $\mathcal{B} \circ \mathcal{A} = \mathcal{I}$, where $\mathcal{I}$ is the identity operator. If $T_{\mathcal{G}}^-$ is injective, the left inverse $\mathcal{B}$ can be applied to $x_-(t)$ to obtain the $u_-(t)$ such that $x_-(t) = (T_{\mathcal{G}}^- u_-)(t)$. That is $u_-(t) = (\mathcal{B}x_-)(t)$.

While we only require $T_{\mathcal{G}}^-$ to be injective, if $T_{\mathcal{G}}^-$ is bijective, or invertible, it is clear that $u_-(t) = \left(\left(T_{\mathcal{G}}^-\right)^{-1} x_-\right)(t)$.

The invertibility of Toeplitz operators on the unit circle has been studied extensively. We highlight several useful theorems.

Theorem 3.1. *Let $\phi, \psi \in L_\infty(\mathbb{T})$. Then*

$$T_{\phi\psi} = T_\phi T_\psi \tag{31}$$

if and only if $\phi \in H_\infty(\mathbb{D})$ or $\overline{\psi} \in H_\infty(\mathbb{D})$

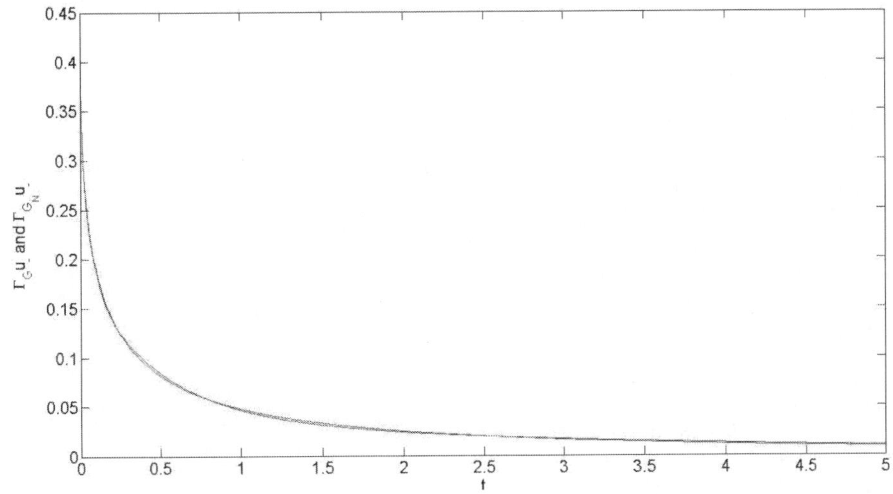

Figure 3. $(\Gamma_{\mathcal{G}} u_{2-})(t)$ and $(\Gamma_{\mathcal{G}_N} u_{2-})(t)$ for u_{2-} an exponential.

Proof. See [13].

Theorem 3.1. states that Toeplitz operator whose symbol is a product of two symbols is the composition of the Toeplitz operators associated with each symbol if and only if one of the multiplicand symbols is in $H_\infty(\mathbb{D})$.

Theorem 3.2. *Let $\phi \in H_\infty(\mathbb{D})$. The Toeplitz operator T_ϕ is invertible if and only if ϕ is invertible in $H_\infty(\mathbb{D})$.*

Proof. See [13].

Theorem 3.2. relates the invertibility of a Toeplitz operator to the invertibility of its symbol.

We now begin to bring together the necessary arguments to show that the Toeplitz operators $T_{\mathcal{G}}^-$ of a certain class of fractional-order systems are invertible. We first require an inequality.

Lemma 3.3. *Let $q \in (0,2)$ and $r \in (0,1]$. Then*

$$p(r,\theta) = \left(\frac{1-re^{i\theta}}{1+re^{i\theta}}\right)^q \neq 1 \tag{32}$$

for any $\theta \in [-\pi, \pi]$.

Proof. Note that the condition $p(r,\theta) = 1$ is equivalent to $\frac{1-re^{i\theta}}{1+re^{i\theta}} = 1^{\frac{1}{q}}$, and that $\left|1^{\frac{1}{q}}\right| = 1$ for $q \in (0,2)$. The magnitude squared of $\frac{1-re^{i\theta}}{1+re^{i\theta}}$ is

$$\left|\frac{1-re^{i\theta}}{1+re^{i\theta}}\right|^2 = \frac{1-r\cos\theta}{1+r\cos\theta}, \tag{33}$$

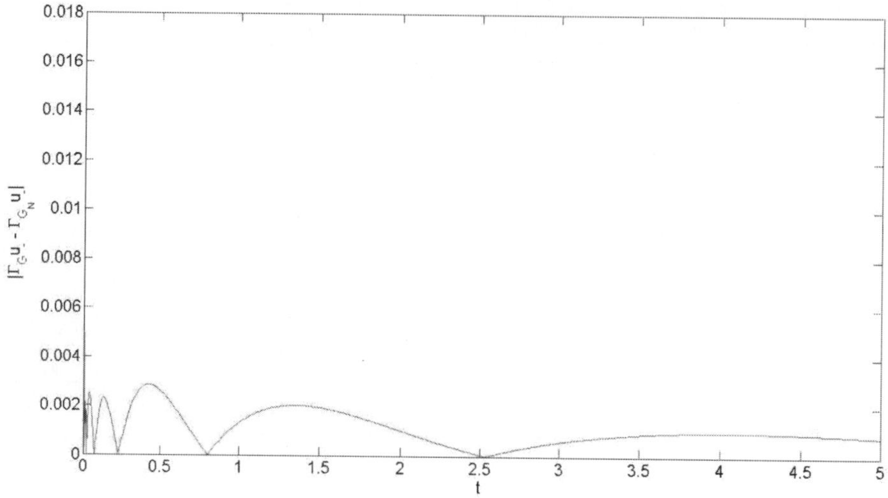

Figure 4. Error between $(\Gamma_{\mathcal{G}} u_{2-})(t)$ and $(\Gamma_{\mathcal{G}_N} u_{2-})(t)$ for u_{2-} an exponential.

which is unity only for $\theta = \pm\frac{\pi}{2}$. Thus, the only $\theta \in [-\pi, \pi]$ for which $p(r, \theta)$ could equal unity are $\theta = \pm\frac{\pi}{2}$. Note that

$$p\left(r, \mp\frac{\pi}{2}\right) = \frac{1 - r^2 \pm i2r}{1 + r^2} = \frac{\sqrt{(1 - r^2)^2 + 4r^2}}{1 + r^2} e^{i\tan^{-1}\left(\frac{\pm 2r}{1-r^2}\right)}. \tag{34}$$

Suppose that $1 = \left(\frac{1-r^2 \mp i2r}{1+r^2}\right)^q$ for some $q \in (0, 2)$. Then,

$$\begin{aligned} 0 + i0 &= q\log\left(\frac{\sqrt{(1 - r^2)^2 + 4r^2}}{1 + r^2} e^{i\tan^{-1}\left(\frac{\pm 2r}{1-r^2}\right)}\right) \\ &= q\log\left(\frac{\sqrt{(1 - r^2)^2 + 4r^2}}{1 + r^2}\right) + iq\tan^{-1}\left(\frac{\pm 2r}{1 - r^2}\right). \end{aligned} \tag{35}$$

Since $q \neq 0$, it must be that $\tan^{-1}\left(\frac{\pm 2r}{1-r^2}\right)$ is zero. But this quantity is only zero when $r = 0$. Therefore $p(r, \theta) \neq 1$ for any $q \in [-\pi, \pi]$ and $r \in (0, 1]$. ∎

We now demonstrate a sufficient condition under which symbols for Toeplitz operators of the form $G(s) = \frac{1}{s^q - a}$ are invertible in $H_\infty(\mathbb{D})$.

Lemma 3.4. *Let $G(s)$ be given by*

$$G(s) = \frac{1}{s^q - a} \tag{36}$$

with $q \in (0, 2)$ and $|\arg a| > q\frac{\pi}{2}$. Then $g(z)$ determined by

$$g(z) = \Pi_+\left(\frac{1}{z} G\left(\frac{1 - z}{1 + z}\right)\right), \tag{37}$$

may be written as

$$g(z) = \frac{1}{z}G\left(\frac{1-z}{1+z}\right) - \frac{1}{1-a}z^{-1}, \tag{38}$$

and $g \in H_\infty(\mathbb{D})$ *and* g *is invertible in* $H_\infty(\mathbb{D})$.

Proof. The conditions on a and q mean $G(s) \in H_\infty(\mathbb{C}^+)$ ([3]). Define $\hat{g}(z)$ by

$$\hat{g}(z) = \frac{1}{z}G\left(\frac{1-z}{1+z}\right). \tag{39}$$

Since $G(s)$ is analytic in $\mathbb{C}^+$, $G\left(\frac{1-z}{1+z}\right)$ is analytic for $\mathbb{D}$ and has a Taylor series expansion $\sum_{k=0}^{\infty} g_k z^k$. Thus, $\hat{g}(z)$ may be written as

$$\begin{aligned} \hat{g}(z) &= \frac{1}{z}\sum_{k=0}^{\infty} g_k z^k = \sum_{k=0}^{\infty} g_k z^{k-1} \\ &= \sum_{k=-1}^{\infty} g_{k+1} z^k = g_0 z^{-1} + \sum_{k=0}^{\infty} g_{k+1} z^k. \end{aligned} \tag{40}$$

Then the only portion of $\hat{g}(z)$ not analytic in $\mathbb{D}$ is $g_0 z^{-1}$. Note that g_0 is given as

$$g_0 = G\left(\frac{1-z}{1+z}\right)\bigg|_{z=0} = G(s)|_{s=1} = \frac{1}{1-a}. \tag{41}$$

Thus we may write $g(z)$ as

$$\begin{aligned} g(z) &= \Pi_+(\hat{g}(z)) = \frac{1}{z}G\left(\frac{1-z}{1+z}\right) - g_0 z^{-1} \\ &= \frac{1}{z}G\left(\frac{1-z}{1+z}\right) - \frac{1}{1-a}z^{-1}. \end{aligned} \tag{42}$$

It is clear that if $G(s) \in H_\infty(\mathbb{C}^+)$, then $g(z) \in H_\infty(\mathbb{D})$ because the mapping $G(s) \mapsto g(z)$ is an isomorphism from $H_\infty(\mathbb{C}^+)$ onto $H_\infty(\mathbb{D})$. It remains to show that $g(z)$ is invertible in $H_\infty(\mathbb{D})$.

We first show that $g(z)$ is nonzero in $\overline{\mathbb{D}}$. We note that as long as $g(z)$ is nonzero on $\mathbb{T}$, $g(z)$ is invertible in $L_\infty(\mathbb{T})$ with inverse $\frac{1}{g(z)}$. Suppose that $g(z)$ is zero for some $z = re^{i\theta} \in \overline{\mathbb{D}}$. That is for this z

$$g(z) = \frac{1}{z\left(\left(\frac{1-z}{1+z}\right)^q - a\right)} - \frac{\frac{1}{1-a}}{z} = 0, \tag{43}$$

or, equivently

$$\frac{1}{\left(\frac{1-z}{1+z}\right)^q - a} - \frac{1}{1-a} = 0. \tag{44}$$

This may be written as

$$\frac{1}{\left(\frac{1-z}{1+z}\right)^q - a} = \frac{1}{1-a}, \tag{45}$$

or

$$\left(\frac{1-z}{1+z}\right)^q - a = 1 - a. \tag{46}$$

This may be written as

$$\left(\frac{1-re^{i\theta}}{1+re^{i\theta}}\right)^q = 1 \tag{47}$$

for some $r \in [0,1]$ and $\theta \in (-\pi, \pi)$. Assume $r \in (0,1]$. Since $q \in (0,2)$, by Lemma 3.3. there is no such θ for $r \in (0,1]$. Thus, $g(z)$ is nonzero on $\overline{\mathbb{D}} - \{0\}$. We now show that $g(0) \neq 0$. Note that

$$\begin{aligned} \lim_{|z|\to 0} g(z) &= \lim_{|z|\to 0} \frac{\left(\frac{1-z}{1+z}\right)^q - 1}{\left(\left(\frac{1-z}{1+z}\right)^q - a\right) z(1-a)} \\ &= \lim_{|z|\to 0} \frac{2q\left(\frac{1-z}{1+z}\right)^q}{\left(\left(\frac{1-z}{1+z}\right)^q (z^2 + 2qz - 1) + a - az^2\right)(1-a)} \\ &= -\frac{2q}{(1-a)^2} \neq 0 \end{aligned} \tag{48}$$

Thus $g(z)$ is nonzero in $\overline{\mathbb{D}}$. In particular, $g(z)$ is nonzero in $\mathbb{T}$, so $g(z)$ is invertible in $L_\infty(\mathbb{T})$ with inverse $\frac{1}{g(z)}$.

Further $g^{-1}(z) = \frac{1}{g(z)}$ is the reciprocal of an analytic function that is nonzero in $\mathbb{D}$, therefore $g^{-1}(z)$ is analytic in $\mathbb{D}$. ∎

We now show sufficient conditions for Toeplitz operators $T^-_{\mathcal{G}}$ whose symbols are the product of single fractional-order poles to be invertible.

Theorem 3.5. *Let $G(s)$ be written as*

$$G(s) = \prod_{k=1}^{N} \frac{1}{s^{q_k} - a_k}, \tag{49}$$

where $q_k \in (0,2)$ and $|\arg a_k| > q_k \frac{\pi}{2}$ for $1 \leq k \leq q$. The Toeplitz operator $T^-_{\mathcal{G}}$ is invertible.

Proof. Define $G_k(s)$ by

$$G_k(s) = \frac{1}{s^{q_k} - a_k} \tag{50}$$

for $1 \leq k \leq N$. Then each $g_k(z)$ is in $H_\infty(\mathbb{D})$, and thus by repeated application of Theorem 3.1.,

$$T^-_{\mathcal{G}} = T^-_{\mathcal{G}_1} T^-_{\mathcal{G}_2} \cdots T^-_{\mathcal{G}_N}. \tag{51}$$

By Lemma 3.4., each $g_k(z)$ is invertible in $H_\infty(\mathbb{D})$, and thus by Theorem 3.2. $T^-_{\mathcal{G}_k}$ is invertible for $1 \leq k \leq N$. It follows that $\left(T^-_{\mathcal{G}}\right)^{-1}$ exists and is given by

$$\left(T^-_{\mathcal{G}}\right)^{-1} = \left(T^-_{\mathcal{G}_N}\right)^{-1} \cdots \left(T^-_{\mathcal{G}_2}\right)^{-1} \left(T^-_{\mathcal{G}_1}\right)^{-1}. \tag{52}$$

∎

4. Output-Based Initialization Examples

We consider the system whose symbol is

$$\begin{aligned} G(s) &= \frac{2s+1}{s^{\frac{3}{2}} + s + 2s^{\frac{1}{2}} + 2} = \frac{2s+1}{(s+2)\left(\sqrt{s}+1\right)} \\ &= \frac{1}{\sqrt{s}+1} + \frac{\sqrt{s}}{s+2} - \frac{1}{s+2}. \end{aligned} \tag{53}$$

To demonstrate that the Toeplitz operator associated with $G(s)$ is invertible, we consider the Toeplitz operators associated with

$$G_2(s) = \frac{1}{\sqrt{s}+1} \tag{54}$$

and

$$G_1(S) = \frac{2s+1}{s+2}. \tag{55}$$

$G_1(s)$ meets the criteria for Theorem 3.5., so $g_1(z) \in H_\infty(\mathbb{D})$, and $T^-_{\mathcal{G}_1}$ is invertible. Using the transformation in Equation 10, $g_2(z)$ is

$$g(z) = -\frac{2}{z+3}, \tag{56}$$

which is clearly invertible in $H_\infty(\mathbb{D})$. Therefore, $T^-_{\mathcal{G}} = T^-_{\mathcal{G}_1} T^-_{\mathcal{G}_2}$ and $T^-_{\mathcal{G}}$ is invertible.

The impulse response for the system with transfer function $G(s)$ is

$$\tilde{g}(t) = \left(\frac{1}{\sqrt{\pi t}} - e^t \mathrm{Erfc}\left(\sqrt{t}\right)\right) + \left(\frac{1}{\sqrt{\pi t}} - \sqrt{2}e^{-2t}\mathrm{Erfi}\left(\sqrt{2t}\right)\right) - \left(e^{-2t}\right), \tag{57}$$

where Erfc $(\cdot)$ is the complementary error function and Erfi $(\cdot)$ is the imaginary error function.

A fifth-order approximation for $G(s)$ is given by

$$G_N(s) = \frac{\hat{N}(s)}{\hat{D}(s)}, \tag{58}$$

where

$$\hat{N}(s) = 33.9s^4 + 3290s^3 + 36040s^2 + 23090s + 2002 \tag{59}$$

and

$$\hat{D}(s) = s^5 + 286.2s^4 + 9866s^3 + 59240s^2 + 64770s + 4679. \tag{60}$$

The Hankel-norm error bounds were calculated using the technique of [2] to be

$$0.008537 \leq \|\Gamma_{G-G_N}\| \leq 0.009944. \tag{61}$$

The impulse response of $G_N(s)$ is

$$\begin{aligned} \tilde{g}_N(t) &= 25.03e^{-247.3t} + 6.561e^{-31.66t} + 2.926e^{-5.888t} \\ &\quad -0.6211e^{-1.3064t} + 0.007609e^{-0.07769t}. \end{aligned} \tag{62}$$

Because $G_N(s)$ is an integer-order approximation to $G(s)$, only a finite number of conditions are required to fully describe $(\Gamma_{\mathcal{G}_N} u_-)(t)$ for any $u_-(t)$. The traditional technique of using a finite number of initial conditions to determine $(\Gamma_{\mathcal{G}_N} u_-)(t)$ because $G_N(s)$ is an integer-order system. To that end, it is convenient to have a state-space realization for $G_N(s)$ such as

$$\begin{aligned} \dot{z} &= \begin{pmatrix} -31.4489 & -2.1604 & 50.2453 & 12.1242 & 25.5252 \\ 2.1604 & -0.0128 & 2.6038 & 0.5238 & 0.8068 \\ 50.2453 & -2.6038 & -140.8286 & -37.1702 & -98.6801 \\ 12.1242 & -0.5238 & -37.1702 & -10.0157 & -28.3958 \\ 25.5252 & -0.8068 & -98.6801 & -28.3958 & -103.9222 \end{pmatrix} z \\ &\quad + \begin{pmatrix} 3.9578 \\ -0.0598 \\ -3.8190 \\ -0.8818 \\ -1.6953 \end{pmatrix} u \\ x &= \begin{pmatrix} 3.9578 & 0.0598 & -3.8190 & -0.8818 & -1.6953 \end{pmatrix} z. \end{aligned} \tag{63}$$

This balanced state-space representation is used because it is more numerically robust than the companion state-space representation.

Because the Hankel norm error between $G(s)$ and $G_N(s)$ is at most 0.009944, the error between any $x_+(t)$ calculated with this state-space representation and the $x_+(t)$ obtained using the full fractional-order system is no greater than $0.009944 \|u_-\|$.

4.1. Output 1

Suppose the system given by Equation 53 is known to have the output

$$x_-(t) = \begin{cases} t + 2 - \frac{2}{\sqrt{\pi}}\sqrt{t+1} - e^{t+1}\text{Erfc}\left(\sqrt{t+1}\right), & t \in [-1, 0] \\ 0, & t \notin [-1, 0] \end{cases}. \tag{64}$$

To obtain the appropriate $u_-(t)$, $w_-(t) = {}_{-\infty}d_t^{\frac{1}{2}}\, x_-(t) + x_-(t)$ is calculated as

$$w_-(t) = \begin{cases} t + 1, & t \in [-1, 0] \\ 0, & t \notin [-1, 0] \end{cases}, \tag{65}$$

and $v_-(t) = \frac{d}{dt}w_-(t) + 2w_-(t)$ is calculated as

$$w_-(t) = \begin{cases} 2t+3, & t \in [-1,0] \\ 0, & t \notin [-1,0] \end{cases}, \tag{66}$$

and $u_-(t)$ is given by

$$\begin{aligned} u_-(t) &= \frac{1}{2}e^{-\frac{t}{2}} \star v_-(t) \\ &= \begin{cases} 2t-1+3e^{-\frac{t}{2}}, & t \in [-1,0] \\ 0, & t \notin [-1,0] \end{cases}, \end{aligned} \tag{67}$$

where $\star$ denotes convolution.

The input $u_-(t)$ from Equation 67 is applied to the Hankel operator $\Gamma_{\mathcal{G}}$ of the fractional-order system and to the Hankel operator $\Gamma_{\mathcal{G}_N}$ of the approximate system. The input $u_-(t)$ has norm $\|u_-\|_2 = 1.8927$, so the error in the responses between $\Gamma_{\mathcal{G}}$ and $\Gamma_{\mathcal{G}_N}$ can be at most

$$\|\Gamma_{\mathcal{G}}u_- - \Gamma_{\mathcal{G}_N}u_-\| \leq 0.009944 \times 1.8927 = 0.0188216. \tag{68}$$

The response $\Gamma_{\mathcal{G}}u_-$ was calculated analytically to be

$$\begin{aligned} (\Gamma_{\mathcal{G}}u_-)(t) &= \int_{-\infty}^{0} \tilde{g}(t-\tau)u_-(\tau)d\tau \\ &= \frac{1}{20}e^{-2t}\left(-6 + (3-i)e^{-3-i} + (3+i)e^{-3+i}\right. \\ &\quad +10i\left(-1+e^{2i}\right)e^{-i+3t}\text{Erfc}\left[\sqrt{1+t}\right] \\ &\quad +(3-11i)\sqrt{-1-i}e^{(3+i)t}\text{Erfi}\left[\sqrt{-1-i}\sqrt{t}\right] \\ &\quad +(3+11i)\sqrt{-1+i}e^{(3-i)t}\text{Erfi}\left[\sqrt{-1+i}\sqrt{t}\right] - 6\sqrt{2}\text{Erfi}\left[\sqrt{2}\sqrt{t}\right] \\ &\quad -(3-11i)\sqrt{-1-i}e^{(3+i)t}\text{Erfi}\left[\sqrt{-1-i}\sqrt{1+t}\right] \\ &\quad -(3+11i)\sqrt{-1+i}e^{(3-i)t}\text{Erfi}\left[\sqrt{-1+i}\sqrt{1+t}\right] \\ &\quad +(3-i)\sqrt{2}e^{-3-i}\text{Erfi}\left[\sqrt{2}\sqrt{1+t}\right] \\ &\quad \left.+(3+i)\sqrt{2}e^{-3+i}\text{Erfi}\left[\sqrt{2}\sqrt{1+t}\right]\right). \end{aligned} \tag{69}$$

The response $\Gamma_{\mathcal{G}_N}u_-$ was calculated analytically to be

$$\begin{aligned} (\Gamma_{\mathcal{G}_N}u_-)(t) &= \int_{-\infty}^{0} \tilde{g}_N(t-\tau)u_-(\tau)d\tau \\ &= 0.1008e^{-247.3t} + 0.2007e^{-31.66t} + 0.4159e^{-5.888t} \\ &\quad -0.2227e^{-1.3064t} + 0.004104e^{-0.07769t}. \end{aligned} \tag{70}$$

The response $\Gamma_{\mathcal{G}_N}u_-$ was also calculated numerically through simulation with the initial conditions

$$z(0) = \begin{pmatrix} 0.2464 & 0.6262 & 0.0188 & 0.0097 & 0.0021 \end{pmatrix}^T \tag{71}$$

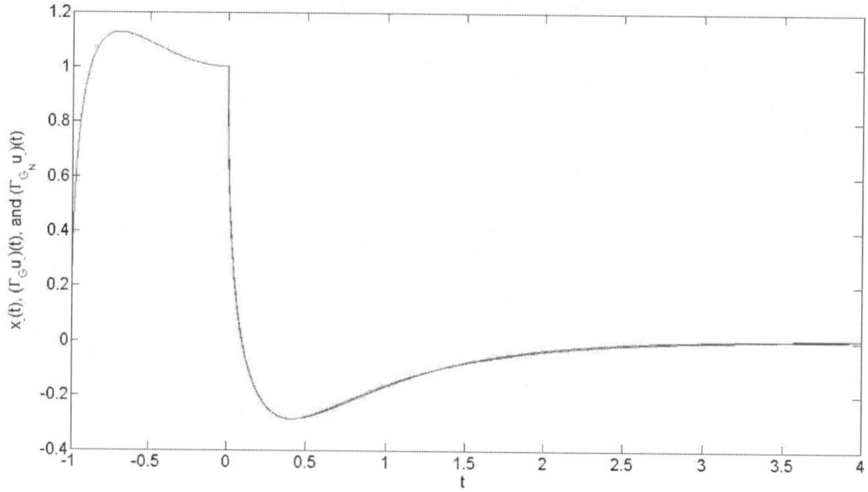

Figure 5. Initial Condition Response for Output-Based Initialization Example.

The past output $x -_(t)$ and the future outputs $x_+(t)$ calculated analytically using $\Gamma_{\mathcal{G}}$ and $\Gamma_{\mathcal{G}_N}$ and the future output $x_+(t)$ calculated numerically using $\Gamma_{\mathcal{G}_N}$ are shown in Figure 5. As expected the responses of $\Gamma_{\mathcal{G}_N} u_-$ calculated analytically and through simulation are exactly the same. The error between $\Gamma_{\mathcal{G}_N} u_-$ and $\Gamma_{\mathcal{G}} u_-$ is shown in Figure 6. The error is sufficiently small for $t \in (0,4)$ that the error bound of 0.01882 is reasonable.

4.2. Output 2

Suppose the system given by Equation 53 is known to have the output

$$\begin{aligned} x_-(t) &= \frac{3\sqrt{3+t}}{\sqrt{\pi}} + \frac{4t\sqrt{3+t}}{3\sqrt{\pi}} + \frac{1}{4}\left(11 - 10t - 2t^2\right) - 4e^{3+t}\mathrm{Erfc}\left[\sqrt{3+t}\right] \\ &\quad -\frac{7}{4}e^{-6-2t}\left(1 + \sqrt{2}\mathrm{Erfi}\left[\sqrt{2}\sqrt{3+t}\right]\right) \end{aligned} \tag{72}$$

for $t \in (-3,0)$. The $u_-(t)$ required to generate $x_-(t)$ is

$$u_-(t) = 9 - t^2, \tag{73}$$

for $t \in (-3,0)$. The $\Gamma_{\mathcal{G}} u_-$ that results from this input is

$$\begin{aligned} \left(\Gamma_{\mathcal{G}} u_-\right)(t) &= \frac{3\sqrt{3+t}}{\sqrt{\pi}} + \frac{4t\sqrt{3+t}}{3\sqrt{\pi}} - \frac{\sqrt{t}(15+4t)}{3\sqrt{\pi}} \\ &\quad + e^t\left(7\mathrm{Erfc}\left[\sqrt{t}\right] - 4e^3\mathrm{Erfc}\left[\sqrt{3+t}\right]\right) \\ &\quad -\frac{1}{4}e^{-6-2t}\left(7 + 17e^6 + 17\sqrt{2}e^6\mathrm{Erfi}\left[\sqrt{2}\sqrt{t}\right] + 7\sqrt{2}\mathrm{Erfi}\left[\sqrt{2}\sqrt{3+t}\right]\right) \end{aligned} \tag{74}$$

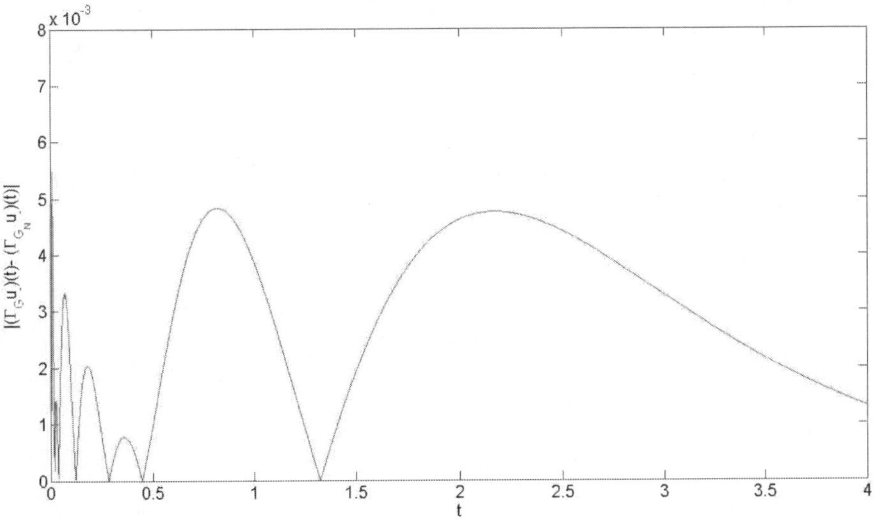

Figure 6. Error for Initial Condition Response for Output-Based Initialization Example.

The response $\Gamma_{\mathcal{G}_N} u_-$ was calculated analytically to be

$$\begin{aligned}(\Gamma_{\mathcal{G}_N} u_-)(t) &= 0.9109e^{-247.3t} + 1.8646e^{-31.66t} + 4.443e^{-5.888t} \\ &\quad -3.776e^{-1.3064t} + 0.1257e^{-0.07769t}.\end{aligned} \tag{75}$$

The response $\Gamma_{\mathcal{G}_N} u_-$ was also calculated numerically with initial conditions

$$z(0) = \begin{pmatrix} 0.7037 & 3.8055 & -0.1954 & 0.5053 & 0.1374 \end{pmatrix}^T. \tag{76}$$

Since the norm of u_- is

$$\|u_-\| = 11.38 \tag{77}$$

the error bound is

$$\|\Gamma_{\mathcal{G}} u_- - \Gamma_{\mathcal{G}_N} u_-\| \leq 0.009944 \times 11.38 = 0.1132. \tag{78}$$

The past output $x_-(t)$ and the future outputs $x_+(t)$ calculated analytically using $\Gamma_{\mathcal{G}}$ and $\Gamma_{\mathcal{G}_N}$ and the future output $x_+(t)$ calculated numerically using $\Gamma_{\mathcal{G}_N}$ are shown in Figure 5. The error between $\Gamma_{\mathcal{G}_N} u_-$ and $\Gamma_{\mathcal{G}} u_-$ is shown in Figure 6. The error is sufficiently small for $t \in (0, 4)$ that the error bound of 0.1132 is reasonable.

4.3. Past Output 3

Suppose that the system given by Equation 53 is known to have the output

$$x_-(t) = \begin{cases} \cos(t), & t \in \left(-\frac{3\pi}{2}, 0\right) \\ 0, & t \notin \left(-\frac{3\pi}{2}, 0\right) \end{cases}. \tag{79}$$

The calculations to analytically determine the $u_-(t)$ that results in this output result in a complicated function that then needs to be multiplied by an equally complicated function

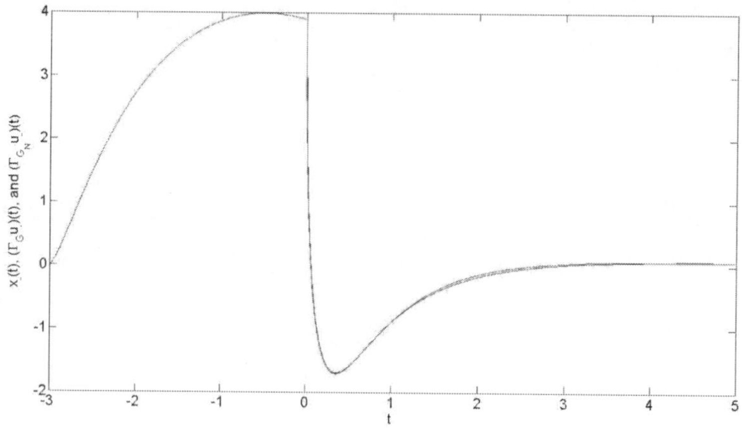

Figure 7. Initial Condition Response for Output-Based Initialization Example.

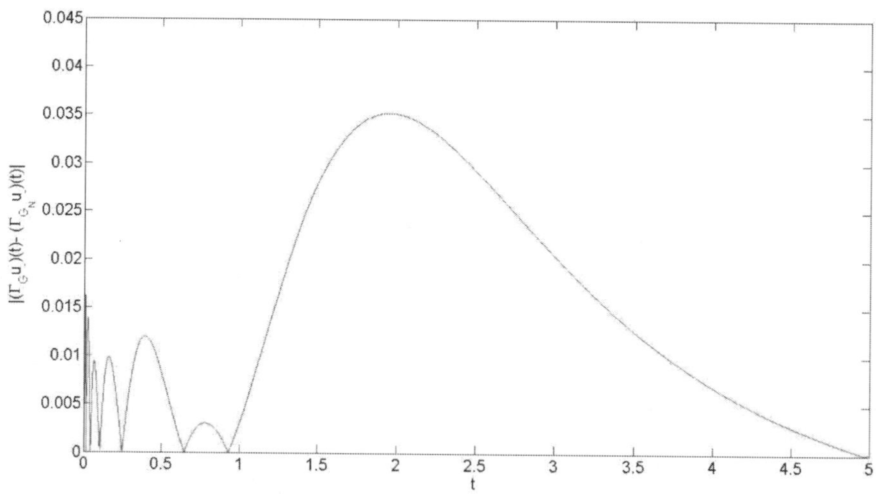

Figure 8. Error for Initial Condition Response for Output-Based Initialization Example.

and integrated. Instead we will calculate the future output numerically using the initial conditions $x(0^-) = 1$, $\dot{x}(0^-) = 0$, $x^{(2)}(0^-) = -1$, $x^{(3)}(0^-) = 0$, and $x^{(4)}(0^-) = 1$, which give the initial conditions

$$z(0^-) = \begin{pmatrix} 0.3363 & -2.6085 & -1.9797 & 10.4523 & -0.8737 \end{pmatrix}^T. \tag{80}$$

Figure 9 shows the results of the numerical calculation of $x_+(t)$. Because we do not know $u_-(t)$ or $\|u_-\|_2$, we can not put a bound on the error between this numerical calculation and the result from the full fractional-order system. However, because the approximation error between $G(s)$ and $G_N(s)$ is small, we can be confident that the error between the simulated response and the actual response is also small.

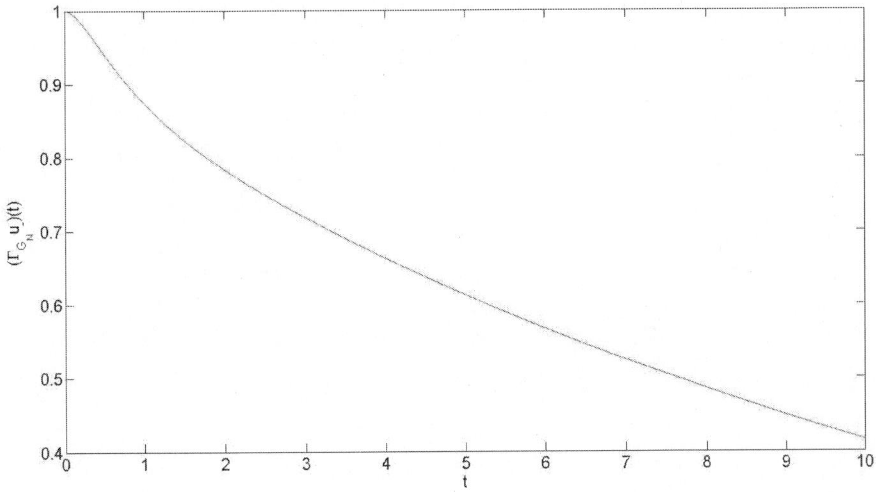

Figure 9. Initial Condition Response for Output-Based Initialization Example.

5. Conclusion

In this chapter the application of the Hankel operator to a fractional-order system was discussed. When the past input $u_-(t)$ to the system is known, the future output $x_+(t)$ may be calculated using Hankel operator Γ_G for the fractional-order system. An approximate future response can also be calculated using the Hankel operator Γ_{G_N} associated with an approximation to the fractional-order system.

When only the past output $x_-(t)$ of the system $G(s)$ is known, it is possible to reconstruct the past input $u_-(t)$ when the Toeplitz operator T_G^- is injective. System with transfer functions of the form

$$G(s) = \prod_{k=1}^{N} \frac{1}{s^{q_k} - a_k}, \tag{81}$$

where $q_k \in (0,2)$ and $|\arg a_k| > q_k \frac{\pi}{2}$ for $1 \leq k \leq q$ were shown to have invertible Toeplitz operators. Output-based initialization using the Hankel operator for a fractional-order system was demonstrated for a fractional-order system.

References

[1] Jay L. Adams and Tom T. Hartley. Hankel operators for fractional-order systems. *Journal Européen des Systèmes Automatisés*, 3(2), 2008.

[2] Jay L. Adams, Robert J. Veillette, and Tom T. Hartley. Hankel-norm estimation for fractional-order systems using the Rayleigh-Ritz method. *Comput. Math. Appl.*, 59:1773–1781, March 2010.

[3] Jay L. Adams, Robert J. Veillette, and Tom T. Hartley. Conjugate-order systems for signal processing: stability, causality, boundedness, compactness. *Signal, Image, and Video Processing*, 6(3):373–380, September 2012.

[4] John B. Conway. *A Course in Functional Analysis*. Springer-Verlag, New York, 2nd edition, 1990.

[5] Anatoly A. Kilbas, Hari M. Srivastava, and Juan J. Trujillo. *Theory and Applications of Fractional Differential Equations*. Elsevier, New York, 2006.

[6] Carl F. Lorenzo and Tom T. Hartley. Initialized fractional calculus. *International Journal of Applied Mathematics*, 3:249–265, 2000.

[7] Carl F. Lorenzo and Tom T. Hartley. Initialization of fractional-order operators and fractional differential equations. *Journal of Computational and Nonlinear Dynamics*, 3(2), 2008.

[8] Keith S. Miller and Bertram Ross. *An Introduction to the Fractional Calculus and Fractional Differential Equations*. Wiley, New York, 1993.

[9] Keith B. Oldham and Jerome Spanier. *The Fractional Calculus, Integrations and Differentiations of Arbitrary Order*. Academic Press, New York, 1974.

[10] Manuel D. Ortigueira. On the initial conditions in continuous time fractional linear systems. *Signal Processing*, 83:2301–2309, 2003.

[11] Alain Oustaloup, F. Levron, B. Mathieu, and F. M. Nanot. Frequency-band complex noninteger differentiator: Characterization and synthesis. *IEEE Transactions on Circuits and Systems I*, 47:25–39, 2000.

[12] Jonathan R. Partington. *An Introduction to Hankel Operators*. Cambridge University Press, Cambridge, 1988.

[13] Vladimir V. Peller. *Hankel Operators and Their Applications*. Springer, New York, 2003.

[14] Igor Podlubny. *Fractional Differential Equations, An Introduction to Fractional Derivatives, Fractional Differential Equations, Some Methods of Their Solution and Some of Their Applications*. Academic Press, New York, 1999.

[15] David Porter and David S. G. Stirling. *Integral Equations, a Practical Treatment from Spectral Theory to Applications*. Cambridge University Press, Cambridge, 1990.

[16] Jean-Claude Trigeassou, Nexha Maamri, and Alain Oustaloup. The pseudo state space model of linear fractional differential systems. In *Proceedings of the 4th IFAC Workshop on Fractional Differentiation and its Applications*. IFAC, 2010.

In: Fractional Calculus: Theory
Editors: R. Abi Zeid Daou and X. Moreau
ISBN: 978-1-63463-002-3

Chapter 8

Fractional Reaction-transport Equations Arising from Evanescent Continuous Time Random Walks

***E. Abad*[1,*]*, S. B. Yuste*[2]*, and K. Lindenberg*[3]**
[1]Departamento de Física Aplicada and
Instituto de Computación Científica Avanzada (ICCAEX)
Centro Universitario de Mérida
Universidad de Extremadura, Mérida, Spain
[2]Departamento de Física and
Instituto de Computación Científica Avanzada (ICCAEX)
Universidad de Extremadura, Badajoz, Spain
[3]Department of Chemistry and Biochemistry and BioCircuits Institute
University of California San Diego, La Jolla, California, US

Abstract

Continuous time random walks (CTRWs) describe a particular class of renewal processes used to model a wide variety of phenomena such as the motion of charge carriers in disordered systems, the dynamics of financial markets, the motion of diffusing particles in crowded environments, and certain anomalous relaxation phenomena in dielectric systems. It is well known that, on long time scales, a CTRW described by a separable probability density function (pdf) for the jump length of a particle and its waiting time between consecutive jumps yields a variety of fractional diffusion equations for suitable (and yet rather general) choices of both pdfs. Such fractional diffusion equations give rise to a broad range of behaviors for the mean square displacement of the particle, ranging from subdiffusive to superdiffusive. In this work we show that fractional equations with interesting solutions may also be obtained when the above CTRW is combined with a rather general class of death processes (evanescence), including a simple first-order process typical of radioactive decay. A general feature of the resulting fractional equations is that the parameters describing the decay process explicitly appear in the transport term, as opposed to heuristic fractional equations lacking a rigorous mesoscopic justification. In the subdiffusive case, we consider

*E-mail address: eabad@unex.es

two applications of interest. First, we compute the survival probability of an immobile target surrounded by one or more subdiffusive traps that may spontaneously disappear in the course of their motion. The second example concerns a biologically relevant problem, namely, the formation of stationary morphogen concentration gradients by means of morphogen synthesis, anomalous diffusion, and degradation.

PACS: 05.40.Jc, 82.20.Wt, 87.18.Pj

Keywords: Reaction-subdiffusion and reaction-superdiffusion equations, target problem, morphogen gradient formation

1. Introduction

The importance of fractional calculus for the theoretical description of anomalous transport has been highlighted by recent research on the connection between renewal processes underlying Continuous Time Random Walks (CTRWs) and fractional diffusion equations. CTRWs are paradigmatic models of stochastic transport that incorporate memory effects reflecting complex features of real-world systems, e.g., transport in complex media [1] and stochastic dynamics of financial markets [2–4]. It is by now a well-known fact that a CTRW with a suitable choice of jump statistics yields fractional diffusion equations that can be solved by standard techniques. The boundary and initial conditions depend on the specific application of interest.

In the simplest case one is interested in the propagator solution of the fractional diffusion equation, which characterizes the diffusive spread of a particle starting at the origin of an infinite volume. In the particular case where the spatial and temporal dependences of the single-step probability density of the CTRW are uncoupled and the corresponding waiting time distribution is long-tailed, the behavior of the mean square displacement (MSD) of the propagator solution depends on the choice of the jump length distribution. If the latter is assumed to follow an inverse power law, sublinear or superlinear behavior of the MSD as a function of time may occur, resulting in the emergence of anomalous subdiffusive or superdiffusive transport in the long time regime.

Researchers working in the field of stochastic transport are quite familiar with the passage from the integral equations describing the arrival and sojourn probability densities of a continuous time random walker to a fractional diffusion equation. However, it is not infrequent to encounter serious misconceptions in the literature when the transported species is allowed to undergo a chemical reaction or a transformation in the course of its motion, as is the case, for instance, for a diffusing radionuclide in a complex medium. In this case, it is tempting to write an evolution equation with separate transport and reaction terms, whereby the transport term is exactly the same as in the absence of the chemical reaction. However, this heuristic approach is plainly wrong, since a rigorous derivation of the equations for particle concentrations from the CTRW probability density equations shows that both terms are mathematically intertwined [5–12], and the reaction rate parameters often appear explicitly in the transport term (even for very simple reaction processes of birth and death type). Failing to take the above into account may result in fractional reaction-transport equations that yield unphysical results, such as negative particle concentrations.

Instead of a heuristic approach, a procedure is required whereby the fractional equations obtained when a reaction is at play are substantiated by the proper CTRW integral equations for the probability densities. The resulting equations may then be shown to fulfill elementary consistency properties that prevent an ill-behaved evolution of the system. In several instances we shall see that the fractional reaction-transport equations may be cast in the form of pure transport equations by a suitable transformation of the concentration variable, much in the spirit of Danckwerts' solution for a Fickian reaction-diffusion equation with a linear loss term [13, 14]. Physically, this means that if one constantly feeds back the amount of reactant lost in the reaction process by multiplying the concentration by a proper time dependent factor, then a pure transport process is obtained (alternatively, one can view this as a rescaling of the concentration in terms of the survival probability of the particle). As it turns out, in the case of anomalous diffusion such a transformation does not work if the transport term is not made to depend on the reaction rate constant.

The aim of this work is to recapitulate previous results obtained in different contexts where the derivation of fractional reaction-transport equations is required. In particular, these results refer to relatively simple reactions, namely, first-order death processes (also called evanescence process). Despite their apparent simplicity, the combination of these reactions with anomalous transport gives rise to fractional equations that are not particularly intuitive. Our focus lies in the derivation of the equations in terms of the Riemann-Liouville fractional derivative, but we also discuss two applications, namely, the survival probability of a particle surrounded by evanescent traps, and the properties of morphogen profiles giving rise to embryogenic cellular differentiation.

The plan of this presentation is as follows. In Sec. 2 we derive an integral equation for the probability density of a particle subject to a rather general class of death processes. In Sec. 3 we consider special cases of the above equation, leading to a fractional reaction-subdiffusion equation and a reaction-superdiffusion equation, as well as to a bifractional reaction diffusion equation. Two applications of the resulting reaction-subdiffusion equation are addressed in Sec. 4. Finally, Sec. 5 is devoted to a summary of our main findings.

2. Integral Equation for a CTRW with Evanescence

Our starting point is a particle performing a CTRW in continuum space (we adopt this type of description for concreteness, but on-lattice approaches do not pose difficulties of principle). The particle jumps take place in negligible time after a randomly distributed waiting time during which the particle remains immobile, as opposed to creeper models where the particle moves continuously and switches direction after a random time after the last change in direction. The probability density that a CTRWer located at $\mathbf{r}_0$ when $t = 0$ is at position $\mathbf{r}$ at time t is denoted by $w(\mathbf{r}, t|\mathbf{r}_0, 0)$. In order to obtain the desired integral equation for this quantity it is necessary to introduce the probability density $\Psi(\mathbf{r} - \mathbf{r}', t - t')$ for the walker to jump in a single step from $\mathbf{r}'$ to $\mathbf{r}$ after having waited a time interval $t - t'$ at position $\mathbf{r}'$. Note that we restrict Ψ to depend only on the difference between the initial position and the final position after the jump, as well as on the difference between the start and the end of the waiting time interval, i.e., we assume translational and temporal invariance of the random walk. The waiting time probability density $\psi(t)$ and the probability density $\chi(\mathbf{r})$

for a single step displacement are respectively obtained by integration, i.e.,

$$\psi(t) = \int d\mathbf{r}' \Psi(\mathbf{r}', t), \tag{1}$$

and

$$\chi(\mathbf{r}) = \int_0^\infty dt' \, \Psi(\mathbf{r}, t'). \tag{2}$$

The specific forms of the above probability densities determine whether the transport is "normal" or "anomalous". In the latter case one can distinguish between subdiffusive and superdiffusive transport. However, in our case we must bear in mind that the walker disappears at a certain rate in the course of its motion (in physical systems, this death process can for instance be the result of a radioactive or photoemissive decay). Therefore, the above jump statistics are conditional on the walker's survival and must be weighted by a suitable survival function. To this end, let us specify the manner in which the death process proceeds by introducing the probability $\Xi(\mathbf{r}, \mathbf{r}'; t, t')$ that the particle has not evanesced during the stepping process described by $\Psi(\mathbf{r} - \mathbf{r}', t - t')$.

As is customary in the derivation of fractional equations in the absence of evanescence, we next introduce $q_n(\mathbf{r}, t|\mathbf{r}_0, 0)$, the probability density that the particle arrives at position $\mathbf{r}$ at time t exactly after making its nth jump, given that it was initially at $\mathbf{r}_0$ [6]. For $n = 0$ one must recover the initial condition, that is,

$$q_0(\mathbf{r}, t|\mathbf{r}_0, 0) = \delta(\mathbf{r} - \mathbf{r}_0)\, \delta(t). \tag{3}$$

The probability density that the particle steps onto position $\mathbf{r}$ exactly at time t is obtained by summing over n:

$$q(\mathbf{r}, t|\mathbf{r}_0, 0) \equiv \sum_{n=0}^{\infty} q_n(\mathbf{r}, t|\mathbf{r}_0, 0). \tag{4}$$

One can then write the following equation for the q_n's:

$$q_{n+1}(\mathbf{r}, t|\mathbf{r}_0, 0) = \int d\mathbf{r}' \int_0^t dt' \, \Psi(\mathbf{r} - \mathbf{r}', t - t')\, q_n(\mathbf{r}', t'|\mathbf{r}_0, 0)\, \Xi(\mathbf{r}, \mathbf{r}'; t, t'). \tag{5}$$

This equation states that the probability density for the particle to arrive at location $\mathbf{r}$ at time t on the $(n + 1)$st step is given by the probability density for the particle to arrive at any point $\mathbf{r}'$ at any earlier time t' on the nth step and then to step from $\mathbf{r}'$ to $\mathbf{r}$ at time t, provided that the particle does not die in the interval between the nth and $(n + 1)$st steps. The usual integral equation without evanescence is immediately recovered upon setting Ξ equal to unity. We now take advantage of the fact that Ψ has been assumed to not depend on n; summing over n we obtain the following integral equation:

$$\begin{aligned} q(\mathbf{r}, t|\mathbf{r}_0, 0) &= \int d\mathbf{r}' \int_0^t dt' \, \Psi(\mathbf{r} - \mathbf{r}', t - t') \\ &\times \; q(\mathbf{r}', t'|\mathbf{r}_0, 0)\, \Xi(\mathbf{r}, \mathbf{r}'; t, t') + q_0(\mathbf{r}, t|\mathbf{r}_0, 0). \end{aligned} \tag{6}$$

The next step in our route to a fractional reaction-transport equation is to write an equation relating w and q. To this end, we introduce $\varXi(\mathbf{r}, \mathbf{r}'; t, t')$ as the probability that the particle does not die between the times t' and t as it moves from $\mathbf{r}'$ to $\mathbf{r}$, with no reference to steps. We can then write the exact relation

$$w(\mathbf{r}, t|\mathbf{r}_0, 0) = \int_0^t dt'\, \Phi(t-t') q(\mathbf{r}, t'|\mathbf{r}_0, 0)\, \mathit{\Xi}(\mathbf{r}, \mathbf{r}; t, t'), \tag{7}$$

where

$$\Phi(t) = 1 - \int_0^t dt'\, \psi(t') = \int_t^\infty dt'\, \psi(t') \tag{8}$$

stands for the probability that the particle does not take a step in the time interval between 0 and t. Equation (7) is intuitively clear. It simply states that the probability density for the particle to be at $\mathbf{r}$ at time t is the probability that it arrived at that location at time $t' < t$ and then neither moved nor died until time t.

Note that $\mathit{\Xi}(\mathbf{r}, \mathbf{r}'; t, t')$ is in general different from the function $\Xi(\mathbf{r}, \mathbf{r}'; t, t')$. We recall that the latter is the probability that the particle does not die in a waiting time interval exactly delineated by two steps (and none in between), one taking it to location $\mathbf{r}'$ and the other to location $\mathbf{r}$. If particles die at a constant rate independent of position and of stepping times, then $\mathit{\Xi}$ and Ξ are equal. However, if, for example, particles die only when they take a step, then these probabilities are different. For this reason, Eq. (6) and Eq. (7) cannot be easily combined with one another to obtain an integral equation for $w(\mathbf{r}, t|\mathbf{r}_0, 0)$ unless one assumes that Ξ and $\mathit{\Xi}$ are equal. We shall do so hereafter.

The other (admittedly debatable) assumption needed to obtain a fractional reaction-transport equation is that the survival function Ξ can be written as follows:

$$\Xi(\mathbf{r}, \mathbf{r}'; t, t') = \frac{\varphi(\mathbf{r}, t)}{\varphi(\mathbf{r}', t')} \tag{9}$$

with a suitably defined function $\varphi(\mathbf{r}, t)$. This assumption is not as limiting as might appear at first glance, since it can be shown to embrace many special cases that have been treated in the literature. The possibilities that φ depends explicitly on the walker density w or that it is prescribed by an external mechanism not related to the intrinsic properties of the walker are not excluded (see discussion in 3.1 and 3.2).

The simplest type of evanescence is an exponential, location-independent decay, $\varphi(\mathbf{r}, t) = \exp(-\lambda t)$, implying that $\Xi(\mathbf{r}, \mathbf{r}'; t, t') = \exp[-\lambda(t - t')]$. Thus, Ξ depends only on the time difference. This case was considered early on in the literature [5], possibly because of its great practical importance, notably for the description of decay processes uncoupled from transport as is often the case for a population of diffusing particles that decay into other species by light or particle emission [15].

We are now in the position to obtain an integral equation for w by taking Eqs. (6) and (7) as a starting point. To begin with, we implement the replacements $t \to t'$ and $t' \to t''$ in Eq. (6). Then we multiply by $\Phi(t - t')\, \Xi(\mathbf{r}, \mathbf{r}; t, t')$, integrate over t', and use Eq. (7) to write

$$w(\mathbf{r},t|\mathbf{r}_0,0) =$$
$$\int_0^t dt'\,\Phi(t-t')\,\Xi(\mathbf{r},\mathbf{r};t,t')\left(\int d\mathbf{r}'\int_0^{t'} dt''\Psi(\mathbf{r}-\mathbf{r}',t'-t'')q(\mathbf{r}',t''|\mathbf{r}_0,0)\,\Xi(\mathbf{r},\mathbf{r}';t',t'')\right)$$
$$+\int_0^t dt'\,\Phi(t-t')\,\Xi(\mathbf{r},\mathbf{r};t,t')\,q_0(\mathbf{r},t'|\mathbf{r}_0,0). \tag{10}$$

Next, we insert the special form given by Eq. (9) and perform a series of simple manipulations, whereby the double convolution structure of the resulting equations with respect to space and time is exploited (details can be found in ref. [10]). Finally, one is left with the desired integral equation:

$$w(\mathbf{r},t|\mathbf{r}_0,0) = \int d\mathbf{r}'\int_0^t dt'\,\Psi(\mathbf{r}-\mathbf{r}',t-t')w(\mathbf{r}',t'|\mathbf{r}_0,0)\frac{\varphi(\mathbf{r},t)}{\varphi(\mathbf{r}',t')}+\Phi(t)\frac{\varphi(\mathbf{r},t)}{\varphi(\mathbf{r},0)}\delta(\mathbf{r}-\mathbf{r}_0). \tag{11}$$

The above equation takes into account all the possible ways in which the particle can be at location $\mathbf{r}$ at time t by carrying out detailed bookkeeping of the positions $\mathbf{r}'$ at prior times t' from which the particle may later arrive at $\mathbf{r}$. The second term on the right hand side is needed to account for the possibility that the particle was initially at $\mathbf{r}$ and has not moved by time t.

Equation (11) is central to the present work; shortly we shall see that it allows one to deduce various fractional reaction-diffusion equations for different forms of $\Psi(\mathbf{r}-\mathbf{r}',t-t')$.

3. Fractional Diffusion Equations with Evanescence

In order to use Eq. (11) to derive various fractional reaction-transport equations, we start by making the simplifying assumption that the waiting time and the jump length distributions are uncoupled, that is,

$$\Psi(\mathbf{r},t) = \psi(t)\chi(\mathbf{r}). \tag{12}$$

The assumption of decoupling is needed at this stage to make Eq. (11) more tractable from a mathematical point of view. Decoupling is admittedly not fulfilled in many real-world systems, e.g., a grasshopper is likely to need a longer recovery time if she is to perform a longer jump. However, dealing with coupled CTRW schemes in the presence of reactions is a very challenging problem beyond the scope of the present work.

With the above assumption, one can use the techniques employed by various authors to obtain fractional diffusion equations for a conventional, non-evanescent CTRWer [16–18]. In particular, when $\psi(t)$ has long tails and $\chi(\mathbf{r})$ does not, one arrives at a fractional subdiffusion equation. When, on the other hand, $\chi(\mathbf{r})$ has long tails but $\psi(t)$ does not, a fractional superdiffusion equation is found. The most "anomalous" case occurs when both have long tails, which leads to a bifractional equation. In what follows we consider a transport process *with a superimposed evanescence process* and find equations that are not trivial superpositions of the reaction and the transport terms, as would be the case for a Markovian walker.

In view of the form of Eq. (11), it turns out to be convenient to work with a ratio introduced earlier in ref. [10], namely,

$$\eta(\mathbf{r}, t|\mathbf{r}_0, 0) = \frac{w(\mathbf{r}, t|\mathbf{r}_0, 0)}{\varphi(\mathbf{r}, t)} \tag{13}$$

rather than with $w(\mathbf{r}, t|\mathbf{r}_0, 0)$ itself. One now proceeds as in the non-evanescent case, namely, one takes the Fourier (for space)-Laplace (for time) transform of Eq. (11) and finds

$$u\hat{\tilde{\eta}}(\mathbf{q}, u) = u\hat{\chi}(\mathbf{q})\tilde{\psi}(u)\hat{\tilde{\eta}}(\mathbf{q}, u) + \left(1 - \tilde{\psi}(u)\right)\frac{e^{i\mathbf{q}\cdot\mathbf{r}_0}}{\varphi(\mathbf{r}_0, 0)}, \tag{14}$$

where $\mathbf{q}$ is the wave vector variable of the Fourier transform and u stands for the Laplace variable (the functions with a hat are Fourier transforms and those with a tilde are Laplace transforms).

3.1. Fractional Subdiffusive Equation with Evanescence

The first example we shall explore involves the waiting time distribution

$$\psi(t) \sim \frac{\kappa}{\tau_D}\left(\frac{t}{\tau_D}\right)^{-\gamma-1}, \tag{15}$$

with $0 < \gamma < 1$. Here κ is a dimensionless constant and τ_D is a characteristic mesoscopic time (but not a first moment). By construction, all the integer moments of the above long-tailed waiting time distribution diverge. When combined with a jump distance distribution with finite moments, $\psi(t)$ gives rise to subdiffusive behavior of the particle. In the limit of small u, the Laplace transform of $\psi(t)$ reads as follows:

$$\tilde{\psi}(u) \sim 1 - \mathcal{A}u^{\gamma}, \tag{16}$$

with $\mathcal{A} = \gamma^{-1}\kappa\Gamma(1-\gamma)\tau_D^{\gamma}$.

In contrast to the waiting time distribution, we shall assume that the jump distance distribution is "normal", i.e., that it has finite moments. In the limit of small $q = |\mathbf{q}|$ we can perform the following expansion:

$$\hat{\chi}(\mathbf{q}) \sim 1 - \frac{\sigma^2\mathbf{q}^2}{2} + \mathcal{O}(\mathbf{q}^4), \tag{17}$$

where

$$\sigma^2 = \int d\mathbf{r}\, \mathbf{r}^2\chi(\mathbf{r}) \tag{18}$$

is the second moment of the jump length distribution. Substitution of the expansions (16) and (17) into Eq. (14) and neglect of a term of $\mathcal{O}(u\mathbf{q}^2)$ yields

$$u\hat{\tilde{\eta}}(\mathbf{q}, u) - \frac{e^{i\mathbf{q}\cdot\mathbf{r}_0}}{\varphi(\mathbf{r}_0, 0)} = -u^{1-\gamma}\frac{\sigma^2\mathbf{q}^2}{2\mathcal{A}}\hat{\tilde{\eta}}(\mathbf{q}, u). \tag{19}$$

The corresponding double inversion with respect to the Fourier and Laplace variables gives

$$\frac{\partial\eta(\mathbf{r}, t|\mathbf{r}_0, 0)}{\partial t} = K_{\gamma}\, {}_0\mathcal{D}_t^{1-\gamma}\nabla_{\mathbf{r}}^2\,\eta(\mathbf{r}, t|\mathbf{r}_0, 0), \tag{20}$$

where the anomalous diffusion coefficient is given by

$$K_\gamma = \frac{\sigma^2}{2\mathcal{A}}. \tag{21}$$

The integro-differential operator ${}_0\mathcal{D}_t^{1-\gamma}$ introduced in Eq. (20) is defined via the relation

$$\mathcal{L}^{-1}_{u\to t}\left\{u^{1-\gamma}\tilde{y}(u)\right\} = {}_0\mathcal{D}_t^{1-\gamma} y(t), \tag{22}$$

where $\mathcal{L}^{-1}_{u\to t}$ denotes the inverse Laplace transform and $y(t)$ is a time-dependent function. The operator ${}_0\mathcal{D}_t^{1-\gamma}$, called the Grünwald-Letnikov fractional derivative, is closely related to the Riemann-Liouville operator ${}_0D_t^{1-\gamma}$, defined as follows:

$${}_0D_t^{1-\gamma} y(t) = \frac{1}{\Gamma(\gamma)}\frac{\partial}{\partial t}\int_0^t dt' \frac{y(t')}{(t-t')^{1-\gamma}}. \tag{23}$$

When applied to sufficiently regular functions $y(t)$ fulfilling the condition $\lim_{t\to 0}\int_0^t d\tau (t-\tau)^{\gamma-1} y(\tau) = 0$, the operators ${}_0\mathcal{D}_t^{1-\gamma}$ and ${}_0D_t^{1-\gamma}$ can be shown to be equivalent (see [19, p. 76]). In practice, this condition is satisfied for all the situations we are interested in here. Thus we shall use the Riemann-Liouville operator.

In terms of the probability density w, Eq. (20) explicitly yields the fractional reaction-subdiffusion equation

$$\frac{\partial w(\mathbf{r},t|\mathbf{r}_0,0)}{\partial t} = \varphi(\mathbf{r},t)\, K_\gamma\, {}_0D_t^{1-\gamma}\nabla^2_{\mathbf{r}}\frac{1}{\varphi(\mathbf{r},t)} w(\mathbf{r},t|\mathbf{r}_0,0) + \frac{\dot{\varphi}(\mathbf{r},t)}{\varphi(\mathbf{r},t)} w(\mathbf{r},t|\mathbf{r}_0,0). \tag{24}$$

Equation (24) describes the combination of subdiffusive transport and a class of evanescence processes that, despite the requirement (9), is much more general than would appear at first glance. It actually includes the possibility of a density-dependent reaction rate constant, or an external feedback mechanism resulting in enhancement of the evanescent reaction. While Eq. (24) has been formulated for a single walker, it is straightforward to show that if one initially has one walker or an ensemble of identical walkers, the concentration or number density $\rho(\mathbf{r},t)$ satisfies Eq. (24). For a one-particle system, there is a trivial connection between $\rho(\mathbf{r},t)$ and $w(\mathbf{r},t|\mathbf{r}_0,0)$, namely

$$\rho(\mathbf{r},t) = \int d\mathbf{r}_0\, p(\mathbf{r}_0) w(\mathbf{r},t;\mathbf{r}_0,0), \tag{25}$$

where $p(\mathbf{r}_0)$ is the distribution of all possible initial positions for the walker.

At this stage, a technical remark is in order. In deriving Eq. (24) we have followed the standard route used for the case without reaction, namely, an expansion of the relevant integral equation in Fourier-Laplace space. In order for this procedure to be valid in the presence of the evanescence process, the latter has been implicitly assumed to be slow enough so as to allow the walker to perform a sufficiently large number of steps and thus have access to the regime of long times and large displacements, corresponding to small wavevectors and Laplace variables in Fourier-Laplace space.

In passing, we mention that it is possible to derive a reaction-transport equation with a Caputo fractional derivative [19, p. 78] as an alternative to equation (24). Both equations are valid under the same conditions. The equation with the Caputo derivative reads

$$\frac{\partial^\gamma}{\partial t^\gamma}\frac{w(\mathbf{r},t|\mathbf{r}_0,0)}{\varphi(\mathbf{r},t)} = K_\gamma \nabla^2_{\mathbf{r}}\frac{w(\mathbf{r},t|\mathbf{r}_0,0)}{\varphi(\mathbf{r},t)}. \tag{26}$$

3.2. Subdiffusion with Specific Types of Evanescence

The simplest type of evanescence contained in Eq. (24) is the exponential case, $\varphi(\mathbf{r},t) \propto \exp(-\lambda t)$. With this choice, the results of Sokolov et al. [5] and Henry et al. [6] are recovered. However, significantly more complex forms of evanescence are also included. For instance, the case of a position dependent reactivity is immediately obtained if one takes $\varphi(\mathbf{r},t) \propto \exp(-\lambda(\mathbf{r})t)$. We consider the following particular choice:

$$\varphi(\mathbf{r},t) = \exp\left\{\int_{\tau}^{t} \Re[\rho(\mathbf{r},t')]dt'\right\}. \tag{27}$$

The time τ in Eq. (27) is an arbitrary reference time. In the language of ref. [8], this corresponds to an evanescence reaction that follows a local law of mass action of the form $\dot{\rho} = \Re(\rho)\rho$. The rate coefficient

$$\Re[\rho(\mathbf{r},t)] = \left[\frac{\dot{\rho}}{\rho}\right]_{\text{Reaction}} < 0 \tag{28}$$

is thus given by the logarithmic derivative of $\varphi(\mathbf{r},t)$. In general, the rate coefficient may depend on the density itself, thus leading to an interesting nonlinear problem. To make the connection with previous work, we note that substitution of Eq. (27) with the specific choice $\tau = 0$ into our general equation (24) for the one-dimensional case, and subsequent integration over the initial condition, yields Eq. (23) in ref. [8].

A simplified version of the above problem arises if the rate coefficient depends *only* on the global density

$$\overline{\rho}(t) = \frac{1}{V}\int d\mathbf{r}\,\rho(\mathbf{r},t), \tag{29}$$

where V stands for the system volume. One then has

$$\Re(\overline{\rho}(t)) = \frac{1}{\overline{\rho}}\frac{d\overline{\rho}}{dt}, \tag{30}$$

and (taking $\tau = 0$)

$$\varphi(\mathbf{r},t) = \varphi(t) = \exp\left\{\int_{0}^{t} \Re(\overline{\rho}(t'))dt'\right\} = \frac{\overline{\rho}(t)}{\overline{\rho}(0)}. \tag{31}$$

In this case, $\varphi(t)$ has a clear physical interpretation. In a single-particle system this can be interpreted as the survival probability of the particle up to time t. In a many-particle system, it can be interpreted as the fraction of surviving particles. As a result of Eq. (9), the probability $\Xi(t,t')$ that a randomly chosen evanescent particle present in the system at time t' survives up to time t is $\overline{\rho}(t)/\overline{\rho}(t')$. This is consistent with what one would expect from the interpretation of $\Xi(t,t')$ as a conditional probability: the survival probability $\overline{\rho}(t)/\overline{\rho}(0)$ of a given particle up to time t must be equal to the probability $\Xi(t,t')$ that it does not die during the interval (t',t) provided that it survived up to time t', times the probability $\overline{\rho}(t')/\overline{\rho}(0)$ for this condition to be fulfilled. Using Eq. (31) in Eq. (24), one finds that the kinetics of individual particles subject to a density-dependent homogeneous rate coefficient is dictated by the reaction-subdiffusion equation

$$\frac{\partial w(\mathbf{r},t|\mathbf{r}_0,0)}{\partial t} = \overline{\rho}(t)\, K_\gamma\, {}_0D_t^{1-\gamma}\nabla_{\mathbf{r}}^2 \frac{1}{\overline{\rho}(t)} w(\mathbf{r},t|\mathbf{r}_0,0) + \frac{\dot{\overline{\rho}}(t)}{\overline{\rho}(t)} w(\mathbf{r},t|\mathbf{r}_0,0). \tag{32}$$

In Sec. 4, we show that Eq. (32) can be used to set up the boundary value problem for the survival probability of an evanescent diffusing particle in the presence of a fully absorbing target. One can subsequently use this equation to compute the survival probability of the target, both in the presence of a single evanescent diffusive particle and of a collection thereof. Such target problems are of great importance in biological systems where diffusive transport is at play.

Finally, we note that, in more general terms, Eq. (24) also allows for the possibility of an *external* control mechanism of the evanescence reaction not necessarily coupled to the instantaneous particle population. Indeed, one can prescribe a reactivity field $\Re(\mathbf{r},t) < 0$ and subsequently set $\varphi(\mathbf{r},t) = \exp\left\{\int_0^t \Re(\mathbf{r},t')dt'\right\}$, which is then the survival probability of the particle up to time t given that it stays at $\mathbf{r}$ all the time (immobile reactant limit). Consequently, our assumption (9) implies that the probability for a particle to vanish during the time interval (t',t) if it is known to be at position $\mathbf{r}'$ at time t' and at position $\mathbf{r}$ at time t depends only on the instantaneous values of the local rate coefficient at the start and end points via the ratio of the associated survival probabilities for particles respectively fixed at $\mathbf{r}'$ and $\mathbf{r}$. The survival probability does not depend on the particular trajectory of the particle between these two points.

3.3. Fractional Superdiffusive Equation with Evanescence

In Section 3.1 we derived a rather general reaction-subdiffusion equation using a waiting time distribution with long tails together with a "normal" distribution of displacements. In contrast, we now consider a "normal" waiting time distribution along with a distribution of displacements that has long tails. In this case, the transport turns out to be superdiffusive.

The waiting time distribution we consider has a finite mean τ, implying that the small u behavior of the Laplace transform (corresponding to long times) is

$$\tilde{\psi}(u) \sim 1 - \tau u. \tag{33}$$

The jump length distribution is assumed to behave as follows:

$$\chi(\mathbf{r}) \sim \frac{\sigma^\mu}{r^{1+\mu}} \tag{34}$$

with $1 \leq \mu \leq 2$ and $r = |\mathbf{r}|$. As a result of this choice, the variance diverges. The Fourier transform of the jump length distribution is

$$\hat{\chi}(\mathbf{q}) \sim 1 - \sigma^\mu q^\mu, \tag{35}$$

where $q = |\mathbf{q}|$. Once again taking Eq. (14) as a starting point, one straightforwardly obtains (upon retention of the leading terms)

$$u\hat{\tilde{\eta}}(\mathbf{q},u) - \frac{e^{i\mathbf{q}\cdot\mathbf{r}_0}}{\varphi(\mathbf{r}_0,0)} = -\frac{\sigma^\mu}{\tau} q^\mu \hat{\tilde{\eta}}(\mathbf{q},u), \tag{36}$$

or, inverting the time Laplace transform,

$$\frac{\partial}{\partial t}\hat{\eta}(\mathbf{q},t) = -\frac{\sigma^\mu}{\tau} q^\mu \hat{\eta}(\mathbf{q},t). \tag{37}$$

Let $\partial^\mu/\partial r^\mu$ be the operator defined by the Fourier transform property,

$$\mathfrak{F}\left\{\frac{\partial^\mu f(\mathbf{r})}{\partial r^\mu}\right\} = -q^\mu f(\mathbf{q}). \tag{38}$$

We can thus write

$$\frac{\partial \eta(\mathbf{r},t|\mathbf{r}_0,0)}{\partial t} = K\frac{\partial^\mu}{\partial r^\mu}\eta(\mathbf{r},t|\mathbf{r}_0,0), \tag{39}$$

where we have introduced the anomalous diffusion coefficient

$$K = \frac{\sigma^\mu}{\tau}. \tag{40}$$

For a one-dimensional system, $\partial^\mu/\partial r^\mu = \partial^\mu/\partial x^\mu$ is identical to the Riesz operator [16–18].

Undoing the transformation (13), we obtain the following fractional reaction-superdiffusion equation:

$$\frac{\partial w(\mathbf{r},t|\mathbf{r}_0,0)}{\partial t} = \varphi(\mathbf{r},t)\, K\frac{\partial^\mu}{\partial r^\mu}\frac{1}{\varphi(\mathbf{r},t)}w(\mathbf{r},t|\mathbf{r}_0,0) + \frac{\dot{\varphi}(\mathbf{r},t)}{\varphi(\mathbf{r},t)}w(\mathbf{r},t|\mathbf{r}_0,0). \tag{41}$$

3.4. Bifractional Equation with Evanescence

Finally, in this subsection we combine subdiffusion and superdiffusion by choosing a waiting time distribution that lacks finite moments (and thus leads to subdiffusion by itself) with a jump distance distribution that also lacks moments (and thus leads to superdiffusion by itself). Our methodology directly lends itself to this combination.

We choose the waiting time distribution of Eq. (15) whose Laplace transform is given in Eq. (16), and the jump distribution Eq. (34) whose Fourier transform is given in Eq. (35). The steps to follow are now essentially the same as in the previous sections, with appropriate care given to the retention of the leading contributions. After some algebra we find

$$\frac{\partial \eta(\mathbf{r},t|\mathbf{r}_0,0)}{\partial t} = K\, {}_0D_t^{1-\gamma}\frac{\partial^\mu}{\partial r^\mu}\eta(\mathbf{r},t|\mathbf{r}_0,0). \tag{42}$$

Undoing the transformation (13), we obtain the fractional reaction-sub/superdiffusion equation,

$$\frac{\partial w(\mathbf{r},t|\mathbf{r}_0,0)}{\partial t} = \varphi(\mathbf{r},t)\, K\, {}_0D_t^{1-\gamma}\frac{\partial^\mu}{\partial r^\mu}\frac{1}{\varphi(\mathbf{r},t)}w(\mathbf{r},t|\mathbf{r}_0,0) + \frac{\dot{\varphi}(\mathbf{r},t)}{\varphi(\mathbf{r},t)}w(\mathbf{r},t|\mathbf{r}_0,0). \tag{43}$$

This is our third mesoscopic result and is unique in that it combines both subdiffusion and superdiffusion in a single equation. The behavior of the rescaled mean square displacement depends on the relative values of γ and μ.

4. Applications

We now turn to two specific examples where the reaction-subdiffusion equation (32) proves to be useful. The first one concerns the survival probability of an immobile target that is surrounded by one or more subdiffusively moving particles (also called traps) that disappear spontaneously in the course of their motion. The second example refers to the formation of morphogen gradients, which are crucial in embryogenesis.

4.1. The Target Problem with Evanescent Traps

4.1.1. The Single-trap Problem

We consider a target of radius R centered at the origin $\mathbf{r} = 0$ of a d dimensional volume V and an evanescent subdiffusive point trap placed at an initial position $\mathbf{r}_0$ inside V. In three dimensions the target is spherical, in two dimensions it is circular, and in one dimension it is a line. Unless otherwise specified, the survival probability of the trap will hereafter be assumed to decay exponentially, $\varphi(t) = e^{-\lambda t}$, in the absence of the target. In other words, the associated (global) number density $\overline{\rho}(t)$ obeys the following first-order differential equation:

$$\dot{\overline{\rho}}(t) \equiv \frac{d\overline{\rho}(t)}{dt} = -\lambda\,\overline{\rho}(t). \tag{44}$$

Since the trap may evanesce as it moves, the behavior of the trap is described by Eq. (32). The target is impenetrable and is killed by the trap as soon as the latter hits the target surface. We are interested in the survival probability $Q_{1,tg}(\mathbf{r}_0, t; R)$ of the target up to time t. The trap will also be assumed to disappear as soon as it hits the target (pairwise annihilation). The survival probability $Q_1(\mathbf{r}_0, t; R) = \int w(\mathbf{r}, t|\mathbf{r}_0; 0)\, d\mathbf{r}$ of the trap can be shown to be related to $Q_{1,tg}(\mathbf{r}_0, t; R)$ by the following equation [20, 21]:

$$Q_{1,tg}(\mathbf{r}_0, t; R) = Q_1(\mathbf{r}_0, t; R) + \lambda \int_0^t Q_1(\mathbf{r}_0, t'; R) dt'. \tag{45}$$

The survival probability of the target is greater than that of the trap, as the trap has a second decay channel due to the spontaneous death process described by Eq. (44). Now, the strategy to obtain $Q_{1,tg}(\mathbf{r}_0, t; R)$ consists of first solving the boundary value problem

$$\begin{aligned}
\frac{\partial Q_1(\mathbf{r}_0, t; R)}{\partial t} &= e^{-\lambda t} K_\gamma\, {}_0D_t^{1-\gamma} e^{\lambda t} \nabla^2_{\mathbf{r}_0} Q_1(\mathbf{r}_0, t; R) \\
&\quad - \lambda Q_1(\mathbf{r}_0, t; R) && \text{(46a)}\\
Q_1(\mathbf{r}_0, 0; R) &= 1, && \text{(46b)}\\
Q_1(R, t; R) &= 0, && \text{(46c)}\\
Q_1(\mathbf{r}_0 \to \infty, t; R) &= e^{-\lambda t}. && \text{(46d)}
\end{aligned}$$

The detailed solution can be found in ref. [21]. In dimensions $d = 1, 3$ one arrives at the following long-time expression:

$$Q_{1,tg}(r_0,\infty;R) - Q_{1,tg}(r_0,t;R) \sim -\left(\frac{R}{r_0}\right)^{(d-1)/2} \frac{1}{\Gamma(1-\frac{\gamma}{2})} \frac{r_0 - R}{\lambda^{1+\frac{\gamma}{2}}\,\ell_\gamma} t^{-1-\frac{\gamma}{2}} e^{-\lambda t}, \quad (47)$$

where $\ell_\gamma = (4K_\gamma/\lambda^\gamma)^{1/2}$ is a length scale associated with the typical distance explored by the trap before decaying spontaneously. The asymptotic survival probability of the target in $d = 1, 3$ is

$$Q_{1,tg}(r_0,\infty;R) = 1 - \left(\frac{R}{r_0}\right)^{(d-1)/2} e^{-2(r_0-R)/\ell_\gamma}. \quad (48)$$

In $d = 2$ the long-time decay is more complex. It has the following form:

$$Q_{1,tg}(r_0,t;R) \sim Q_{1,tg}(r_0,\infty;R) + \frac{2}{\gamma}\ln\left(\frac{r_0}{R}\right)\frac{e^{-\lambda t}}{\lambda t \ln^2(\alpha_\gamma t)} \quad (49)$$

with $\alpha_\gamma = (4K_\gamma/R^2)^{1/\gamma}$ and

$$Q_{1,tg}(r_0,\infty;R) = 1 - \frac{\mathcal{K}_0(2r_0/\ell_\gamma)}{\mathcal{K}_0(2R/\ell_\gamma)}, \quad (50)$$

where $\mathcal{K}_\nu$ stands for a modified Bessel function of the second kind of order ν.

Note that in $d = 1, 2$ the target has a finite survival probability due to the evanescence of the trap. In contrast, if the trap were non-evanescent ($\lambda = 0$), the target would eventually be killed with certainty because of the compact space exploration of a random walker in one and two dimensions.

4.1.2. The Many-trap Problem

Let us now extend the results in Section 4.1.1 to the case where the target centered at $\mathbf{r} = 0$ is surrounded by a collection of N_0 point traps moving independently of one another. The traps are assumed to be evanescent and randomly distributed in a spherical volume V (or circular area or line, depending on dimension) around $\mathbf{r} = 0$. At time $t = 0$ they begin to move (and decay) according to Eq. (32). As in the single trap problem, we assume that whenever any of the traps hits the target, both reaction partners are annihilated. Our goal is once again to compute the survival probability $Q_{tg}(t)$ of the target, that is, the probability that none of the traps has collided with the target surface up to time t.

Since the traps are statistically independent, we can write the solution in terms of the single-trap problem discussed previously. One has

$$Q_{tg}(t;R) = \left[\frac{1}{V}\int_{r_0>R} Q_{1,tg}(\mathbf{r}_0,t;R)\,d\mathbf{r}_0\right]^{N_0}. \quad (51)$$

We focus on the thermodynamic limit $N_0 \to \infty$, $V \to \infty$, which implies fixing the initial global trap density $\overline{\rho}_0 \equiv \overline{\rho}(0) = \lim_{N_0,V\to\infty} N_0/V$. We already know the solution for $Q_{1,tg}$ in the infinite volume case, so in that limit we find

$$Q_{tg}(t;R) = \exp\left\{-\overline{\rho}_0 R^d \sigma(t,R)\right\}, \tag{52}$$

where we have introduced the auxiliary quantity

$$\sigma(t;R) \equiv \frac{1}{R^d}\int_{r_0>R} [1 - Q_{1,tg}(\mathbf{r}_0, t; R)]\, d\mathbf{r}_0. \tag{53}$$

We note that the quantity $R^d\sigma(t,R)$ is directly related to the average volume explored by a *single* evanescent trap up to time t, see e.g., refs. [22] and [23].

One can now employ the explicit expressions of the Laplace transform of $Q_{1,tg}$, perform the integration over $\mathbf{r}_0$, and finally make use of Tauberian theorems to find the long-time behavior of Q_{tg}. In terms of $\sigma(t;R) = -(\overline{\rho}_0 R^d)^{-1} \ln Q_{tg}(t;R)$ one has

$$\sigma(\infty;R) - \sigma(t;R) \propto \begin{cases} t^{\gamma/2-1} e^{-\lambda t}, & d=1, \\ \ln^{-1}(\alpha_\gamma t)\, t^{\gamma-1} e^{-\lambda t}, & d=2, \\ t^{\gamma-1} e^{-\lambda t}, & d=3, \end{cases} \tag{54}$$

with $\sigma(\infty;R) = \ell_\gamma/R$ for $d=1$, $\sigma(\infty;R) = (2\pi\ell_\gamma/R)\,(1+\ell_\gamma/2R)$ for $d=3$ and $\sigma(\infty;R) = 2\pi(\ell_\gamma/2R)\,\mathcal{K}_1(2R/\ell_\gamma)/\mathcal{K}_0(2R/\ell_\gamma)$ for $d=2$ (details can be found in refs. [20] and [21]). These are particular cases of the general d-dimensional expression

$$\sigma(\infty;R) = S_d \frac{\ell_\gamma}{2R} \frac{\mathcal{K}_{d/2}(2R/\ell_\gamma)}{\mathcal{K}_{d/2-1}(2R/\ell_\gamma)}, \tag{55}$$

where $S_d = 2\pi^{d/2}/\Gamma(d/2)$ denotes the surface of a d-dimensional hypersphere of unit radius. Thus, the asymptotic survival probability of the target is *finite* in all dimensions despite the infinite initial number of traps. The physical reason for this is the evanescent character of the traps, in the absence of which the final survival probability of a target surrounded by a collection of uncorrelated traps is zero in all dimensions. In the absence of evanescence, the target is eventually killed in all spatial dimensions [24].

Interestingly, the survival probability $Q_{tg}(t;R)$ obtained from Eq.(55) displays non-monotonic behavior as a function of dimensionality. Figure 1 displays a plot of $\sigma(\infty;R)$ as a function of d for three different values of the dimensionless parameter $z = 2R/\ell_\gamma$. It is seen that $\sigma(\infty;R)$ first increases and then decreases with increasing d, implying the opposite behavior of $Q_{tg}(t;R)$. We note the behavior in the limit $d \to \infty$. Using the asymptotic expansion

$$\mathcal{K}_\alpha(z) \sim \frac{\Gamma(\alpha)}{2}\left(\frac{2}{z}\right)^\alpha, \tag{56}$$

valid for $\alpha > 0$ and $0 < z \ll \sqrt{\alpha+1}$, one sees that $\sigma(\infty;R)$ tends to zero in this limit, implying $Q_{tg}(t;R) \to 1$. For large values of d, the traps are more likely to perform long excursions and die in the course of these excursions before they are able to hit the target.

It is also possible to obtain the long time expression for $\sigma(\infty;R)$ in the case of a general global density decay $\overline{\rho}(t)$. To this end, Eq. (45) must be replaced by the more general equation

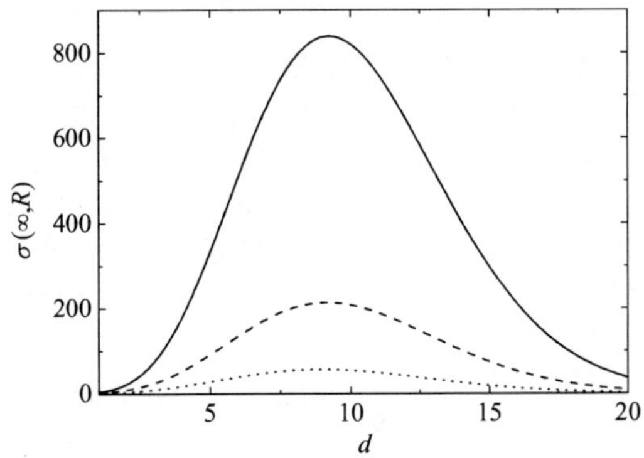

Figure 1. Behavior of $\sigma(\infty; R)$ as a function of d $z = 0.5$ (solid line), $z = 1$ (dashed line) and $z = 3$ (dotted line).

$$Q_{1,tg}(r,t;R) = Q_1(r,t;R) - \int_0^t Q_1(r,t';R)\frac{\dot{\overline{\rho}}(t')}{\overline{\rho}(t')}\,dt'. \tag{57}$$

Using this equation in the definition (53) in combination with (32), one finds

$$\sigma(t;R) = S_d \frac{K_\gamma}{R}\int_0^t \left[{}_0D_{t'}^{1-\gamma} \frac{1}{\overline{\rho}(t')} \frac{\partial Q_1(r,t';R)}{\partial r}\bigg|_{r=R}\right] \overline{\rho}(t')\,dt', \tag{58}$$

where the Gauss theorem was used to convert the volume integral resulting from Eq. (53) into a surface integral. From the specific form of Q_1 for $d = 1$ one finds [20,21]

$$\sigma(t;R) = \frac{1}{\overline{\rho}_0 R}\frac{2\sqrt{K_\gamma}}{\Gamma(\gamma/2)}\int_0^t \overline{\rho}(t')\,t'^{\gamma/2-1}\,dt'. \tag{59}$$

Likewise, using the expression for Q_1 for $d = 3$ we obtain [20,21]

$$\sigma(t;R) = \frac{4\pi\sqrt{K_\gamma}}{\overline{\rho}_0 R\Gamma(\gamma/2)}\int_0^t \overline{\rho}(t')\,t'^{\gamma/2-1}\,dt' + \frac{4\pi K_\gamma}{\overline{\rho}_0 R^2\Gamma(\gamma)}\int_0^t \overline{\rho}(t')\,t'^{\gamma-1}\,dt'. \tag{60}$$

In contrast, for $d = 2$ only a long-time expression is available (provided that $\overline{\rho}(t)$ decays sufficiently fast):

$$\sigma(\infty;R) - \sigma(t;R) \sim \frac{4\pi K_\gamma}{\overline{\rho}_0 R^2\Gamma(\gamma+1)}\int_t^\infty \frac{\overline{\rho}(t')\,t'^{\gamma-1}}{\ln(\alpha_\gamma t')}\,dt', \tag{61}$$

where to the best of our knowledge the exact expression for $\sigma(\infty;R)$ is in general unknown.

4.2. Morphogen Gradient Formation

Morphogens are special signaling molecules that are secreted at a specific location in the embryo and subsequently disseminate through the tissue. One current understanding of the morphogenesis process suggests that different target genes in an embryonic cell are activated above certain concentration thresholds, resulting in a different development and fate of different cells depending on the local morphogen concentration that they sense [25, 26]. According to this mechanism, cell differentiation in embryos thus requires the preexistence of a morphogen gradient controlled by the morphogen secretion rate, the transport properties of morphogens, and their degradation pathways. Traditional synthesis-diffusion-degradation models are based on the classical reaction-diffusion equation

$$\frac{\partial \rho(x,t)}{\partial t} = D\frac{\partial^2}{\partial x^2}\rho(x,t) - \lambda\rho(x,t), \tag{62}$$

where $\rho(x,t)$ is the local concentration of morphogens, λ is the degradation rate constant, and D is the diffusion coefficient in the case of ordinary Brownian motion (because of the high degree of symmetry in experiments, one often restricts the analysis to an effectively one-dimensional setting as above). However, the dissemination process of morphogens is more complex than a Fickian diffusion process. Indeed, some of the transport routes are likely to involve binding and unbinding to surface receptors and transcytosis in the complex intracellular environment [25], resulting in long characteristic waiting times. There is also reason to believe that the distribution of those waiting times displays significant dispersion, a rather common feature in biological systems. Under these circumstances a phenomenological description in terms of a CTRW with a long-tailed waiting time distribution and a concomitant degradation process becomes a plausible picture. In ref. [27], a mesoscopic reaction-transport model was developed that becomes equivalent to Eq. (32) in the case of a constant degradation rate λ. In this case, one can go beyond the ordinary diffusion case described by Eq. (62) by using the equation [1]

$$\frac{\partial \rho(x,t)}{\partial t} = e^{-\lambda t} K_\gamma \, {}_0D_t^{1-\gamma} \frac{\partial^2}{\partial x^2} e^{\lambda t}\rho(x,t) - \lambda\rho(x,t). \tag{63}$$

The above equation is solved assuming a constant injection j_0 of secreted morphogens at, say, $x = 0$. Typically, one neglects other boundary effects and assumes that one is dealing with an infinite system. The solution of the corresponding boundary value problem can be expressed in terms of the propagator solution $G(x,t)$ corresponding to the deterministic initial condition $\rho(x,t) = \delta(x)$. The solution reads

$$\rho(x,t) = j_0 \int_0^t G(x, t-t')dt'. \tag{64}$$

Our formulation yields stationary concentration profiles $\rho_s(x) = \lim_{t\to\infty} \rho(x,t)$. Note that this is by no means a universal feature; in our model, we have assumed that degradation may occur at any time. Models assuming that morphogen degradation occurs only at the time of a jump have also been implemented [28, 29], but in general they do not lead to a steady state solution [28]. Returning to our case, the stationary profile can easily be found from the solution of the corresponding boundary value problem in Laplace space and the subsequent application of the final value theorem. Details can be found in refs. [11] and [27]. One finds exponentially decaying stationary profiles of the form [27]

---

[1]Note the interesting feature that Eq. (63) can be cast into a pure transport equation by means of a rescaling of the concentration, i.e., $u(x,t) = e^{\lambda t}\rho(x,t)$.

$$\rho_s(x) = \frac{j_0}{2} \frac{\lambda^{\gamma/2-1}}{\sqrt{K_\gamma}} \exp\left[-|x|/\Lambda\right], \qquad 0 < \gamma < 1, \tag{65}$$

where the decay length is $\Lambda = \sqrt{K_\gamma/\lambda^\gamma}$.

Equation (65) generalizes the well-known stationary exponential profile found for normal diffusion [26], which is simply this result if one takes $\gamma = 1$ and sets $K_1 \equiv D$. This shows that the same characteristic decay length may be obtained with different combinations of the parameters K_γ, λ and γ, which are difficult to measure individually. We thus conclude that the exponentially decaying experimentally observed profiles are not only compatible with Fickian diffusion, but also with non-Markovian generalizations thereof.

Moreover, in ref. [27] it has been shown that if the characteristic degradation time λ^{-1} is large in comparison with the typical stepping time τ_D of the underlying CTRW, the robustness of the profile against fluctuations in the morphogen secretion rate j_0 increases with decreasing γ. In the small γ regime, it would appear that subdiffusion enhances the stability of the concentration profile. Needless to say, a mechanism for dampening the effects of accidental fluctuations in the morphogen secretion rate is crucial for a proper development of the embryo.

A question of interest in the above context is whether similar results are obtained when the rate constant of the evanescence process becomes density-dependent, a non-linear problem that is also described by our equations. Recent research has focused on the form of the morphogen concentration profiles in this case [30, 31]. These interesting studies concern specific realizations of the microscopic reaction and transport processes. The next step will be to ascertain the influence of the model details on the robustness of the concentration profiles by performing comparative studies.

5. Summary and Outlook

In this work we have presented the derivation of rather general reaction-subdiffusion and reaction-superdiffusion equations starting from a CTRW description of particles that die as they walk, or as they wait to walk. The reaction term describing the death or evanescence of the particles may in general depend on both position and time. An explicit dependence of the rate constant on the particle density is not excluded either, hence the problem is in general a nonlinear one. The final form of the resulting fractional reaction-transport equations is not particularly intuitive, as the reaction rate constant appears explicitly in the transport term.

Two interesting applications of our reaction-subdiffusion equation have been considered, namely, the survival probability of a target surrounded by one or many decaying subdiffusive particles, and the formation of stationary morphogen gradients. However, our fractional equations can be applied to many other interesting problems by considering different initial conditions and boundary conditions.

Acknowledgments

This work was partially funded by the Ministerio de Ciencia y Tecnología (Spain) through Grant No. FIS2010-16587 (partially financed by FEDER funds), by the Junta de Extremadura through Grant No. GRU10158, and by the US National Science Foundation under Grant No. PHY-0855471.

References

[1] H. Scher and M. Lax, *Phys. Rev. B* 7, 4491 (1973); H. Scher and M. Lax, *Phys. Rev. B* 7, 4501 (1973).

[2] E. Scalas, R. Gorenflo, and F. Mainardi, *Phys. Rev. E* 69, 011107 (2004).

[3] J. Masoliver, M. Montero, J. Perelló, and G. H. Weiss, *J. of Economic Behaviour and Org.* 61, 577 (2006).

[4] J. Masoliver, M. Montero, J. Perelló, and G. H. Weiss, *Physica A* 379, 151 (2007).

[5] I. M. Sokolov, M. G. W. Schmidt, and F. Sagués, *Phys. Rev. E* 73, 031102 (2006); M. G. W. Schmidt, F. Sagués, and I. M. Sokolov, *J. Phys.: Condens. Matter*, 19, 065118 (2007).

[6] B. I. Henry, T. A. M. Langlands, and S. L. Wearne, *Phys. Rev. E* 74, 031116 (2006).

[7] D. Campos and V. Méndez, *Phys. Rev. E* 80, 021133 (2009).

[8] S. Fedotov, *Phys. Rev. E* 81, 011117 (2010).

[9] V. Mendez, S. Fedotov, and W. Horsthemke, *Reaction-Transport Systems: Mesoscopic Foundation, Fronts, and Spatial Instabilities* (Springer-Verlag, Berlin, 2010).

[10] E. Abad, S. B. Yuste, and K. Lindenberg, *Phys. Rev. E* 81, 031115 (2010).

[11] S. B. Yuste, E. Abad, and K. Lindenberg, *Reactions in Subdiffusive Media and Associated Fractional Equations*, in *Fractional Dynamics. Recent Advances*, J. Klafter, S. C. Lim, and R. Metzler (Eds.) (World Scientific, Singapore, 2011).

[12] C. N. Angstmann, I. C. Donnelly, and B. I. Henry, *Math. Model. Nat. Phenom.* 8, 32 (2013).

[13] P. V. Danckwerts, *Trans. Far. Soc.* 47, 1014 (1951).

[14] J. Crank, *The mathematics of diffusion*, 2nd edition reprinted (Oxford University Press,New York, 1993), p. 329-332.

[15] A. Zoia, *Phys. Rev. E* 77, 041115 (2008).

[16] R. Metzler and J. Klafter, *Phys. Rep.* 339, 1 (2000)

[17] R. Metzler and J. Klafter, *J. Phys. A: Math. Gen.* 37, R161 (2004).

[18] R. Klages, G. Radons, and I. M. Sokolov (Eds.), *Anomalous Transport: Foundations and Applications* (Wiley-VCH, Weinheim, 2008).

[19] I. Podlubny, *Fractional Differential Equations* (Academic Press, San Diego, 1998).

[20] E. Abad, S. B. Yuste, and K. Lindenberg, *Phys. Rev. E* 86, 061120 (2012).

[21] E. Abad, S. B. Yuste, and K. Lindenberg, *Math. Model. Nat. Phenom.* 8, 100 (2013).

[22] S. B. Yuste, E. Abad, and K. Lindenberg, *Arrival statistics and exploration properties of mortal walkers*, in *First-Passage Phenomena and Their Applications*, R. Metzler, G. Oshanin, and S. Redner (Eds.) (World Scientific, Singapore, 2014).

[23] E. Abad, S. B. Yuste, and K. Lindenberg, *Phys. Rev. E* 88, 062110 (2013).

[24] S. B. Yuste and K. Lindenberg, *Phys. Rev. E* 76, 051114 (2007).

[25] M. Ibáñes and J. C. Izpisúa, *Molecular Systems Biology* 4, 176 (2008).

[26] O. Wartlick, A. Kicheva, and M. González-Gaitán, *Cold Spring Harb. Perspect. Biol.* 1, a001255 (2009).

[27] S. B. Yuste, E. Abad, and K. Lindenberg, *Phys. Rev. E* 82, 061123 (2010).

[28] G. Hornung, B. Berkowitz, and N. Barkai, *Phys. Rev. E* 72, 041916 (2005).

[29] V. P. Shkilev, *J. Exp. Theor. Phys.* (Springer) 112, 1071 (2011).

[30] J. P. Boon, J. F. Lutsko, and C. Lutsko, *Phys. Rev. E* 85, 021126 (2012).

[31] S. Fedotov and S. Falconer, *Phys. Rev. E* 89, 012107 (2014).

In: Fractional Calculus: Theory
Editors: R. Abi Zeid Daou and X. Moreau
ISBN: 978-1-63463-002-3

Chapter 9

EXPONENTIAL INTEGRATORS FOR FRACTIONAL DIFFERENTIAL EQUATIONS

***Roberto Garrappa*[1] * *and Marina Popolizio*[2] †**
[1]Department of Mathematics
University of Bari, Bari, Italy
[2]Department of Mathematics and Physics "Ennio De Giorgi"
University of Salento, Lecce, Italy

Abstract

Exponential integrators are a well–established class of effective methods for the numerical integration of systems of differential equations of large dimension, especially in the presence of stiffness. Problems of this kind are common in several applications and usually come from semi–discretization of partial differential equations.

The main idea behind exponential integrators is to solve in an exact way the stiff term of the problem and hence apply a time–step integration to the non–stiff term, thus to allow to use less expensive explicit schemes and obtain, at the same time, good stability properties.

Recently, the investigation of exponential integrators has been generalized to differential equations of fractional order (FDEs).

In this context exponential integrators turn out to be very effective since they overcome some of the typical weak points of numerical methods for FDEs: indeed, it is possible to derive methods with excellent stability properties and, when working with linear problems, to surmount some limitations in accuracy and order of convergence which represent a severe constriction for common methods for FDEs.

In this Chapter we illustrate the main aspects concerning the derivation of some families of exponential integrators for FDEs and we study the convergence properties.

Moreover, we address their numerical implementation with special attention to some peculiar aspects as, for example, the evaluation of Mittag–Leffler functions, with scalar and/or matrix arguments; this is a central feature in the implementation of exponential integrators for fractional order problems. The application to some time–fractional partial differential equations is also presented by means of some numerical examples.

*E-mail address: roberto.garrappa@uniba.it
†E-mail address: marina.popolizio@unisalento.it

Keywords: fractional differential equation, exponential integrator, product integration, Mittag–Leffler function, time–fractional partial differential equations

AMS Subject Classification: 34A08, 65L06, 33E12, 26A33, 65F60

1. Introduction

Differential equations are a powerful mathematical tool to model real–life systems in several fields such as biology, economics, engineering and physics. Most of the equations describing fundamental phenomena in physics and engineering are indeed differential equations as, for example, the Newton's Second Law, Maxwell's and Schrödinger equations, etc. Economists use differential equations to describe equilibrium and stability in economic markets and investment returns. Biological scientists use them for the population growth in species while medical scientists apply them to determine rates of growth of tumors, just to give some examples.

In general the most used and sophisticated models are described by functions of several variables and the governing equations involve the presence of derivatives along the time variable and the spatial coordinates. These are called *partial differential equations* (PDEs) and often model multidimensional systems.

Historically the definition of classical derivative has been shortly followed by that of *fractional* derivative, since Leibnitz, L'Hospital and other important mathematicians started to intellectualize about this concept at the end of the seventeenth century. For long time the fractional derivatives have been considered just as a theoretical subject and only in the last decades their importance in applications has been recognized. Indeed, in many fields they turn out to be adequate for the description of various real materials or situations, more than classical derivatives. This is particularly true for polymers or viscoelastic materials, as well as in bioengineering, control theory, finance, pharmacokinetics, rheology, signal processing and many others (e.g., see [1, 11, 43, 53] and references therein).

The main difference between fractional and integer derivatives is the capability of the former to better describe processes with memory and hereditary properties. On the other hand, this feature represents one of the main challenges facing the numerical solution of FDEs due to the huge computational effort involved by the presence of the persistent memory.

As in the case of ordinary differential equations (ODEs), resorting to the numerical solution of FDEs is necessary in several situations since most of the systems involving derivatives of fractional order do not usually have solutions in closed form. Unfortunately, the numerical solution of FDEs is much more difficult than that of classical ODEs not only for the memory preservation but also for the non-smooth nature of the solutions which in general forces some limitations to the order of the solvers one can define.

The aim of this chapter is to discuss an innovative approach for the numerical solution of FDEs which essentially grounds on the *exponential integrators* developed in the context of ODEs. It turns out to be effective for linear and semilinear systems of FDEs with large size and in presence of stiffness related to the linear term. We will also see that this approach can be profitable when applied to PDEs with time fractional derivatives.

This approach basically consists on the development of difference schemes grounded on an alternative integral formulation of the exact solution of the FDEs.

The structure of this Chapter is the following. We first discuss the main theoretical aspects concerning fractional derivatives and FDEs and we review the basics of exponential integrators for ODEs. Subsequently we present two different approaches for deriving exponential integrators for FDEs with an accurate description of some specific methods and their main properties. We then deal with an important numerical issue related to these methods, namely the computation of the Mittag–Leffler function and its generalizations. This topic strongly influences the effectiveness of the exponential product integration and thus deserves great attention. Finally we consider the application of exponential product integration to the solution of fractional PDEs and we present some numerical tests to better explain the concepts involved in the analysis of the chapter.

2. Linear Systems of FDEs: Theoretical Aspects

In this chapter we address the numerical computation of stiff systems of FDEs. In general, the definition of *stiffness* is puzzling even for ODEs (see e.g., the careful description in [28]). Here we restrict our analysis to linear systems of FDEs of the form

$$ {}_0^C D_t^\alpha y(t) + Ay(t) = f(t) \tag{1} $$

where $f, y : [0, T] \to \mathbb{R}^M$, f is smooth and the matrix $A \in \mathbb{R}^{M\times M}$ has eigenvalues with large positive real parts. To this case the oldest definition of stiffness holds, that is, for these equations certain implicit methods perform better, usually tremendously better, than explicit ones [8]. Equation (1) is usually combined with the initial conditions

$$ y^{(\nu)}(0) = y_{0,\nu}, \quad \nu = 0, 1, \ldots, m-1, \tag{2} $$

where $m = \lceil \alpha \rceil$ is the smallest integer such that $\alpha < m$.

Throughout this chapter ${}_0^C D_t^\alpha$ denotes the derivative of fractional order $\alpha > 0$ according to the Caputo's definition introduced in 1967 [11, 53]

$$ {}_0^C D_t^\alpha y(t) = \frac{1}{\Gamma(m-\alpha)} \int_0^t (t-s)^{m-\alpha-1} y^{(m)}(s) ds. $$

In addition to be a classical test equation of theoretical relevance, the solution of (1-2) has more practical motivations, as we will discuss hereinafter, since important problems modeled in terms of PDEs can be reformulated as linear problems of this type. Anyway, some aspects concerning more general problems than (1-2), in which the source term f is a non-linear function of the solution y, will be also discussed.

In the fully linear case, i.e., when $f(t) \equiv 0$, it is well–known that the exact solution of (1) can be written as $y(t) = Y_\alpha(t; A)$ with

$$ Y_\alpha(t; A) = \sum_{\nu=0}^{m-1} e_{\alpha,\nu+1}(t; A) y_{0,\nu} \tag{3} $$

and

$$ e_{\alpha,\beta}(t; A) = t^{\beta-1} E_{\alpha,\beta}(-t^\alpha A), $$

where $E_{\alpha,\beta}(z)$ is the Mittag–Leffler (ML) function in two parameters defined as

$$E_{\alpha,\beta}(z) = \sum_{\ell=0}^{\infty} \frac{z^{\ell}}{\Gamma(\alpha\ell+\beta)}, \quad z \in \mathbb{C}, \tag{4}$$

with $\Gamma(\cdot)$ the Euler's gamma function, α and β complex parameters and $\Re(\alpha) > 0$.

In the more general case to determine an explicit representation of the solution of (1) it is useful to use the Laplace transform. To this purpose we first recall that for any real value $\nu > -1$ the Laplace transform of the real power t^{ν} is

$$\mathcal{L}\left(t^{\nu}\,;\,s\right) = \frac{\Gamma(\nu+1)}{s^{\nu+1}}, \Re(s) > 0 \tag{5}$$

and for any $\alpha > 0$ the Laplace transform of the Caputo's fractional derivative of a function $g(t) \in \mathcal{C}^m\big([0,T]\big)$ is

$$\mathcal{L}\left({}^{C}_{0}D^{\alpha}_{t} g(t)\,;\,s\right) = s^{\alpha}\hat{g}(s) - \sum_{\nu=0}^{m-1} s^{\alpha-\nu-1} g^{\nu}(0),$$

where $\hat{g}$ denotes the Laplace transform of g and $g^{(\nu)}$ classical integer-order derivatives [53].

We can now reformulate equation (1) in the Laplace transform domain as

$$s^{\alpha}\hat{y}(s) - \sum_{\nu=0}^{m-1} s^{\alpha-\nu-1} y^{(\nu)}(0) + A\hat{y}(s) = \hat{f}(s) \tag{6}$$

from which, after multiplying both sides by $s^{-\alpha}$ and imposing the initial conditions (2), we get

$$\hat{y}(s) = \sum_{\nu=0}^{m-1} s^{-\nu-1} y_{0,\nu} + \frac{1}{s^{\alpha}}\left(-A\hat{y}(s) + \hat{f}(s)\right).$$

We can now transform back to the time–domain and, by means of (5), we obtain the following classical integral representation of the solution of (1)

$$y(t) = \sum_{\nu=0}^{m-1} \frac{t^{\nu}}{\nu!} y_{0,\nu} + \frac{1}{\Gamma(\alpha)} \int_0^t (t-\tau)^{\alpha-1}\big(-Ay(\tau) + f(\tau)\big)\, d\tau. \tag{7}$$

The integral formulation (7) is a widely used starting point for the development of numerical methods for FDEs. An alternative formulation can however be derived. Starting from (6), we can indeed solve with respect to $\hat{y}(s)$ in order to obtain

$$\hat{y}(s) = \sum_{\nu=0}^{m-1} s^{\alpha-\nu-1}\left(s^{\alpha}I + A\right)^{-1} y_{0,\nu} + \left(s^{\alpha}I + A\right)^{-1}\hat{f}(s)$$

where I denotes the $M \times M$ identity matrix. Now, since $e_{\alpha,\beta}(t; A)$ is the inverse Laplace transform of $s^{\alpha-\beta}\left(s^{\alpha}I + A\right)^{-1}$ [43], by turning back to the temporal domain we obtain

$$y(t) = Y_{\alpha}(t; A) + \int_0^t e_{\alpha,\alpha}(t-\tau; A) f(\tau)\, d\tau. \tag{8}$$

The function $e_{\alpha,\beta}(t; A)$, also known as the *generalized* ML function, has been investigated by many authors (see e.g [43, 53]). It is immediate to verify that $e_{\alpha,\beta}(t; A)$ is a generalization of the exponential function. Indeed, for $\alpha = \beta = 1$ it is $e_{1,1}(t; A) = e^{-tA}$.

The following properties of the generalized ML function will be useful later.

Lemma 1. *Let* $\alpha, \beta > 0$ *and* $h > 0$. *For any* $t \geq 0$ *it is*

$$e_{\alpha,\beta}(t; A) = h^{\beta-1} e_{\alpha,\beta}(t/h; h^{\alpha} A). \tag{9}$$

Proof. The Lemma can be easily proved by using the series representation (4) of the ML function; we omit the details. □

Lemma 2. *Let* $\alpha, \beta > 0$ *and* $r \in \mathbb{R}$ *such that* $r > -1$. *For any* $t > a$ *it is*

$$\frac{1}{\Gamma(r+1)} \int_a^t e_{\alpha,\beta}(t-s; A)(s-a)^r ds = e_{\alpha,\beta+r+1}(t-a; A).$$

Proof. The result can be proved by simply operating in the Laplace transform domain (see also [53]). □

3. Exponential Integrators: A Short Review

The history of exponential integrators is not easy to unravel, since many contributions overlapped over the years, some of them reinvented several times. Moreover, seminal ideas can be found in various papers and, as described in [34], probably the oldest one is due to Hersch dating back to 1958. The intention of this section is just to recall the basics of this theory, while for a full account on its history we refer to [46, 34].

To provide introductory material on exponential integrators we consider a semilinear system of ODEs in the form

$$\frac{d}{dt}U(t) + AU(t) = f(U(t)), \quad U(0) = U_0 \tag{10}$$

where $U(t)$ is a vector valued function, A a real matrix of the appropriate dimension and $f(U)$ the non–linear part of the vector field. In some situations, problem (10) can be the result of the linearization of a fully non linear problem. In general, exponential integrators apply to the solution of PDEs which, after a space discretization, can be brought to the form (10). Examples are Allen–Cahn, Burgers, Navier–Stokes, nonlinear Schrödinger equations. Thus, in an abstract formulation, $U(t)$ describes the time evolution of a system in some function space, A is a linear (typically unbounded) differential operator (for instance, the Laplace operator) and f is a nonlinear operator. We just use formulation (10) as a useful and elegant framework comprising both ODEs and PDEs.

Typically the linear part of the problem is stiff and the nonlinear part nonstiff. In these situations classical numerical methods fail: indeed, the explicit ones require extremely small step–sizes, probably making the computation unfeasible. At the same time implicit methods, which overcome the stiffness problem, are generally extremely costly.

There exist many exponential schemes, all sharing the same peculiarity: the linear part of (10), which is usually the source of stiffness, is solved exactly (or, at least, with a very high accuracy) and explicit time integrators are used to process the nonlinear term without severe restrictions on the step–size. This approach turns out to be very efficient, especially when the size and the complexity of the problem advise against the use of implicit methods.

Among the various exponential schemes the Lawson ones, or *Integrating Factor* methods, deserve a mention for the simple effective idea they ground on, that is, the use of an explicit solver for an equivalent problem deriving from a change of variables. Unfortunately these schemes have only linear stiff order [2]. J. Certaine can be considered one of the most important contributors to the subject: indeed, in 1967, he proposed the *Exponential Time Differencing* (ETD) methods, which are still the most common class of exponential integrators. The starting point is a variation–of–constants formula for the exact solution of (10)

$$U(t) = e^{-tA}U_0 + \int_0^t e^{-(t-s)A} f(U(s))ds$$

which, for a given grid of points $\{t_n\}_{n\geq 0}$ on the integration interval, can be written as

$$U(t_{n+1}) = e^{-(t_{n+1}-t_n)A}U(t_n) + \int_{t_n}^{t_{n+1}} e^{-(t_{n+1}-s)A} f(U(s))ds. \tag{11}$$

One of the simplest method to numerically compute (11) is to interpolate the nonlinearity at the known value $f(U_n)$ only, leading to the *explicit ETD Euler method*

$$U_{n+1} = e^{-hA}U_n + h\varphi_1(-hA)f(U_n) \tag{12}$$

with $\varphi_1(z) = (e^z - 1)/z$; observe that, since $\lim_{z\to 0}\varphi_1(z) = 1$, when $A = 0$ the above formula reduces to the classical forward Euler scheme. The use of different and more sophisticated schemes for evaluating the integral in (11) leads to a variety of methods such as exponential multistep methods, exponential Runge–Kutta methods, exponential general methods and so on (e.g., see [4, 5, 7, 33, 50, 52]).

All the methods in the class of exponential integrators involve matrix functions, like the exponential and φ–functions defined by means of the recursion

$$\varphi_\ell(z) = \frac{1}{(\ell-1)!}\int_0^1 e^{(1-s)z}s^{\ell-1}ds, \quad \ell \geq 1.$$

For several years, however, this feature made exponential integrators pointless for the main difficulty related to the evaluation of functions with matrix arguments, usually of large dimension. It is only the recent spread of efficient methods to perform the numerical computation of matrix functions that has renewed the interest in exponential integrators. For a wide literature on matrix functions we refer to the monograph by Higham [30] and references therein. The application of these recent results has recently led to the design of robust software packages implementing exponential integrators for a wide class of problems (see for instance the Matlab package EXPINT [2] and `exp4` by Hochbruck and coauthors [32]).

4. Exponential Product Integration for FDEs

Product Integration (PI) is a widely used technique to derive numerical methods for FDEs. It has been originally developed in the context of numerical solution of Volterra Integral Equations [65] and its application to the FDE (1) basically consists in replacing just the term $-Ay(t) + f(t)$ in the weighted integral (7) by some piecewise polynomial approximation. PI can hence be considered as a generalization of classical Adams multistep methods for ODEs [28].

In the recent past several authors have investigated methods for FDEs grounded on PI rules; the most renowned application is the predictor–corrector algorithm proposed by Diethelm and some coauthors [13] for which a robust code is available on the Matlab central repository [18].

Unfortunately, the use of PI rules is not completely satisfactory for FDEs. The derivatives of the solution of (1) are, indeed, usually unbounded at the origin [39] and polynomials do not accurately fit non–smooth functions (it is worthwhile to underline that the possible smoothness of $f(t)$ is not sufficient to assure the smoothness of the solution $y(t)$ of (1) and, hence, of the integrand in (7)).

As a consequence, the convergence order of PI rules does not agree, contrary to what happens with ODEs, with the degree of the polynomial approximation. Just to clarify, consider a FDE with order $0 < \alpha < 1$; while a constant piecewise interpolation generates, as expected, methods converging with order 1, a convergence order not greater than $1 + \alpha$ is instead in general achieved by piecewise polynomials of any higher degree.

Dixon was one of the first authors to study in details this drop in the convergence order [14] and, successively, similar convergence properties were observed in the analysis of the predictor–corrector method for FDEs [12].

Some authors have proposed the use of nonuniform meshes to regain the order expected from the polynomials degree, with a distribution of nodes depending on the smoothness of the solution at the endpoints [58]; in [3] it has been showed that graded meshes are particularly suitable to deal with the weak singularity in (7). The computation of weights over nonuniform meshes is however more expensive and, moreover, in several applications the source term can be only sampled at specific given instants, thus making the use of nonuniform meshes unfeasible.

Besides, PI rules share with classical Adams methods for ODEs the same stability issues; for instance, it has been proved [40] that for $\alpha = \frac{1}{2}$ unconditional stability is guaranteed only for methods generated on the basis of polynomials of degree not greater than 2; in other words, the well–known second Dahlquist barrier holds also for PI rules and FDEs.

As a result of all these downsides, only PI rules developed on the basis of very low degree polynomials are actually employed in the numerical treatment of FDEs.

An alternative approach which allows, at the same time, to overcome stability issues and generate higher order methods (at least for linear problems) consists in applying the PI approach in the integral formulation (8) instead of (7). This means that the polynomial replacement applies just to the source term $f(t)$ of (1) which, in several applications, turns out to be smooth enough to be approximated in a satisfactory way by higher degree polynomials.

As it will be clear later, a polynomial replacement of $f(t)$ in (8) leads to special quadrature rules with some ML functions in the weights. Since these functions are generalization of the exponential function, hereafter we will use the same term *exponential integrator* adopted in the context of ODEs to denote methods of this type for FDEs.

Exponential integrators for FDEs were first described in [21] and [22]. In the following we describe two alternative approaches to generate methods of this type and we analyze their main properties.

4.1. Fractional Exponential Adams Methods

Let us consider, on the interval $[0,T]$, a grid $t_n = nh$ which, just for simplicity, we assume to be equispaced.

For an integer $k \geq 1$, on each subinterval $[t_n, t_{n+1}]$ the source term $f(\tau)$ can be approximated by the polynomial $p_n(t_n + h\theta)$ of degree $k-1$ interpolating the data

$$(t_{n-k+2}, f_{n-k+2}), \ldots, (t_n, f_n), (t_{n+1}, f_{n+1}),$$

where for shortness we denoted $f_n = f(t_n)$. It is a basic result that these polynomials can be represented in terms of classical backward differences

$$\begin{aligned} \nabla^0 f_n &= f_n \\ \nabla^\ell f_n &= \nabla^{\ell-1} f_n - \nabla^{\ell-1} f_{n-1}, \quad \ell = 1, 2, \ldots \end{aligned}$$

according to the formula

$$p_n(t_n + h\theta) =: P_n(\theta) = \sum_{\ell=0}^{k-1} (-1)^\ell \binom{-\theta+1}{\ell} \nabla^\ell f_{n+1},$$

where the binomial coefficients are defined in the usual way

$$\binom{-\theta+1}{\ell} = \begin{cases} 1, & \ell = 0 \\ \dfrac{(-\theta+1)(-\theta)\cdots(-\theta-\ell+2)}{\ell!}, & \ell \geq 1. \end{cases}$$

To derive a numerical scheme for (1-2), in the integral formulation (8) we replace $f(\tau)$ by its polynomial interpolation and in each interval we operate the change of variable $\tau = t_j + \theta h$. Thus, for each $t = t_n$ we obtain an approximation y_n to the solution $y(t_n)$ in the form

$$\begin{aligned} y_n &= Y_\alpha(t_n; A) + h \int_{-k+2}^{1} e_{\alpha,\alpha}(t_n - t_{k-2} - h\theta; A) P_{k-2}(\theta)\, d\theta \\ &\quad + h \sum_{j=k-1}^{n-1} \int_0^1 e_{\alpha,\alpha}(t_n - t_j - h\theta; A) P_j(\theta)\, d\theta \end{aligned}$$

for $Y_\alpha(t_n; A)$ defined in (3); notice that the second summation vanishes when $k < 2$ (when $k \geq 2$, in order to properly initialize the computation, the same polynomial $P_{k-2}(\theta)$ is used on $[0, t_{k-1}]$).

After putting, for convenience,

$$\begin{aligned}\hat{e}_{\alpha,\ell}(t;A) &= (-1)^{\ell} h^{1-\alpha} \int_{-k+2}^{1} e_{\alpha,\alpha}(t-h\theta;A) \binom{-\theta+1}{\ell} d\theta \\ \tilde{e}_{\alpha,\ell}(t;A) &= (-1)^{\ell} h^{1-\alpha} \int_{0}^{1} e_{\alpha,\alpha}(t-h\theta;A) \binom{-\theta+1}{\ell} d\theta\end{aligned} \tag{13}$$

we obtain for $n \geq k-1$ the following convolution quadrature rule

$$y_n = Y_\alpha(t_n;A) + h^\alpha \sum_{\ell=0}^{k-1} \left(\hat{e}^{[k]}_{\alpha,\ell}(t_n - t_{k-2};A)\nabla^\ell f_{k-1} + \sum_{j=k-1}^{n-1} \tilde{e}_{\alpha,\ell}(t_n - t_j;A)\nabla^\ell f_{j+1} \right) \tag{14}$$

where, again, the first term in the summation must be omitted when $k < 2$.

Methods (14) are the natural extension to non–integer order equations of the exponential multistep methods of Adams–type investigated in [35]; they are therefore named as k–step *fractional exponential Adams methods*.

The convolution weights $\tilde{e}_{\alpha,\ell}(t;A)$ in (13) can be determined in a simpler and algebraic way by means of the following result.

Proposition 3. *There exist some sequences of coefficients $c_{\nu,\ell}$, $c^{[k]}_{\nu,\ell}$ and $d_{\nu,\ell}$, $\nu = 0,\ldots,\ell$, such that the weights of the convolution quadrature (14) can be expressed as*

$$\tilde{e}_{\alpha,\ell}(t_n - t_j;A) = \sum_{\nu=0}^{\ell} \left(c_{\nu,\ell} e_{\alpha,\alpha+\nu+1}(n-j;h^\alpha A) - d_{\nu,\ell} e_{\alpha,\alpha+\nu+1}(n-j-1;h^\alpha A) \right)$$

and

$$\hat{e}^{[k]}_{\alpha,\ell}(t_n - t_{k-2};A) = \sum_{\nu=0}^{\ell} \left(c^{[k]}_{\nu,\ell} e_{\alpha,\alpha+\nu+1}(n;h^\alpha A) - d_{\nu,\ell} e_{\alpha,\alpha+\nu+1}(n-k+1;h^\alpha A) \right).$$

Proof. Primarily expand the binomial coefficients as polynomials with respect to θ according to

$$(-1)^\ell \binom{-\theta+1}{\ell} = \sum_{\nu=0}^{\ell} b_{\nu,\ell}\theta^\nu,$$

for some coefficients $b_{\nu,\ell}$ which can be easily determined in an algebraic way. Moreover, since $t_n - t_j = (n-j)h$, by using (9) we can write from (13)

$$\tilde{e}_{\alpha,\ell}(t_n - t_j;A) = \sum_{\nu=0}^{\ell} b_{\nu,\ell} \int_0^1 e_{\alpha,\alpha}(n-j-\theta;h^\alpha A)\theta^\ell \, d\theta. \tag{15}$$

After splitting the above integral as $\int_0^1 = \int_0^{n-j} - \int_1^{n-j}$, the proof for $\tilde{e}_{\alpha,\ell}(t_n - t_j;A)$ easily follows thanks to the application of Lemma 2. Similarly for $\hat{e}^{[k]}_{\alpha,\ell}(t_n - t_{k-2};A)$. □

The explicit formulation for the coefficients of Proposition 3 can be easily computed by means of standard algebraic manipulations; in Table 1, for $\ell \in \{0,1,2,3,4\}$ we present the first instances of $c_{\nu,\ell}$ and $d_{\nu,\ell}$, $\nu = 0,\ldots,\ell$ and the first instances of $\{c^{[k]}_{\nu,\ell}\}_{\nu=0,\ell}$ for $\ell = 0,\ldots,k-1$ and $k \in \{2,3,4\}$ are shown in Table 2.

Table 1. First instances of the coefficients $\{c_{\nu,\ell}\}_{\nu=0,\ell}$ and $\{d_{\nu,\ell}\}_{\nu=0,\ell}$ for $\ell = 0,1,2,3,4$

ν	$c_{\nu,0}$	$c_{\nu,1}$	$c_{\nu,2}$	$c_{\nu,3}$	$c_{\nu,4}$
0	1	-1	0	0	0
1		1	$-\frac{1}{2}$	$-\frac{1}{6}$	$-\frac{1}{12}$
2			1	0	$-\frac{1}{12}$
3				1	$\frac{1}{2}$
4					1

ν	$d_{\nu,0}$	$d_{\nu,1}$	$d_{\nu,2}$	$d_{\nu,3}$	$d_{\nu,4}$
0	1	0	0	0	0
1		1	$\frac{1}{2}$	$\frac{1}{3}$	$\frac{1}{4}$
2			1	1	$\frac{11}{12}$
3				1	$\frac{3}{2}$
4					1

Table 2. First instances of the coefficients $\{c^{[k]}_{\nu,\ell}\}_{\nu=0,\ell}$ for $\ell = 0,\ldots,k-1$ and $k \in \{2,3,4\}$

ν	$c^{[2]}_{\nu,0}$	$c^{[2]}_{\nu,1}$
0	1	-1
1		1

ν	$c^{[3]}_{\nu,0}$	$c^{[3]}_{\nu,1}$	$c^{[3]}_{\nu,2}$
0	1	-2	1
1		1	$-\frac{3}{2}$
2			1

ν	$c^{[4]}_{\nu,0}$	$c^{[4]}_{\nu,1}$	$c^{[4]}_{\nu,2}$	$c^{[4]}_{\nu,3}$
0	1	-3	3	-1
1		1	$-\frac{5}{2}$	$\frac{11}{6}$
2			1	-2
3				1

Example 4. Fractional exponential Euler method. *The simplest fractional exponential Adams method is a 1–step method which reads as*

$$y_n = Y_\alpha(t_n; A) + h^\alpha \sum_{j=0}^{n-1} \left(e_{\alpha,\alpha+1}(n-j; h^\alpha A) - e_{\alpha,\alpha+1}(n-j-1; h^\alpha A) \right) f(t_{j+1}).$$

When $A = 0$, since it is $e_{\alpha,\alpha+1}(n;0) = n^\alpha/\Gamma(\alpha+1)$, the method corresponds to the classical product rectangular rule studied, for instance, in [6, 12, 19].

The case $\alpha = 1$ corresponds to the classical exponential Euler method. Thanks indeed to the recursive relation [26]

$$e_{\alpha,\alpha+\beta}(t;\lambda) = \frac{e_{\alpha,\beta}(t;\lambda) - e_{\alpha,\beta}(t;0)}{-\lambda}, \quad e_{\alpha,\beta}(t;0) = \frac{t^{\beta-1}}{\Gamma(\beta)},$$

which holds for any $\beta \geq 1$, for $\alpha = \beta = 1$ it is

$$e_{1,2}(t;\lambda) = \frac{e_{1,1}(t;\lambda) - 1}{-\lambda} = \frac{e^{-t\lambda} - 1}{-\lambda} = t\varphi_1(-t\lambda)$$

and a standard manipulation allows to obtain (12) (with the only difference of the source term evaluated at t_{n+1} instead of t_n).

Example 5. 2–step fractional exponential trapezoidal method. *The use of a linear piece-wise interpolation to approximate $f(t)$ leads to the exponential counterpart of the trape-zoidal product integration rule*

$$y_n = Y_\alpha(t_n; A) + h^\alpha \tilde{W}_n f_0 + h^\alpha \sum_{j=1}^{n} W_{n-j} f(t_j),$$

where

$$\begin{aligned}
\tilde{W}_n &= e_{\alpha,\alpha+1}(n; h^\alpha A) - e_{\alpha,\alpha+2}(n; h^\alpha A) + e_{\alpha,\alpha+2}(n-1; h^\alpha A)\\
W_0 &= e_{\alpha,\alpha+2}(1; h^\alpha A)\\
W_n &= e_{\alpha,\alpha+2}(n-1; h^\alpha A) - 2e_{\alpha,\alpha+2}(n; h^\alpha A) + e_{\alpha,\alpha+2}(n+1; h^\alpha A), \quad n = 1, 2, \ldots
\end{aligned}$$

(it is immediate to verify that $A = 0$ leads to the classical trapezoidal product integration rule).

The way in which stability issues are overcome is straightforward. Indeed, in most of the applications in which system (1) is involved, the source of the stiffness is in the linear term $Ay(t)$ whose contribution to the solution is dealt in an almost exact way (we will discuss this topic more in–depth later) by means of the ML function in (8), (13) and (15).

Methods (14) have also the remarkable feature to allow the computation of the solution $y(t)$ directly at any time instant t_n without the need to previously compute the solution over all the grid–points. This is an obvious consequence of recasting the differential equation into the integral (8) but it is of extreme importance in reducing the computational cost.

Another capability of the exponential approach lies on the improvement of the conver-gence properties. To this purpose we must first provide a preliminary result which will be used also later.

Lemma 6. *Let $\omega > -1$, $g(t) = (1-t)^\omega \psi(t)$, where $\psi(t) \in C^1([0,1])$, and $t_j = j\tilde{h}$, $j = 0, 1, \ldots, n$ an equispaced grid of $n+1$ points on $[0,1]$ with stepsize $\tilde{h} = 1/n$. For $|\gamma| < 1$ consider the quadrature rule*

$$R^{[n,\gamma]} g = \tilde{h} \sum_{j=0}^{n-1} g\left(t_j + \frac{\gamma+1}{2}\tilde{h}\right)$$

for approximating the integral $Ig = \int_0^1 g(t)dt$. The error $E^{[n,\gamma]}g = R^{[n,\gamma]}g - Ig$ is

$$E^{[n,\gamma]} g = -\frac{\gamma}{2}\psi(0)\tilde{h} + \bar{\xi}\left(-\omega, -\frac{1+\gamma}{2}\right)\psi(1)\tilde{h}^{\omega+1} + \mathcal{O}(\tilde{h}^2)$$

with $\bar{\xi}(s,a)$ a periodic generalization of the Hurwitz zeta function $\xi(s,a)$ [51] defined by $\bar{\xi}(s,a) = \xi(s,\bar{a})$, with $a - \bar{a} =$ integer, $0 < \bar{a} \le 1$.

Proof. The proof follows from [41, Eq. (7.3)] where the first coefficient $-\frac{\gamma}{2}$ in the expan-sion of $E^{[n,\gamma]}g$ has been obtained by applying formula 1.441(3) in [27]. □

We point out that coefficients in the term $\mathcal{O}(\tilde{h}^2)$ in the above Lemma depend on the derivative of $\psi(t)$ in the endpoints of $[0,1]$ and this justifies the requirement of continuity of $\psi(t)$ on $[0,1]$.

To study the converge properties of the method (14) we denote the error at the step t_n with $E_n = y(t_n) - y_n$ and we present the following result.

Theorem 7. *Let $f \in C^k([0,T])$. The error of the k–step fractional exponential Adams method (14) is given by*

$$E_n = C_k h^k e_{\alpha,\alpha+1}(t_n; A) f^{(k)}(\eta_n) + \mathcal{O}(h^{k+\alpha})$$

for $\eta_n \in [0, t_n]$ and C_k a constant which does not depend on h and n.

Proof. By classical results on the remainder of polynomial interpolation we preliminarily observe that

$$\begin{aligned} E_n &= \sum_{j=0}^{k-2} \int_{t_j}^{t_{j+1}} e_{\alpha,\alpha}(t_n - \tau; A) \frac{\omega_{k-2,k-1}(\tau)}{k!} f^{(k)}(t_{k-1} - \xi_{k-2}(\tau)h) d\tau + \\ & \sum_{j=k-1}^{n-1} \int_{t_j}^{t_{j+1}} e_{\alpha,\alpha}(t_n - \tau; A) \frac{\omega_{j,k-1}(\tau)}{k!} f^{(k)}(t_{j+1} - \xi_j(\tau)h) d\tau, \end{aligned}$$

where $\omega_{j,k-1}(\tau) = (\tau - t_{j+1})(\tau - t_j) \cdots (\tau - t_{j-k+2})$, $\xi_{k-2}(\tau) \in [0, k-1]$ and $\xi_j(\tau) \in [0,1]$. The first of the two terms in E_n vanishes when $k < 2$ and it is otherwise clearly proportional to $\mathcal{O}(h^{k+1+\alpha})$. Thus, by means of the change of variable $\tau = t_j + \theta h$ and using the scaling (9) together with the Taylor's formula, we obtain

$$E_n = h^{k+\alpha} \sum_{j=0}^{n-1} \int_0^1 e_{\alpha,\alpha}(n - j - \theta; h^\alpha A) \frac{\omega_{k-1}(\theta)}{k!} f^{(k)}(t_j + \theta h) d\theta + \mathcal{O}(h^{k+1+\alpha})$$

with $\omega_{k-1}(\theta) = (\theta - 1)\theta \cdots (\theta + k - 2)$. We denote now for shortness $\bar{f}(s) = f^{(k)}(sh)$ and, after using the series expansion defining the ML function (4), we can write the error as

$$E_n = \frac{h^{k+\alpha}}{k!} \sum_{\ell=0}^{\infty} \frac{h^{\alpha\ell} A^\ell}{\Gamma(\alpha\ell + \alpha)} \int_0^1 \omega_{k-1}(\theta) \sum_{j=0}^{n-1} (n - j - \theta)^{\alpha(\ell+1)-1} \bar{f}(j + \theta) d\theta + \mathcal{O}(h^{k+1+\alpha}).$$

We observe that for $\beta > -1$ the definite integral in $[0,1]$ of the function $g(t) = (1 - t)^\beta \bar{f}(nt)$

$$Ig = \int_0^1 g(t) dt = \frac{1}{n^{\beta+1}} \int_0^n (n - \tau)^\beta \bar{f}(\tau) d\tau$$

can be approximated by the general trapezoidal quadrature rule $R^{[n,\gamma]}g$ of Lemma 6

$$R^{[n,\gamma]}g = \frac{1}{n^{\beta+1}} \sum_{j=0}^{n-1} (n - j - \theta)^\beta \bar{f}(j + \theta)$$

for $\delta = 2\theta - 1$. By means of the error expansion in Lemma 6 it is immediate to verify

$$\sum_{j=0}^{n-1}(n-j-\theta)^{\beta}\bar{f}(j+\theta) = \int_0^n (n-\tau)^{\beta}\bar{f}(\tau)d\tau + n^{\beta}\Big(\frac{1}{2}-\theta\Big)\bar{f}(0) + C_{\beta},$$

for C_{β} a constant depending on β and on the value of $\bar{f}(0)$ but not on n. We can use this result in the expansion of the error E_n in order to write

$$\begin{aligned} E_n &= \frac{h^{k+\alpha}}{k!}\sum_{\ell=0}^{\infty}\frac{h^{\alpha\ell}A^{\ell}}{\Gamma(\alpha\ell+\alpha)}\bar{C}_k\int_0^n (n-\tau)^{\alpha(\ell+1)-1}\bar{f}(\tau)d\tau \\ &\quad + \frac{h^{k+\alpha}}{k!}\sum_{\ell=0}^{\infty}\frac{h^{\alpha\ell}A^{\ell}}{\Gamma(\alpha\ell+\alpha)}n^{\alpha(\ell+1)-1}\bar{f}(0)\tilde{C}_k + \mathcal{O}\big(h^{k+\alpha}\big) \end{aligned}$$

where for shortness we put

$$\bar{C}_k = \int_0^1 \omega_{k-1}(\theta)d\theta, \quad \tilde{C}_k = \int_0^1 \omega_{k-1}(\theta)\Big(\frac{1}{2}-\theta\Big)d\theta.$$

By the mean value theorem we deduce the existence of $\eta \in (0,n)$ such that

$$E_n = \frac{h^k}{k!}\bar{C}_k\sum_{\ell=0}^{\infty}\frac{h^{\alpha\ell+\alpha}n^{\alpha\ell+\alpha}A^{\ell}}{\Gamma(\alpha\ell+\alpha+1)}\bar{f}(\eta) + \frac{h^{k+\alpha}}{k!}\sum_{\ell=0}^{\infty}\frac{h^{\alpha\ell}n^{\alpha(\ell+1)-1}A^{\ell}}{\Gamma(\alpha\ell+\alpha)}\bar{f}(0)\tilde{C}_k + \mathcal{O}\big(h^{k+\alpha}\big)$$

from which it is immediate to see that

$$E_n = \frac{h^k}{k!}\bar{C}_k e_{\alpha,\alpha+1}(t_n;A)\bar{f}(\eta) + \frac{h^{k+1}}{k!}\tilde{C}_k e_{\alpha,\alpha}(t_n;A)\bar{f}(0) + \mathcal{O}\big(h^{k+\alpha}\big)$$

which concludes the proof after putting $C_k = \bar{C}_k/k!$. □

As we clearly see, under the assumption of regularity for the source term $f(t)$, it is possible to obtain high order methods by simply increasing the number of the steps as in classical multistep methods for ODEs.

This is a remarkable feature of the exponential integrators which makes them especially suited and effective for the application to FDEs. Whilst exponential integrators were introduced for ODEs with the only aim to overcome stability issues, for FDEs the further aim of increasing the convergence order is also achieved.

Also the computational cost is different for the exponential integrators for ODEs and those for FDEs: for the latter, indeed, the required computation of matrix ML functions is even more difficult than the evaluation of the exponential matrix required by the former. Nevertheless, the advantage in terms of convergence properties make the exponential integrators for FDEs even more fruitful than the ODEs' counterpart.

4.2. Fractional Exponential Quadrature Rules

One of the distinctive features of the fractional exponential Adams methods presented in the Subsection 4.1 is that the forcing term is evaluated just at well–established time–steps. This feature is necessary in several applications when the input f of the system is known, or can be evaluated, only at some given, and often uniformly distributed, points in the time interval.

On the contrary, when the values of f are available at any time, some slightly different methods can be devised in order to further improve, in some cases, the convergence properties. This alternative approach mainly consists in considering interpolant polynomials on intermediate nodes with respect to the grid $t_n = nh$.

Given k distinct points $0 \le c_1 < c_2 < \cdots < c_k \le 1$ we consider on each subinterval $[t_n, t_{n+1}]$ the polynomial of degree $k-1$ interpolating the source term $f(t)$ at the nodes $t_n + c_\ell h$, $\ell = 1, \dots, k$. This polynomial can be formulated in terms of the classical Lagrangian basis

$$p_n(t_n + \theta h) =: P_n(\theta) = \sum_{\ell=1}^{k} L_\ell^{[k]}(\theta) f(t_n + c_\ell h), \quad L_\ell^{[k]}(\theta) = \prod_{\substack{\nu=1,k \\ \nu\neq\ell}} \frac{\theta - c_\ell}{c_\nu - c_\ell}.$$

The piecewise replacement of $f(t)$ by $p_n(t_n + \theta h)$ in the integral representation (8) together with the change of variable $\tau = t_j + \theta h$ and the use of Lemma 1 lead to the approximation

$$y_n = Y_\alpha(t_n; A) + h^\alpha \sum_{j=0}^{n-1} \sum_{\ell=1}^{k} \omega_{\alpha,\ell}^{[k]}(n-j; h^\alpha A) f(t_j + c_\ell h), \tag{16}$$

where

$$\omega_{\alpha,\ell}^{[k]}(n-j; h^\alpha A) = \int_0^1 e_{\alpha,\alpha}(n-j-\theta; h^\alpha A) L_\ell^{[k]}(\theta)\, d\theta. \tag{17}$$

In [34] it was introduced the term *exponential quadrature* for rules obtained on the basis of the same principle in the field of linear ODEs. This is the reason for which we propose the same name also for fractional–order systems

The convolution weights $\omega_{\alpha,\ell}^{[k]}(n; h^\alpha A)$ are more difficult to be represented by means of algebraic coefficients as in the case of fractional exponential Adams methods. We discuss their derivation on a case by case basis.

Example 8. 1-step exponential quadrature rule. *In the simplest case in which $k = 1$, since $L_1^{[1]}(\theta) \equiv 1$ we readily have*

$$\int_0^1 e_{\alpha,\alpha}(n-\theta; h^\alpha A)\, d\theta = e_{\alpha,\alpha+1}(n; h^\alpha A) - e_{\alpha,\alpha+1}(n-1; h^\alpha A)$$

and the weights $\omega_{\alpha,1}^{[1]}(n-j; h^\alpha A)$ are the same of the fractional Euler exponential method of Example 4. Anyway, we will show later how the capability of changing the points $t_n + c_1 h$ in which f is evaluated can be exploited to improve the accuracy of the method.

Example 9. **2-step exponential quadrature rule.** *When two distinct intermediate points c_1 and c_2 are selected, the weights of the resulting rule can be readily evaluated as*

$$\begin{aligned}\omega^{[2]}_{\alpha,1}(n;h^\alpha A) &= \frac{1}{c_1-c_2}\Big(e_{\alpha,\alpha+2}(n;h^\alpha A) - e_{\alpha,\alpha+2}(n-1;h^\alpha A) - c_2 e_{\alpha,\alpha+1}(n;h^\alpha A) \\ &\quad - (1-c_2)e_{\alpha,\alpha+1}(n-1;h^\alpha A)\Big) \\ \omega^{[2]}_{\alpha,2}(n;h^\alpha A) &= \frac{1}{c_2-c_1}\Big(e_{\alpha,\alpha+2}(n;h^\alpha A) - e_{\alpha,\alpha+2}(n-1;h^\alpha A) - c_1 e_{\alpha,\alpha+1}(n;h^\alpha A) \\ &\quad - (1-c_1)e_{\alpha,\alpha+1}(n-1;h^\alpha A)\Big).\end{aligned}$$

In the special case $c_1 = 0$ and $c_2 = 1$ we obtain the 2–step fractional exponential trapezoidal method presented in the Example 5.

The convergence properties of the exponential quadrature rule (16) can be studied in a similar way as in Theorem 7. To be thorough we present here the main result.

Theorem 10. *Let $\alpha > 0$ and $f(t) \in C^k([0,T])$. The error of the fractional exponential quadrature rule (16-17) on k nodes is given by*

$$E_n = h^k \frac{C^{[k]}_0}{k!} e_{\alpha,\alpha+1}(t_n;A) f^{(k)}(\eta_n) + R_n,$$

where $\eta_n \in [0,t_n]$ and

$$R_n = \frac{h^{k+1}}{k!} C^{[k]}_1 e_{\alpha,\alpha}(t_n;A) f^{(k)}(0) + \frac{h^{k+\alpha}}{k!} \sum_{\nu=0}^{\infty} \frac{h^{\alpha\nu} C^{[k]}_{\alpha+\alpha\nu} A^\nu}{\Gamma(\alpha\nu+\alpha)} f^{(k)}(t_n) + \mathcal{O}(h^{k+2})$$

with all the constant $C^{[k]}_\nu$, $\nu \in \{0,1,\alpha,2\alpha,\ldots\}$ depending only on the nodes c_ℓ and not on n and h.

Proof. We first express the error E_n in terms of the remainders $r_j(\tau) = f(\tau) - p_j(\tau)$ of the piecewise polynomial interpolation

$$E_n = \sum_{j=0}^{n-1} \int_{t_j}^{t_n} e_{\alpha,\alpha}(t_n-\tau;A) r_j(\tau)\, d\tau$$

and hence, by using the series expansion (4) in the definition of the ML function, we can provide the following formulation

$$E_n = \sum_{\nu=0}^{\infty} \frac{A^\nu}{\Gamma(\alpha\nu+\alpha)} E^{[\nu]}_n, \quad E^{[\nu]}_n = \sum_{j=0}^{n-1} \int_{t_j}^{t_{j+1}} (t_n-\tau)^{\alpha-1+\alpha\nu} r_j(\tau)\, d\tau \qquad (18)$$

with $E^{[\nu]}_n$ the error of the quadrature rule based on the same piecewise polynomial interpolation but applied to the integral of the function $(t_n-\tau)^{\alpha-1+\alpha\nu} f(\tau)$ on the same interval

$[0, t_n]$. By the change of variable $\tau = ut_n$ we can scale this integration from $[0, t_n]$ to the interval $[0, 1]$ and hence

$$E_n^{[\nu]} = t_n^{\alpha+\alpha\nu} \sum_{j=0}^{n-1} \int_{\frac{j}{n}}^{\frac{j+1}{n}} (1-u)^{\alpha-1+\alpha\nu} \tilde{r}_j(u)\, du$$

with $\tilde{r}_j(u) = r_j(ut_n)$ the remainder between $\tilde{f}(u) = f(ut_n)$ and its piecewise polynomial interpolation on the subinterval $[j/n, j/(n+1)]$. The application of Lemma 2.1 in [9] leads to

$$\sum_{j=0}^{n-1} \int_{\frac{j}{n}}^{\frac{j+1}{n}} (1-u)^{\alpha-1+\alpha\nu} \tilde{r}_j(u)\, du = \frac{\tilde{h}^k}{k!} \int_0^1 \omega_k(u) \tilde{R}(u)\, du + \mathcal{O}(\tilde{h}^{k+1})$$

where $\tilde{h} = 1/n$, $\omega_k(u) = (u - c_1)(u - c_2) \cdots (u - c_k)$ and

$$\tilde{R}(u) = \tilde{h} \sum_{j=0}^{n-1} \left(1 - \frac{j}{n} - u\tilde{h}\right)^{\alpha-1+\alpha\nu} \tilde{f}^{(k)}\left(\frac{j}{n} + u\tilde{h}\right). \tag{19}$$

For a fixed $u \in [0, 1]$, the summation formula (19) is a generalized Euler approximation to $\int_0^1 (1-s)^{\alpha-1+\alpha\nu} \tilde{f}^{(k)}(s) ds$ and the application of Lemma 6, with $\gamma = 2u - 1$, provides the asymptotic estimation

$$\begin{aligned} \tilde{R}(u) = & \int_0^1 (1-s)^{\alpha-1+\alpha\nu} \tilde{f}^{(k)}(s) ds + \left(\frac{1}{2} - u\right) \tilde{f}^{(k)}(0) n^{-1} \\ & + \bar{\xi}(-\alpha + 1 - \alpha\nu, -u) n^{-\alpha(\nu+1)} \tilde{f}^{(k)}(1) + \mathcal{O}(n^{-2}) \end{aligned}$$

and hence, after observing that $\bar{\xi}(1-\nu, -u) = \xi(1-\nu, 1-u)$, we are able to write

$$\begin{aligned} \sum_{j=0}^{n-1} \int_{\frac{j}{n}}^{\frac{j+1}{n}} (1-u)^{\alpha-1+\alpha\nu} \tilde{r}_j(u)\, du = & \frac{\tilde{h}^k}{k!} \Big(C_0^{[k]} \int_0^1 (1-s)^{\alpha\nu+\alpha-1} \tilde{f}^{(k)}(s) ds \\ & + C_1^{[k]} n^{-1} \tilde{f}^{(k)}(0) + C_{\alpha+\alpha\nu}^{[k]} n^{-\alpha(\nu+1)} \tilde{f}^{(k)}(1) + \mathcal{O}(n^{-2}) \Big) + \mathcal{O}(\tilde{h}^{k+1}) \end{aligned}$$

where for shortness we denoted

$$C_0^{[k]} = \int_0^1 \omega_k(u) du, \quad C_1^{[k]} = \int_0^1 \omega_k(u) \left(\frac{1}{2} - u\right) du,$$

$$C_\nu^{[k]} = \int_0^1 \omega_k(u) \xi(1-\nu, 1-u) du, \quad \nu \in \mathbb{R}.$$

We can now turn back to the change of variable $u = \tau/t_n$ and, after observing that $\tilde{f}^{(k)}(0) = t_n^k f^{(k)}(0)$, $\tilde{f}^{(k)}(1) = t_n^k f^{(k)}(t_n)$ and $\tilde{h} t_n = h$, we can express $E_n^{[\nu]}$ as

$$\begin{aligned} E_n^{[\nu]} = & \frac{h^k}{k!} \Big(C_0^{[k]} \int_0^{t_n} (t_n - \tau)^{\alpha-1+\alpha\nu} f^{(k)}(\tau)\, d\tau + \\ & + n^{-1} t_n^{\alpha+\alpha\nu} C_1^{[k]} f^{(k)}(0) + n^{-\alpha-\alpha\nu} t_n^{\alpha+\alpha\nu} C_{\alpha+\alpha\nu}^{[k]} f^{(k)}(t_n) + \mathcal{O}(n^{-2}) \Big) + \mathcal{O}(\tilde{h}^{k+1}) \end{aligned}$$

and hence

$$E_n^{[\nu]} = \frac{h^k}{k!}\Big(C_0^{[k]} \int_0^{t_n} (t_n - \tau)^{\alpha-1+\alpha\nu} f^{(k)}(\tau)\, d\tau +$$
$$+ h t_n^{\alpha-1+\alpha\nu} C_1^{[k]} f^{(k)}(0) + h^{\alpha+\alpha\nu} C_{\alpha+\alpha\nu}^{[k]} f^{(k)}(t_n)\Big) + \mathcal{O}\big(h^{k+1}\big);$$

by inserting in (18) and using the mean value theorem for integrals we can conclude the existence of a value $\eta_n \in [0, t_n]$ such that

$$E_n = \frac{h^k}{k!} C_0^{[k]} f^{(k)}(\eta_n) \int_0^{t_n} e_{\alpha,\alpha}(t_n - \tau; A)\, d\tau + R_n$$

from which the proof immediately follows after applying Lemma 2. □

As expected, the convergence properties of the fractional exponential quadrature rules are comparable to those of the fractional exponential Adams methods as stated in Theorem 7. The more detailed formulation for the $\mathcal{O}\big(h^{k+\alpha}\big)$ term R_n in the error suggests that the accuracy of the method could be further improved; the coefficients $C_\nu^{[k]}$ indeed depend only on the nodes and thus a focused choice of them can zero out some of these coefficients to increase the order. To characterize these methods we need to distinguish some different cases.

Case 1) $0 < \alpha < \frac{1}{2}$**:** the first terms in the expansion of the error E_n are

$$E_n = h^k \frac{C_0^{[k]}}{k!} e_{\alpha,\alpha+1}(t_n; A) f^{(k)}(\eta_n) + h^{k+\alpha} \frac{C_\alpha^{[k]}}{k!\Gamma(\alpha)} f^{(k)}(t_n) + \mathcal{O}\big(h^{k+2\alpha}\big)$$

and hence by selecting nodes such that $C_0^{[k]} = C_\alpha^{[k]} = 0$ it is possible to obtain methods with an error proportional to $\mathcal{O}\big(h^{k+2\alpha}\big)$.

Case 2) $\frac{1}{2} < \alpha < 1$**:** the first terms in the expansion of E_n are

$$E_n = h^k \frac{C_0^{[k]}}{k!} e_{\alpha,\alpha+1}(t_n; A) f^{(k)}(\eta_n) + h^{k+\alpha} \frac{C_\alpha^{[k]}}{k!\Gamma(\alpha)} f^{(k)}(t_n) + \mathcal{O}\big(h^{k+1}\big)$$

and choosing nodes such that $C_0^{[k]} = C_\alpha^{[k]} = 0$ will provide a method with an error proportional to $\mathcal{O}\big(h^{k+1}\big)$.

Case 3) $\alpha > 1$**:** the first terms in the expansion of E_n are

$$E_n = h^k \frac{C_0^{[k]}}{k!} e_{\alpha,\alpha+1}(t_n; A) f^{(k)}(\eta_n) + h^{k+1} \frac{C_1^{[k]}}{k!} e_{\alpha,\alpha+1}(t_n; A) f^{(k)}(0) + \mathcal{O}\big(h^{k+\alpha}\big)$$

and the selection of nodes with $C_0^{[k]} = C_1^{[k]} = 0$ leads to methods with an error proportional to $h^{k+\alpha}$.

The most convenient way to make $C_0^{[k]} = 0$ for odd k is to choose symmetric nodes with respect to the middle of the interval $[0, 1]$. With even k it is instead necessary to solve some simple algebraic equations. Thus, classical examples of methods with $C_0^{[k]} = 0$ are

$$\begin{aligned} k = 1: &\quad c_1 = 1/2 \\ k = 2: &\quad c_1 \in [0,1],\ c_2 = \tfrac{3c_1-2}{3(2c_2-1)} \\ k = 3: &\quad c_1 \in [0,1],\ c_2 = 1/2,\ c_3 = 1 - c_1. \end{aligned}$$

By means of an algebraic computation, we can see that the combined requirement of having $C_0^{[k]} = C_1^{[k]} = 0$ is obtained for $k = 2$ by using $c_1 = (3 - \sqrt{3})/6$ and $c_2 = (3 + \sqrt{3})/6$ while when $k = 3$ it is necessary to select $c_1 = (1 - \sqrt{3/5})/2$, $c_2 = 1/2$ and $c_3 = (1 + \sqrt{3/5})/2$.

The task of setting $C_\alpha^{[k]}$ to zero can be quite difficult. Not only the points depend on α but their computation involves the knowledge of some properties of the Hurwitz zeta function. We refer to [22] where this topic is discussed in more details and examples of nodes for some values of α are also provided.

However selecting nodes with the only aim of increasing the convergence properties is not always the best choice. For $k = 3$, for instance, imposing $c_1 = 0$ and $c_3 = 1$ allows to reduce the computational effort since the forcing term is evaluated on a smaller number of points and the computation can be also organized to reduce redundant operations. Thus it is preferable to avoid imposing also $C_\alpha^{[k]} = 0$ (when $0 < \alpha < 1$) or $C_1^{[k]} = 0$ (when $\alpha > 1$) and hence an optimal selection is $c_1 = 0$, $c_2 = 1/2$ and $c_3 = 1$.

4.3. Methods for Semi–Linear Systems

Exponential integrators for fractional–order systems can be in general employed, with minor changes, also to solve semilinear problems in the form

$$ {}_0^C D_t^\alpha y(t) + Ay(t) = f(t, y(t)), \tag{20}$$

where, as usual, the non linear term $f(t, y(t))$ is assumed non stiff.

It is worthwhile to note that the application of the exponential quadrature rules of Subsection 4.2 to (20) involves some difficulties, related to the fact that the approximation of the source term on non–grid points needs the knowledge of the solution at these points. Indeed, when applied to semilinear problems the exponential quadrature rules actually give rise to a counterpart of the exponential Runge–Kutta methods with the same nontrivial issues which are typical of all Runge–Kutta methods.

On the other side, the use of fractional exponential Adams methods is more direct since it involves only the knowledge of the solution at the grid points t_n and hence it is preferable for semilinear problems.

It is obviously necessary to slightly modify the way in which the rules are devised since it is advisable to generate explicit formulas with the aim of overcoming the computational complexity due to the nonlinear term.

The use of exponential integrators for semilinear systems presents however a not negligible shortcoming. The usual conditions of smoothness for f and (10) are not sufficient to assure the smoothness of the solution $y(t)$ and hence that of the nonlinear term $f(t, y(t))$;

as a consequence, the same drop in the order of convergence of classical product integration will be exhibited by exponential integrators. We do not cover in details convergence properties in this case; we just refer to some related results in [21].

It is important to stress that, due to this restriction on the possible convergence order, just methods based on polynomial approximation of low degree are used in practice with semilinear systems [20].

5. Evaluation of the Matrix Mittag–Leffler Function

When in 1903 the Swedish mathematician Gösta Mittag–Leffler introduced the functions (4) that bear his name, he probably ignored the important role they would have gained in the years. Indeed, although the theoretical properties of these functions have been largely analyzed since then, they have become crucial in the last decades on account of their applications in the analysis and solution of FDEs.

We refer, for instance, to [29, 38, 43, 44, 45, 10, 53] and references therein for a thorough review of the main properties and applications of these functions.

The numerical evaluation of the ML functions is a delicate aspect whose importance goes hand in hand with the development of efficient solvers for FDEs. As described in Section 4, exponential integrators for FDEs require a massive computation of ML functions and the efficiency of this computation affects the effectiveness of the whole procedure.

Recently Podlubny and Kacenak, starting from the work in [25], have offered the very accurate and robust Matlab `mlf` code [54] which evaluates the classical ML function in any region of the complex plane at any prescribed tolerance; anyway this code is not suitable for practical usage in exponential integrators since it works very slowly for some of the parameters; moreover, it does not apply to evaluate the function for matrix arguments but just for the scalar case.

Here we describe two approaches for the computation of $e_{\alpha,\beta}$: the first grounds on the numerical inversion of the Laplace transform; it turns out to be effective and accurate and easily extends to matrix arguments. The second one relies on a rational approximation of the ML function and is tailored for the matrix case.

5.1. The Inversion of the Laplace Transform

The numerical evaluation of special functions by means of the inversion of their Laplace transform has been deeply discussed, for instance, in [37, 63, 64, 62].

This strategy is fruitful whenever the function of interest has a Laplace transform which is easier to compute than the function itself. It thus fits the computation of $e_{\alpha,\beta}(t;\lambda)$, as deepened in [62, 24], since its Laplace transform is

$$F_{\alpha,\beta}(s;\lambda) = \frac{s^{\alpha-\beta}}{s^\alpha + \lambda}, \quad \Re(s) > \lambda^{1/\alpha}$$

(e.g., see [43, 53]). To uniquely define $F_{\alpha,\beta}(s;\lambda)$, which is a multivalued function for the presence of non integer powers, we introduce a branch cut along the negative real semi–axis.

To compute $e_{\alpha,\beta}(t;\lambda)$ the inversion Laplace transform formula holds

$$e_{\alpha,\beta}(t;\lambda) = \frac{1}{2\pi i}\int_{\mathcal{C}} e^{st} F_{\alpha,\beta}(s;\lambda) ds. \tag{21}$$

The computation of $e_{\alpha,\beta}(t;\lambda)$ is thus reduced to the numerical approximation of the contour integral in (21). The contour $\mathcal{C}$ is a suitable deformation of the Bromwich line and common examples are parabola, hyperbola or cotangent contours; it should be as much as possible to the left to hold the factor e^{st} down and avoid instability phenomena; at the same time, if $\mathcal{C}$ is too close to the singularities of $F_{\alpha,\beta}$ the integrand function can blow up and consequently the quadrature error increases exceedingly. It is therefore fundamental to select $\mathcal{C}$ in such a way to maximize the region of analyticity surrounding it.

Once the geometric shape of $\mathcal{C}$ has been fixed, a common strategy is to parametrize the contour $\mathcal{C}$ by a complex function $z(u)$, $-\infty < u < \infty$, and rewrite the integral (21) as

$$e_{\alpha,\beta}(t;\lambda) = \frac{1}{2\pi i}\int_{-\infty}^{\infty} e^{z(u)t} \frac{z(u)^{\alpha-\beta}}{z(u)^{\alpha}+\lambda} z'(u) du .$$

A standard numerical quadrature, as, for example, the trapezoidal rule can be applied to approximate the integral above; thus, given a uniform mesh–grid $u_k = kh$, with step size h, the resulting approximation is

$$e_{\alpha,\beta}^{[h,N]}(t;\lambda) = \frac{h}{2\pi i}\sum_{k=-N}^{N} e^{z(u_k)t} \frac{z(u_k)^{\alpha-\beta}}{z(u_k)^{\alpha}+\lambda} z'(u_k). \tag{22}$$

The quadrature parameters h and N and those defining $z(u)$ are chosen according to an accurate analysis of the error $e_{\alpha,\beta}(t;\lambda) - e_{\alpha,\beta}^{[h,N]}(t;\lambda)$ with the aim of achieving a fixed accuracy ε.

To this purpose, the approach described in [62] gives useful insights. It is grounded on the analysis of the three main components of the error: DE_{+} is the analytical error of the quadrature rule associated with the width of the region of analyticity of the integrand at the left of the contour $\mathcal{C}$; DE_{-} the counterpart of DE_{+} with respect to the region of analyticity of the integrand at the right of the contour; TE is the truncation error due to the fact that in (22) just a finite number of quadrature nodes is considered on the infinite length contour. The procedure proposed in [62] aims to obtain the free parameters by asymptotically balancing the three error factors with the required accuracy ε, that is,

$$DE_{+} = DE_{-} = TE = \varepsilon.$$

We refer to [62] for a detailed description on the way in which the parameters can be chosen in a simple, elegant and computationally inexpensive way. For the ML function the analysis carried out in [62] applied to a restrict range of the parameter β, while [24] (which focuses on parabolic contours) extended the analysis to the more general case.

Thanks to its computational simplicity, this approach works well also with matrix arguments, as required in the implementation of the exponential integration methods we are treating. Indeed, formula (22) can be easily generalized to a matrix argument A. Moreover, generally one needs to approximate $e_{\alpha,\beta}(t;A)y$ for some vector y, as encountered in the description in Section 4. This approximation easily reads as

$$e^{[h,N]}_{\alpha,\beta}(t;A)y = \frac{h}{2\pi i}\sum_{k=-N}^{N} z'(u_k)e^{z(u_k)t}z(u_k)^{\alpha-\beta}(z(u_k)^{\alpha}I + A)^{-1}y$$

which involves the solution of $2N+1$ linear systems.

5.2. Rational Approximation

In the context of matrix functions evaluation, say $f(A)$, an effective approach is often to approximate the function f itself. Moreover, rational approximations are usually desired, especially when vectors like $f(A)y$ are required, since they allow an easier formulation of the problem by resorting to a partial fraction expansion (PFE). In particular, in the case the poles are simple, if $\tau_0,\ldots,\tau_\nu,\xi_1,\ldots,\xi_\nu$ denote the residues and the poles of a rational function $\mathcal{R}_\nu$ then

$$f(A)y \approx \mathcal{R}_\nu(A)y = \tau_0 y + \sum_{j=1}^{\nu}\tau_j(A-\xi_j I)^{-1}y \tag{23}$$

which requires solving ν linear systems with shifted complex matrices of the form $A-\xi_j I$.

Depending on the dimension of the matrix A direct or iterative solvers may be applied to solve the linear systems in (23) whose number, moreover, can be almost halved when the A is real.

The tough part is to determine a good rational approximation and this considerable problem goes back to Chebyshev. Indeed, given a function f continuous on a closed interval, the well known *Chebyshev approximation problem* consists in finding a rational approximation, say $r^* \in R_{\mu,\nu}$, such that

$$\|f-r^*\|_\infty \le \|f-r\|_\infty, \quad \forall\, r \in R_{\mu,\nu},$$

where $R_{\mu,\nu}$ is the set of rational functions of the form

$$r(z) = \left(\sum_{k=0}^{\mu} a_k z^k\right) \Big/ \left(\sum_{k=0}^{\nu} b_k z^k\right).$$

Classical theorems state the existence of r^* while uniqueness is not guaranteed, unless $\nu = 0$; moreover, the characterization of a best approximation remains an issue, even though this is a crucial part for the applications.

A more constructive approach follows from the Carathéodory-Fejér (CF) theory, presented in 1917. The previous problem is slightly gone around since the best approximation $\tilde{r}^*$ to f is now searched in the set $\tilde{R}_{\mu,\nu}$ whose elements are of the form

$$\tilde{r}(z) = \left(\sum_{k=-\infty}^{\mu} a_k z^k\right) \Big/ \left(\sum_{k=0}^{\nu} b_k z^k\right).$$

The CF approximation r^{cf} is then obtained by dropping all terms of negative degree in the numerator of $\tilde{r}^*$.

For this problem existence and uniqueness of the solution are guaranteed; moreover a standard technique can actually construct the function $\tilde{r}^*$. Compared with r^* the CF approximation $\tilde{r}^*$ gives only a *near-best* rational approximation. However, Trefethen [59] showed the closeness of $\tilde{r}^*$ and r^{cf} for smooth functions, thus justifying the use of r^{cf} for actual computations. For the exponential function, for example, Magnus [42] found that, for $\mu = \nu$, r^{cf} and r^* differ only for a factor of $\mathcal{O}(56^{-\nu})$.

In the actual implementation, by a slightly modified version of the algorithm presented in [60], we are able to compute the poles and the residues of the CF approximation in a fast and effective way [23].

5.3. Krylov Subspace Methods

Both methods based on the Laplace transform and on rational approximations require the solution of many linear systems, usually complex and of large dimension. A possible way to ride over this issue is to resort to the *Krylov subspace methods*.

We now describe the basics of this method and we make reference to [56, 31, 57] for the details and further research material.

As their name suggests, the essential of the methods are the Krylov subspaces

$$\mathcal{K}_m(A, y) = \text{span}\{y, Ay, \ldots, A^{m-1}y\}$$

defined as the spaces of dimension m spanned by a given vector $y \in \mathbb{C}^M$, which for simplicity we assume with $\|y\|_2 = 1$, and a matrix $A \in \mathbb{C}^{M\times M}$. The matrix A is projected in these spaces and the strength of this approach is that a small dimension m usually suffices to get an appropriate approximation, as we will see. Two matrices are strictly connected with these spaces: $V_m \in \mathbb{C}^{M\times m}$, whose columns form an orthonormal basis of $\mathcal{K}_m(A, y)$, and an unreduced Hessenberg matrix $H_m \in \mathbb{C}^{m\times m}$ whose entries $h_{i,j}$ are the coefficients of the orthonormalization process. The standard Krylov recurrence formula holds

$$AV_m = V_m H_m + h_{m+1,m} v_{m+1} e_m^T, \quad V_m^* V_m = I_m$$

from which the following approximation derives

$$f(A)y \approx V_m f(H_m) e_1. \tag{24}$$

The ratio of this approximation can be explained by the following two results from Saad [57]. The first states that the approximation is exact for any f which is a polynomial of sufficiently low degree.

Lemma 11 ([57]). *Let $A \in \mathbb{C}^{M\times M}$ and V_m, H_m be the results of m steps of the Arnoldi process on A. Then for any polynomial p_j of degree $j \leq m-1$ the following equality holds*

$$p_j(A)y = V_m p_j(H_m) e_1.$$

The approximation (24) can then be seen as the polynomial interpolation of f on the spectrum of H_m rather than the spectrum of A, as the following theorem states.

Theorem 12 ([57]). *Let V_m, H_m be the results of m steps of the Arnoldi process on $A \in \mathbb{C}^{M\times M}$. Then*

$$V_m f(H_m)e_1 = \tilde{p}_{m-1}(A)y$$

where $\tilde{p}_{m-1}$ is the unique polynomial of degree at most $m-1$ that interpolates f on the spectrum of H_m.

As said before, usually a small dimension m suffices to give an accurate approximation; this makes the evaluation of $f(H_m)$ in general simpler than that of $f(A)$, both for the reduced dimension and for the special structure of H_m. Moreover, when the matrix A is symmetric the resulting approach, known as the *Lanczos* method, further simplifies and the matrix H_m is tridiagonal.

Among the first application of (24) probably the exponential function is the most analyzed [15, 56, 31] but the approach has been extended to more general functions [15, 16]. Recently Moret and Novati [48] have considered the use of Krylov subspace methods for the computation of the Mittag–Leffler matrix function applied to vectors.

The convergence of Krylov subspace methods has been analyzed extensively (we just cite [56, 31] and references therein), however for general A and f it is poorly understood. In some situations, like for the exponential for A Hermitian positive semi–definite or skew–Hermitian, the convergence is superlinear but the number of iterations needed to reach the regime of superlinear convergence strictly depends on the field of values of the matrix argument and, as seen in [31], the convergence of Krylov subspace methods can be quite slow when the spectrum of A is large. In these cases enhancements have been proposed to the standard method, like the *restart* version [17, 49] or the *Rational Arnoldi* method (RAM) we describe in the following section.

5.3.1. Rational Krylov Method

The *Rational Krylov* method was first proposed in [47] and, independently, in [61]. The idea of this method, known as *Shift and Invert* in the context of eigenvalue problems, is to fix a parameter γ to consider the Krylov subspaces generated by $Z = (I + \gamma A)^{-1}$, rather than A, and y. The *Arnoldi* relation for Z and y reads

$$ZU_k = U_k S_k + s_{k+1,k} u_{k+1} e_k^T.$$

Thus the *Rational Arnoldi* approximation to $f(A)y$ is

$$U_k f(B_k)e_1,$$

with B_k defined by the relation

$$(I + \gamma B_k)S_k = I.$$

To construct $\mathcal{K}_m(Z, y)$ requires the computation of $m-1$ vectors of the form $(I + \gamma A)^{-1}w$, that is, it requires to solve $m-1$ linear systems with the same shifted coefficient matrix.

The parameter γ determines the convergence of the method and so its choice deserves attention. Its selection in [47] was related to empirical motivations, while in [55] a criterion was suggested, independent on the problem, for the exponential operator and symmetric positive semidefinite matrices. It was based on the strict connection between the RAM method and rational approximations for the exponential function. Indeed, if a rational approximation like (23) is used for $\exp(A)y$, then the RAM method corresponds to precondition all the involved systems by means of the same preconditioner for all poles, which depends on γ.

Also in the context of the ML function the RAM method overcomes the problems related to slow convergence, as shown in [48].

6. Application to Time–Fractional Partial Differential Equations

One of the most attractive fields in which the advanced capabilities of exponential integrators are fully exploited is the solution of partial differential equations. Many complex problems modeling real–life systems are formulated in terms of evolutionary equations involving derivatives with respect both to the spatial coordinates and to the time variable. When anomalous phenomena are involved, it is frequent to use time–fractional partial differential equations (FPDEs).

To this purpose we consider as a first test problem a quite general two dimensional subdiffusion equation with constant diffusion coefficient and variable source term

$$
{}_0^C D_t^\alpha u(t,x,y) + \rho \nabla^2 u(t,x,y) = r(t,x,y), \quad (x,y) \in \Omega, \quad t \geq 0 \tag{25}
$$

onto the unitary square domain $\Omega = [0,1] \times [0,1]$. This equation is subject to initial and boundary conditions

$$
\begin{aligned}
u(0,x,y) &= u_0(x,y), \quad (x,y) \in \Omega \\
u(t,x,y) &= g(t,x,y), \quad (x,y) \in \partial\Omega.
\end{aligned}
$$

A well–established strategy to numerically solve equations of this kind is first to discretize the spatial derivatives by leaving the time variable continuous; in this way the original PDE is transformed into a system of ODEs and, hence, suitable step by step methods can be applied to solve this resulting semi–discrete system.

The spatial discretization can be pursued by means of several methods, such as finite differences, finite elements or spectral methods. For simplicity of presentation, in this work we focus on semi-discrete methods obtained on the basis of finite differences but most of the results can be readily extended to the other aforementioned techniques.

On the domain Ω we consider an equispaced mesh–grid (x_i, y_j), $i = 0, \ldots, N_x + 1$, $j = 0, \ldots, N_y + 1$, with $x_i = i\Delta x$ and $y_j = j\Delta y$, $\Delta x = 1/(N_x + 1)$ and $\Delta y = 1/(N_y + 1)$. A straightforward way to discretize (25) along the spatial directions is to replace the second order spatial derivatives

$$
\nabla^2 u(t,x,y) = \frac{\partial^2}{\partial x^2} u(t,x,y) + \frac{\partial^2}{\partial y^2} u(t,x,y)
$$

by means of central differences

$$\begin{aligned}\frac{\partial^2}{\partial x^2}u(t,x_i,y_j) &= \frac{u(t,x_{i-1},y_j)-2u(t,x_i,y_j)+u(t,x_{i+1},y_j)}{\Delta x^2}+\mathcal{O}(\Delta x^2),\\ \frac{\partial^2}{\partial y^2}u(t,x_i,y_j) &= \frac{u(t,x_i,y_{j-1})-2u(t,x_i,y_j)+u(t,x_i,y_{j+1})}{\Delta y^2}+\mathcal{O}(\Delta y^2),\end{aligned}$$

where the second–order accuracy is achieved under the assumption of sufficiently smoothness of u_{xxxx} and u_{yyyy} (we refer to any classical textbooks on the numerical treatment of PDEs, for instance [36]).

The resulting system of FDEs will be of the form (1) with the state variable $y(t) \in \mathbb{R}^{N_x N_y}$ containing the approximation $u_{i,j}(t) \approx u(x_i, y_j, t)$ ordered in a suitable manner, the source term $f(t)$ generated by $r(t,x,y)$ of (25) and by the boundary function $g(t,x,y)$ and the initial value y_0 given by the initial function $u_0(x,y)$.

The matrix A is in general sparse and large. Moreover, the eigenvalues of A are strictly positive and the largest eigenvalue is proportional to $8N_x^2$. This means that the problem becomes stiffer as the discretization becomes finer [28].

The solution in closed form is not available for this test equation, thus we consider as reference solution the one obtained by the 4–step fractional exponential Adams method with a very small time–step. In Figure 1 we present the solution obtained at time $t = 1$.

For our experiment we consider the source term $r(t,x,y) = 10x(1-x)y(1-y)t^2$ coupled with the boundary condition $g(t,x,y) = (\sin(\pi x)+\sin(\pi y))e^{-2t}/2$ and the initial function $u_0(x,y) = g(0,x,y)$. The diffusion coefficient is $\rho = 0.1$ and the fractional order is $\alpha = 0.7$.

All the numerical simulations are carried out in Matlab, version 7.9.0.529, on a computer equipped with the Intel Dual Core E5400 processor running at 2.70 GHz under the Windows XP operating system.

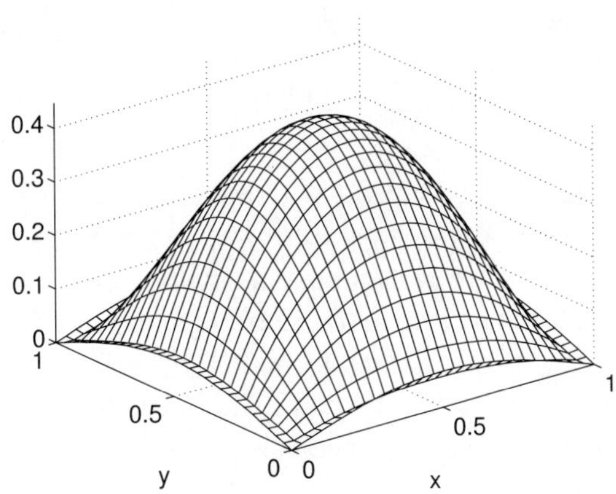

Figure 1. Reference solution for test problem (25) at $t = 1$.

To compare different methods and verify the corresponding convergence properties, for a decreasing sequence of step–sizes h we evaluate the error $E(h)$ with respect to the reference solution and we show the estimated order of convergence (EOC) obtained as $\log_2(E(h)/E(h/2))$.

In Table 3 for a spatial discretization given by $N_x = N_y = 10$ grid-points we compare the classical rectangular product integration rule with the 1–step fractional exponential Adams method. Both methods are expected to converge with order 1 as the numerical experiments clearly confirm.

Table 3. Errors and EOC by first order methods for test problem (25) $N_x = N_y = 10$ spatial grid points

	1-PI		1-Exp. Adams	
h	Error	EOC	Error	EOC
2^{-2}	5.46(−2)		3.63(−2)	
2^{-3}	2.88(−2)	0.922	2.03(−2)	0.839
2^{-4}	1.48(−2)	0.960	1.09(−2)	0.898
2^{-5}	7.52(−3)	0.978	5.69(−3)	0.935
2^{-6}	3.79(−3)	0.988	2.93(−3)	0.959

In Table 4 the same comparison is made between the classical trapezoidal (2-step) product integration rule and fractional exponential Adams methods of higher order. Also in this case the numerical results confirm in a clear way the theoretical prediction from Theorem 7.

Table 4. Errors and EOC by higher order methods for test problem (25) with $N_x = N_y = 10$ spatial grid points.

	2-PI		2-Exp. Adams		3-Exp. Adams		4-Exp. Adams	
h	Error	EOC	Error	EOC	Error	EOC	Error	EOC
2^{-2}	5.74(−3)		2.72(−3)		3.51(−5)		1.35(−5)	
2^{-3}	1.64(−3)	1.804	7.09(−4)	1.942	4.81(−6)	2.867	8.64(−7)	3.962
2^{-4}	4.88(−4)	1.750	1.81(−4)	1.968	6.37(−7)	2.917	5.42(−8)	3.995
2^{-5}	1.47(−4)	1.734	4.59(−5)	1.981	8.28(−8)	2.943	3.41(−9)	3.992
2^{-6}	4.44(−5)	1.726	1.16(−5)	1.989	1.06(−8)	2.970	2.12(−10)	4.002

A more interesting comparison concerns with the computational effort required by the different methods. To this purpose, we plot the error versus the cpu time required to evaluate the solution with the same sequence of step-sizes as in Tables 3 and 4.

As we can see in Figure 2 the computational effort required by exponential methods is remarkably smaller than that of the classical PI rules. This is not a completely expected result. Although the exponential methods allow to evaluate the solution directly at the desired instant t, the more involved computation necessary to evaluate the convolution weights in terms of matrix ML functions would expect to adversely affect the overall efficiency of the method.

It is instead true that in general the exponential methods turn out to be more efficient in spite of the more involved computation of the matrix ML functions. This happens not only with systems of dimension 100 as in Figure 2 but it is more evident when problems of larger size are involved.

In Figure 3 we consider a mesh with $N_x = N_y = 20$ spatial grid points and hence the

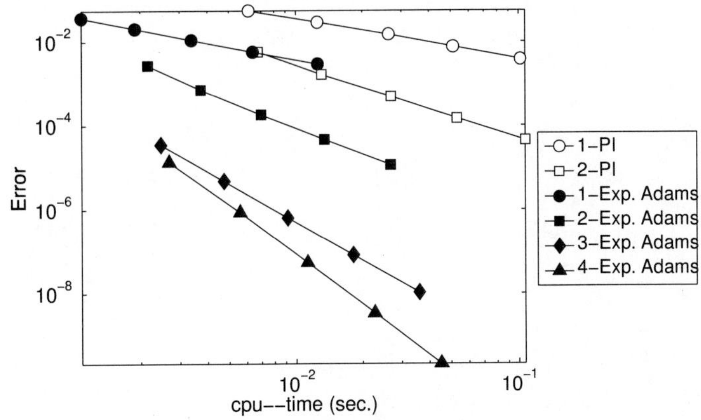

Figure 2. Errors versus cpu-time for test problem (25) with $N_x = N_y = 10$ spatial grid points.

resulting system is of dimension 400. The advantage of the exponential approach is very remarkable in this case. The results for errors and EOC in this case are similar to those in Tables 3 and 4 and therefore we omit them.

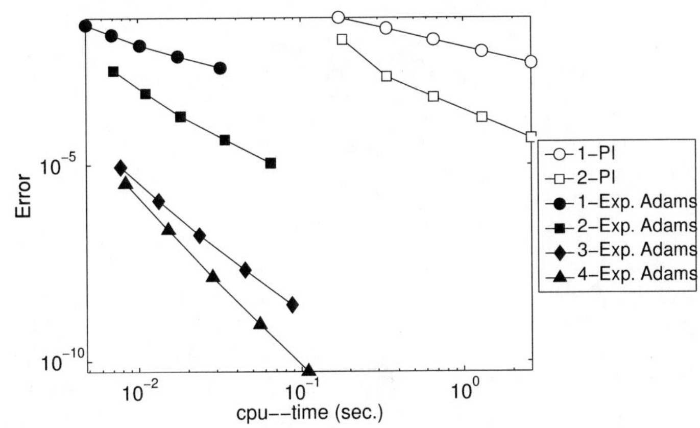

Figure 3. Errors versus cpu-time for test problem (25) with $N_x = N_y = 20$ spatial grid points.

A finer mesh of $N_x = N_y = 30$ spatial grid points, resulting in a system dimension equal to 900, leads to very similar results, as shown in Figure 4. The qualitative behavior is almost the same of Figure 3, even if the cpu-time is one order of magnitude larger than that for $N_x = N_y = 20$, thus confirming the increase of computational complexity as the dimension gets larger.

Although the generalization of exponential integrators to fractional–order systems appears to be more suitable for linear problems, we however investigate their potentials also in the nonlinear case. We anticipate that the results are less conspicuous than the linear case, but are anyhow prominent.

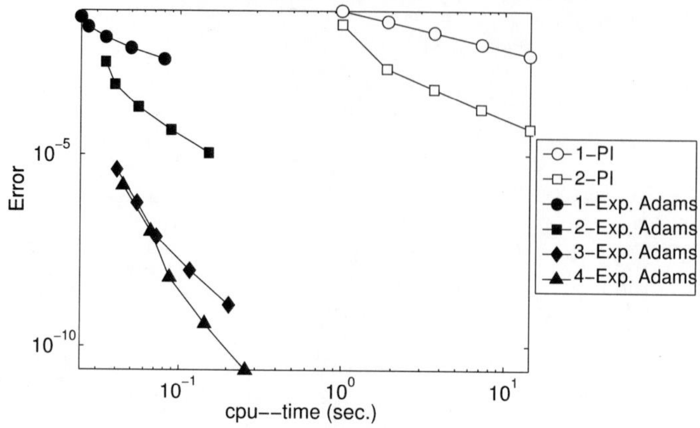

Figure 4. Errors versus cpu-time for test problem (25) with $N_x = N_y = 30$ spatial grid points.

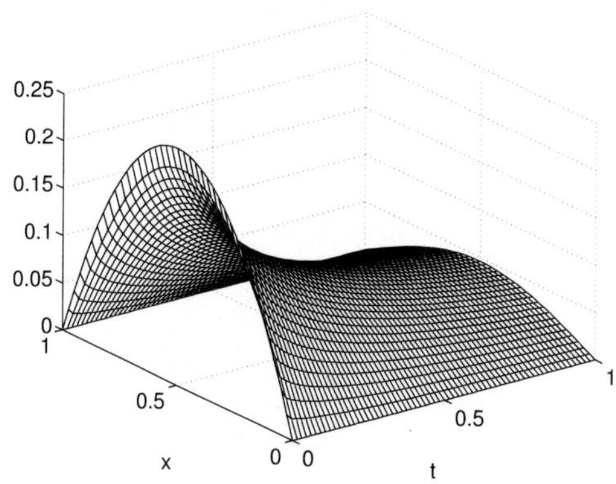

Figure 5. Reference solution for test problem (26) with $\alpha = 0.7$ and $t \in [0, 1]$.

To this aim we consider a generic prototype time–fractional diffusion–reaction equation

$$\frac{\partial^\alpha}{\partial t^\alpha} u - \nu \frac{\partial^2}{\partial x^2} u = \rho f(u), \quad x \in [0,1], \quad t \geq 0, \tag{26}$$

subject to initial and boundary conditions

$$\begin{aligned} & u(0,x) = u_0(x), \quad x \in [0,1] \\ & u(t,0) = g_1(t), \quad u(t,1) = g_2(t), \quad t \in [0,1]. \end{aligned}$$

As in the previous case the second order spatial derivative is discretized by means of central differences, thus giving rise to a semilinear system of FDEs in the form (20).

In our experiment it is assumed a diffusion coefficient $\nu = 0.1$, the reaction term $f(u) = \frac{1}{2}u(u - \frac{1}{2})(1-u) - u^\alpha$, the initial solution $u_0(x) = x(1-x)$ and the homogeneous boundary conditions $g_1(t) = g_2(t) = 0$. The reference solution for $\alpha = 0.7$, obtained in a similar way as in the linear case, is plotted in Figure 5.

As before, we compare the PI rules and fractional exponential Adams methods of 1-step and 2-step type. The comparison is in term of accuracy, order of convergence and cpu time, for different values of the problem dimension. The problem (26) is clearly one dimensional; its discretization thus leads to a system of the same dimension of the mesh refinement. For this reason we start from the small dimension $N_x = 20$, to get to $N_x = 80$ and $N_x = 200$. As in the linear case, in all cases the exponential Adams method is faster than the analogue PI rule in reaching the same accuracy, as shown in Figures 6-7-8. The difference is less striking than in the linear case but anyhow noticeable.

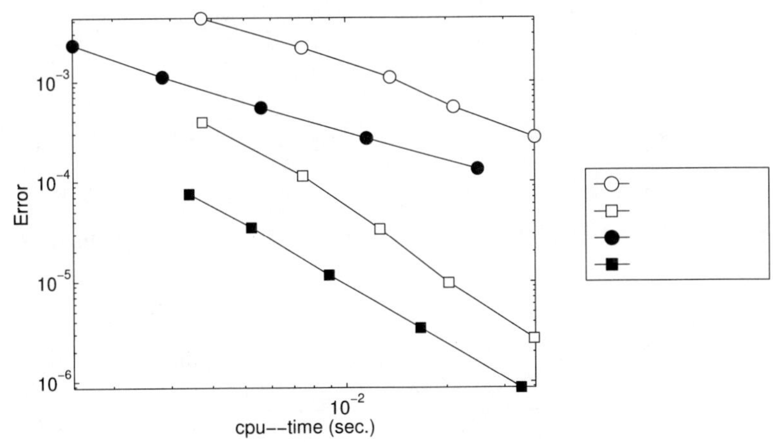

Figure 6. Errors versus cpu-time for test problem (26) with $N_x = 20$ spatial grid points.

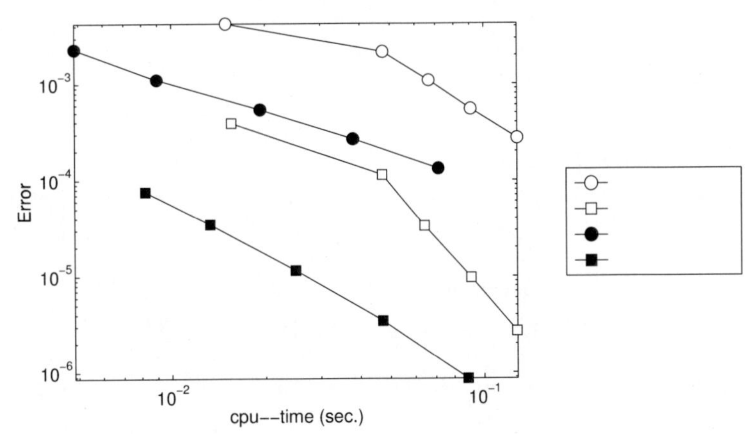

Figure 7. Errors versus cpu-time for test problem (26) with $N_x = 80$ spatial grid points.

Also the results from Table 5 for the accuracy and the order of convergence clearly agree with the theoretical results described in this chapter.

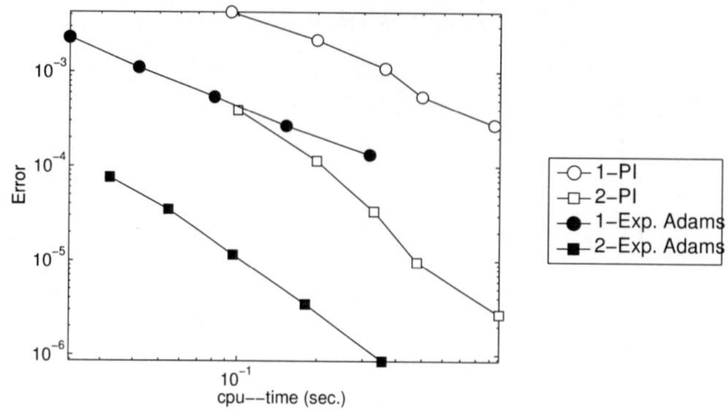

Figure 8. Errors versus cpu-time for test problem (26) with $N_x = 200$ spatial grid points.

Table 5. Errors and EOC by first order methods for the non linear test problem (26) with a mesh of $N_x = 20$ spatial grid points.

	1-PI		1-Exp. Adams		2-PI		2-Exp. Adams	
h	Error	EOC	Error	EOC	Error	EOC	Error	EOC
2^{-3}	$4.28(-3)$		$2.31(-3)$		$3.91(-4)$		$7.55(-5)$	
2^{-4}	$2.17(-3)$	0.979	$1.11(-3)$	1.053	$1.14(-4)$	1.781	$3.48(-5)$	1.116
2^{-5}	$1.09(-3)$	0.991	$5.43(-4)$	1.034	$3.34(-5)$	1.769	$1.16(-5)$	1.590
2^{-6}	$5.48(-4)$	0.995	$2.68(-4)$	1.021	$9.67(-6)$	1.789	$3.44(-6)$	1.749
2^{-7}	$2.74(-4)$	0.997	$1.33(-4)$	1.014	$2.70(-6)$	1.843	$8.67(-7)$	1.990

References

[1] Dumitru Baleanu, Kai Diethelm, Enrico Scalas, and Juan J. Trujillo. *Fractional calculus. Models and numerical methods*, volume 3 of *Series on Complexity, Nonlinearity and Chaos*. World Scientific Publishing Co. Pte. Ltd., Hackensack, NJ, 2012.

[2] Hvard Berland, Brd Skaflestad, and Will M. Wright. EXPINT - A MATLAB package for exponential integrators. *ACM Trans. Math. Softw.*, 33(1), 2007.

[3] Hermann Brunner. The numerical solution of weakly singular Volterra integral equations by collocation on graded meshes. *Math. Comp.*, 45(172):417–437, 1985.

[4] Marco Caliari and Alexander Ostermann. Implementation of exponential Rosenbrock-type integrators. *Appl. Numer. Math.*, 59(3-4):568–581, 2009.

[5] M. P. Calvo and C. Palencia. A class of explicit multistep exponential integrators for semilinear problems. *Numer. Math.*, 102(3):367–381, 2006.

[6] R. F. Cameron and S. McKee. The analysis of product integration methods for Abel's equation using discrete fractional differentiation. *IMA J. Numer. Anal.*, 5(3):339–353, 1985.

[7] S. M. Cox and P. C. Matthews. Exponential time differencing for stiff systems. *J. Comput. Phys.*, 176(2):430–455, 2002.

[8] C. F. Curtiss and J. O. Hirschfelder. Integration of stiff equations. *Proc. Nat. Acad. Sci. U. S. A.*, 38:235–243, 1952.

[9] Frank de Hoog and Richard Weiss. Asymptotic expansions for product integration. *Math. Comp.*, 27:295–306, 1973.

[10] E. Capelas de Oliveira, F. Mainardi, and Jr. Vaz, J. Models based on Mittag–Leffler functions for anomalous relaxation in dielectrics. *The European Physical Journal Special Topics*, 193(1):161–171, 2011.

[11] K. Diethelm. *The analysis of fractional differential equations*, volume 2004 of *Lecture Notes in Mathematics*. Springer-Verlag, Berlin, 2010.

[12] K. Diethelm, N. J. Ford, and A. D. Freed. Detailed error analysis for a fractional Adams method. *Numer. Algorithms*, 36(1):31–52, 2004.

[13] Kai Diethelm and Alan D. Freed. The FracPECE subroutine for the numerical solution of differential equations of fractional order. In S.Heinzel and T.Plesser, editors, *Forschung und wissenschaftliches Rechnen 1998*, Gessellschaft für wissenschaftliche Datenverarbeitung, pages 57–71, Göttingen, 1999.

[14] J. Dixon. On the order of the error in discretization methods for weakly singular second kind Volterra integral equations with nonsmooth solutions. *BIT*, 25(4):624–634, 1985.

[15] V. Druskin and L. Knizhnerman. Two polynomial methods of calculating functions of symmetric matrices. *U.S.S.R. Comput. Math. Math. Phys.*, 29:112–121, 1989.

[16] Vladimir Druskin and Leonid Knizhnerman. Extended Krylov subspaces: approximation of the matrix square root and related functions. *SIAM J. Matrix Anal. Appl.*, 19(3):755–771, 1998.

[17] Michael Eiermann and Oliver G. Ernst. A restarted Krylov subspace method for the evaluation of matrix functions. *SIAM J. Numerical Analysis*, 44(6):2481–2504, 2006.

[18] R. Garrappa. Predictor-corrector PECE method for fractional differential equations. MATLAB Central File Exchange, 2011. File ID: 32918.

[19] Roberto Garrappa. On linear stability of predictor–corrector algorithms for fractional differential equations. *Int. J. Comput. Math.*, 87(10):2281–2290, 2010.

[20] Roberto Garrappa. Exponential integrators for time–fractional partial differential equations. *Eur. Phys. J. Special Topics*, 222(8):1915–1927, 2013.

[21] Roberto Garrappa and Marina Popolizio. Generalized exponential time differencing methods for fractional order problems. *Comput. Math. Appl.*, 62(3):876–890, 2011.

[22] Roberto Garrappa and Marina Popolizio. On accurate product integration rules for linear fractional differential equations. *J. Comput. Appl. Math.*, 235(5):1085–1097, 2011.

[23] Roberto Garrappa and Marina Popolizio. On the use of matrix functions for fractional partial differential equations. *Math. Comput. Simulation*, 81(5):1045–1056, 2011.

[24] Roberto Garrappa and Marina Popolizio. Evaluation of generalized Mittag–Leffler functions on the real line. *Adv. Comput. Math.*, 39(1):205–225, 2013.

[25] R. Gorenflo, J. Loutchko, and Yu. Luchko. Computation of the Mittag-Leffler function and its derivatives. *Fractional Calculus and Applied Analysis*, 5(4):491–518, 2002.

[26] R. Gorenflo and F. Mainardi. Fractional calculus: integral and differential equations of fractional order. In *Fractals and fractional calculus in continuum mechanics (Udine, 1996)*, volume 378 of *CISM Courses and Lectures*, pages 223–276. Springer, Vienna, 1997. Reprinted with revisions in http://arxiv.org/abs/0805.3823.

[27] I. S. Gradshteyn and I. M. Ryzhik. *Table of integrals, series, and products*. Fourth edition prepared by Ju. V. Geronimus and M. Ju. Ceĭtlin. Translated from the Russian by Scripta Technica, Inc. Translation edited by Alan Jeffrey. Academic Press, New York, 1965.

[28] E. Hairer and G. Wanner. *Solving ordinary differential equations. II. Stiff and differential-algebraic problems*, volume 14 of *Springer Series in Computational Mathematics*. Springer-Verlag, Berlin, second edition, 1996.

[29] Hans Joachim Haubold, Arak M. Mathai, and Ram K. Saxena. Mittag–Leffler functions and their applications. *Journal of Applied Mathematics*, pages Art. ID 298628, 51, 2011.

[30] N. J. Higham. *Functions of matrices*. Society for Industrial and Applied Mathematics (SIAM), Philadelphia, PA, 2008.

[31] M. Hochbruck and C. Lubich. On Krylov subspace approximations to the matrix exponential operator. *SIAM J. Numer. Anal.*, 34(5):1911–1925, 1997.

[32] M. Hochbruck, C. Lubich, and H. Selhofer. Exponential integrators for large systems of differential equations. *SIAM J. Sci. Comput.*, 19(5):207–213, 1998.

[33] Marlis Hochbruck and Alexander Ostermann. Explicit exponential Runge-Kutta methods for semilinear parabolic problems. *SIAM J. Numer. Anal.*, 43(3):1069–1090 (electronic), 2005.

[34] Marlis Hochbruck and Alexander Ostermann. Exponential integrators. *Acta Numer.*, 19:209–286, 2010.

[35] Marlis Hochbruck and Alexander Ostermann. Exponential multistep methods of Adams-type. *BIT*, 51(4):889–908, 2011.

[36] Willem Hundsdorfer and Jan Verwer. *Numerical solution of time-dependent advection-diffusion-reaction equations*, volume 33 of *Springer Series in Computational Mathematics*. Springer-Verlag, Berlin, 2003.

[37] María López-Fernández, César Palencia, and Achim Schädle. A spectral order method for inverting sectorial Laplace transforms. *SIAM J. Numer. Anal.*, 44(3):1332–1350 (electronic), 2006.

[38] Carl F. Lorenzo and Tom T. Hartley. Generalized functions for the fractional calculus. *Crit. Rev. Biomed. Eng.*, 36(2):39–55, 2008.

[39] Christian Lubich. Runge-Kutta theory for Volterra and Abel integral equations of the second kind. *Math. Comp.*, 41(163):87–102, 1983.

[40] Christian Lubich. A stability analysis of convolution quadratures for Abel-Volterra integral equations. *IMA J. Numer. Anal.*, 6(1):87–101, 1986.

[41] J. N. Lyness and B. W. Ninham. Numerical quadrature and asymptotic expansions. *Math. Comp.*, 21:162–178, 1967.

[42] A. P. Magnus. Asymptotics and super asymptotics for best rational approximation error norms to the exponential function (the "$1/9$" problem) by the Carathéodory-Fejér method. *Nonlinear numerical methods and rational approximation*, 296(II):173–185, 1994.

[43] Francesco Mainardi. *Fractional calculus and waves in linear viscoelasticity*. Imperial College Press, London, 2010.

[44] Francesco Mainardi and Rudolf Gorenflo. On Mittag-Leffler-type functions in fractional evolution processes. *J. Comput. Appl. Math.*, 118(1-2):283–299, 2000.

[45] Kenneth S. Miller and Stefan G. Samko. A note on the complete monotonicity of the generalized Mittag-Leffler function. *Real Anal. Exchange*, 23(2):753–755, 1997/98.

[46] Borislav V. Minchev and Will M. Wright. A review of exponential integrators for first order semi-linear problems, 2005.

[47] I. Moret and P. Novati. RD-rational approximations of the matrix exponential. *BIT, Numerical Mathematics*, 44(3):595–615, 2004.

[48] I. Moret and P. Novati. On the convergence of Krylov subspace methods for matrix Mittag-Leffler functions. *SIAM J. Numer. Anal.*, 49(5):2144–2164, 2011.

[49] Igor Moret and Marina Popolizio. The restarted shift-and-invert Krylov method for matrix functions. *Numerical Linear Algebra with Appl.*, 21(1):68–80, 2014.

[50] Syvert P. Nørsett. An A-stable modification of the Adams-Bashforth methods. In *Conf. on Numerical Solution of Differential Equations (Dundee, 1969)*, pages 214–219. Springer, Berlin, 1969.

[51] Frank W. J. Olver, Daniel W. Lozier, Ronald F. Boisvert, and Charles W. Clark, editors. *NIST handbook of mathematical functions*. U.S. Department of Commerce National Institute of Standards and Technology, Washington, DC, 2010.

[52] A. Ostermann, M. Thalhammer, and W. M. Wright. A class of explicit exponential general linear methods. *BIT*, 46(2):409–431, 2006.

[53] I. Podlubny. *Fractional differential equations*, volume 198 of *Mathematics in Science and Engineering*. Academic Press Inc., San Diego, CA, 1999.

[54] Igor Podlubny and Martin Kacenak. The Matlab mlf code. MATLAB Central File Exchange, 2001–2009. File ID: 8738.

[55] M. Popolizio and V. Simoncini. Acceleration techniques for approximating the matrix exponential. *SIAM J. Matrix Analysis Appl.*, 30(2):657–683, 2008.

[56] Y. Saad. Analysis of some Krylov subspace approximations to the matrix exponential operator. *SIAM J. Numer. Anal.*, 29:209–228, 1992.

[57] Y. Saad. *Iterative Methods for Sparse Linear Systems, 2nd edition*. SIAM, Philadelpha, PA, 2003.

[58] Claus Schneider. Product integration for weakly singular integral equations. *Math. Comp.*, 36(153):207–213, 1981.

[59] L. N. Trefethen. Rational Chebyshev approximation on the unit disk. *Numer. Math.*, 37(2):297–320, 1981.

[60] L. N. Trefethen, J. A. C. Weideman, and T. Schmelzer. Talbot quadratures and rational approximations. *BIT. Numerical Mathematics*, 46(3):653–670, 2006.

[61] J. van den Eshof and M. Hochbruck. Preconditioning Lanczos approximations to the matrix exponential. *SIAM J. Sci. Comput.*, 27(4):1438–1457, 2006.

[62] J. A. C. Weideman and L. N. Trefethen. Parabolic and hyperbolic contours for computing the Bromwich integral. *Math. Comp.*, 76(259):1341–1356, 2007.

[63] J. Andrè C. Weideman. Computing special functions via inverse Laplace transforms. In T.E. Simos, G. Psihoyios, and Ch. Tsitouras, editors, *International Conference on Numerical Analysis and Applied Mathematics 2005 (Rhodes)*, pages 702–704. Wiley-VCH, 2005.

[64] J. Andrè C. Weideman. Improved contour integral methods for parabolic PDEs. *IMA J. Numer. Anal.*, 30(1):334–350, 2010.

[65] Andrew Young. Approximate product-integration. *Proc. Roy. Soc. London Ser. A.*, 224:552–561, 1954.

In: Fractional Calculus: Theory
Editors: R. Abi Zeid Daou and X. Moreau

ISBN: 978-1-63463-002-3

Chapter 10

NON-FRAGILE TUNING OF FRACTIONAL-ORDER PD CONTROLLERS FOR INTEGRATING AND DOUBLE INTEGRATING TIME-DELAY SYSTEMS

MirSaleh Bahavarnia and Mohammad Saleh Tavazoei*
Electrical Engineering Department,
Sharif University of Technology, Tehran, Iran

Abstract

In this chapter, non-fragile algebraic tuning rules are proposed for fractional-order PD controllers in order to be used in control of integrating time-delay systems. It is assumed that these systems are described by Integrating Plus Time Delay (IPTD) or Double Integrating Plus Time Delay (DIPTD) models. The tuning rules are derived based on a recently introduced approach, known as the centroid approach, which leads to designing non-fragile controllers. Considering this approach, it is found out that the centroids of two-parameter admissible regions or the center of mass of three-parameter admissible regions in the controller parameter space provide non-fragile choices for choosing controller parameters.

Keywords: Fractional-Order PD Controller, Integrating Plus Time Delay (IPTD) Process, Double Integrating Plus Time Delay (DIPTD) Process, Centroid Approach, Non-Fragility, Tuning Rules

1. Introduction

Time-Delay models have been extensively used for modeling of chemical systems, electrical systems, mechanical systems, and biological systems [1]. One of the most popular models in this class is the Integrating Plus Time Delay (IPTD) model [2]. Till now, different PI, PD and

---

* E-mail address: tavazoei@sharif.edu.

A part of this chapter has been presented in the 6th IFAC Workshop on Fractional Differentiation and its Applications, Grenoble, France, February 4-6, 2013.

PID tuning methods have been suggested for controlling IPTD-modeled processes. For example, in [3] a method for tuning PD controllers has been presented which minimizes the ISE criterion and results in a desired slope of the Nyquist curve at the gain crossover frequency. As another example, a tuning method has been presented to achieve an optimal PI controller for IPTD-modeled processes in [4]. Additional tuning methods for IPTD-modeled processes can be found in [2]. Another popular time-delay model is Double Integrating Plus Time Delay (DIPTD) model [2]. Similar to IPTD-modeled processes, different PD and PID tuning methods have been proposed for control of DIPTD-modeled processes [3]. On the other hand, fractional-order PID controllers and their subsets, i.e., fractional-order PD controllers and fractional-order PI controllers, as a generalization for classic PID controllers have attained growing attention from control system community in recent years. In this chapter, we focus on controlling IPTD-modeled processes and DIPTD-modeled processes by fractional-order PD controllers. As we know, up to now different methods have been offered for designing or tuning fractional-order PD controllers [5, 6]. But, to the best of the authors' knowledge non-fragile algebraic tuning rules have not been introduced for these controllers.

This chapter deals with proposing such tuning rules based on the centroid approach which has been recently suggested in [7, 8] for designing non-fragile two-parameter controllers. This approach suggests the centroids of the admissible regions in the controller parameter space as appropriate choices for selecting the controller parameters. In [7, 8], it has been proved that such an approach results in a non-fragile controller. In fact, the point proposed by this approach in the controller parameter space is placed away from the boundaries of the admissible region. Thus, the controller tuned based on such a point will be non-fragile. As mentioned in [8], the centroid approach can be extended for tuning a three-parameter controller by choosing the center of mass of the three-dimensional admissible region where such a region is considered as a mass. Based on such an extension, non-fragile PID controllers have been presented in [9] to control IPTD- and First-Order Pluse Time Delay (FOPTD)- modeled processes.

The chapter is organized as follows. Section 2 deals with investigating the stability region where the process is modeled by IPTD or DIPTD models, and the controller is a fractional-order PD controller. Sections 3 and 4 as the main parts of the chapter present non-fragile tuning rules for fractional-order PD controllers. In Section 3, it is supposed that the differentiator order is not tuneable, whereas in the Section 4 such an order is considered as a tuneable parameter. Finally, conclusions in Section 5 close the chapter.

2. Positive Stability Region

2.1. IPTD-Modeled Processes

Consider the Integrating Plus Time Delay (IPTD) model

$$G(s) = K\frac{e^{-Ls}}{s},\ (K > 0), \tag{1}$$

where K and L express the gain and the time delay, respectively. Suppose that for control of system (1), a fractional-order PD controller (PD^{μ} controller) in the form

$$C(s) = K_p + K_d s^{\mu}, \tag{2}$$

where $\mu \in (0,2)$ is the differentiator order, is used. In this case, for a fixed μ the positive stability region[1] could be obtained by the results which exist in [10]. On the basis of these results, the boundaries

$$K_p = 0, \tag{3}$$

$$K_d = 0, \tag{4}$$

and

$$\begin{cases} K_p = \dfrac{1}{KL} \dfrac{\varphi \cos(\varphi - \frac{\mu\pi}{2})}{\sin(\frac{\mu\pi}{2})} \\ K_d = \dfrac{-1}{KL^{1-\mu}} \dfrac{(\varphi)^{1-\mu} \cos(\varphi)}{\sin(\frac{\mu\pi}{2})}, \end{cases} \tag{5}$$

where $\varphi = \omega L$ and ω varies from 0 to ∞, can be used to specify the positive stability region. The stability regions have been depicted in Figure 1 for different values of μ. It can be simply checked that the part of the boundary (5) which is effective in the boundary of the positive stability region is the part given by

$$\varphi \in [\pi/2, (1+\mu)\pi/2]. \tag{6}$$

This effective part starts from the point $(K_d, K_p) = \left(0, \dfrac{\pi}{2KL}\right)$ in the $K_d - K_p$ plane and ends at the point $(K_d, K_p) = \left(\dfrac{1}{K}(\dfrac{\pi}{2L}(1+\mu))^{1-\mu}, 0\right)$. Two different cases are possible for this part. This part can have a monotonic behavior in K_d when φ is increasing (For instance, the parts related to the orders $\mu = 0.7, 0.8, 0.9, 1$ in Figure 1) or can have a non-monotonic behavior in K_d when φ is increasing (For instance, the parts related to the orders

[1] The positive stability region means the set of all positive controller parameters yielding in a stable closed-loop system.

$\mu = 1.1, 1.2, 1.3$ in Figure 1). The following lemma specifies the range in which the mentioned part has a monotonic behavior in K_d when φ is increasing.

Lemma 1. Consider the positive stability region in control of the IPDT process (1) by the fractional-order PD controller (2) in the $K_d - K_p$ plane. The part in the boundary of this positive stability region which is specified by (5) has a monotonic behavior in K_d when φ is increasing if and only if $0 < \mu \leq \bar{\mu}$ where $\bar{\mu} = 1$.

Proof. First, assume that $\bar{\mu} = 1$. According to (5),

$$\frac{dK_d}{d\varphi} = \frac{-1}{KL^{1-\mu}} \frac{((1-\mu)\varphi^{-\mu}\cos(\varphi) - \varphi^{1-\mu}\sin(\varphi))}{\sin(\frac{\mu\pi}{2})}. \tag{7}$$

To achieve a monotonic behavior in K_d when φ is increasing, we should have

$$\frac{dK_d}{d\varphi} \geq 0. \tag{8}$$

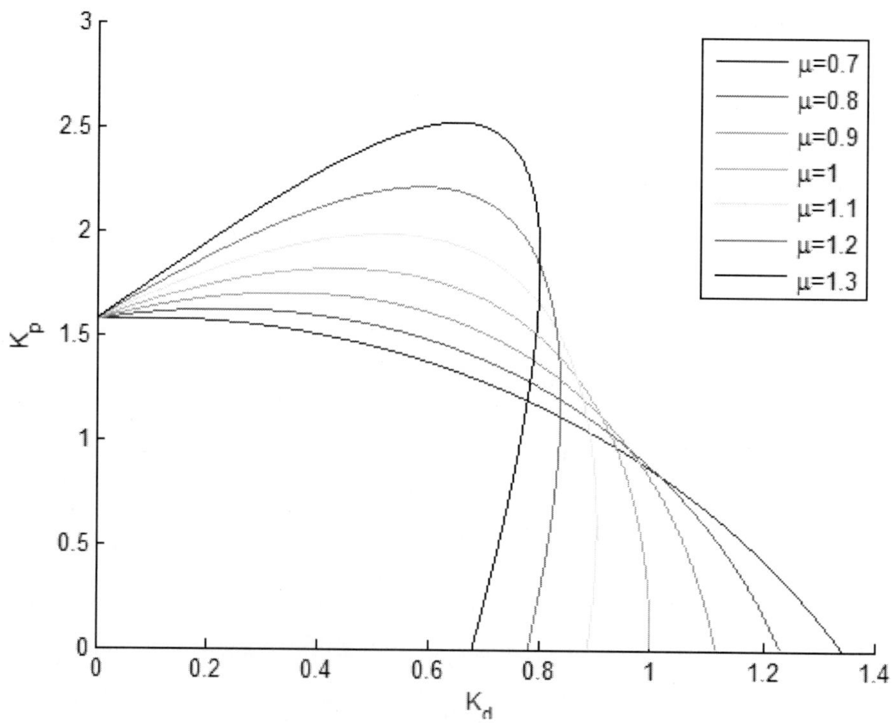

Figure 1. Boundaries specifying the positive stability region for different value of μ where $K = 1$ and $L = 1$ in model (1).

From (7) and (8), it is deduced that

$$(1-\mu)\cos(\varphi)-\varphi\sin(\varphi)\leq 0. \quad (9)$$

Considering the range of φ in (6), it can be verified that $(1-\mu)\cos(\varphi)\leq 0$ and $-\varphi\sin(\varphi)\leq 0$ for all $\mu\in(0,\bar{\mu}]$ where $\bar{\mu}=1$. This means that (9) holds for all $\mu\in(0,\bar{\mu}]$ where $\bar{\mu}=1$. Hence, in this case the aforementioned part of the positive stability region has a monotonic behavior in K_d when φ is increasing. Now, we show that if this part has a monotonic behavior in K_d, $\bar{\mu}$ cannot be greater than 1. If $\bar{\mu}>1$, define $\varphi=(1+\mu)\pi/2$ where $1<\mu\leq\bar{\mu}$. For this value of φ inequality (9) cannot hold because $(1-\mu)\cos(\varphi)>0$ and $-\varphi\sin(\varphi)>0$. This means that $\bar{\mu}$ cannot be greater than 1. ■

By an approach similar to that presented in proof of Theorem 1, it can be shown that the part in the boundary of the positive stability region which is specified by (5) has a monotonic behavior in K_p when φ is increasing if $0<\mu\leq\bar{\mu}$ where $\frac{2}{\pi}(\frac{\pi}{2}-\tan^{-1}(\frac{2}{\pi}))\approx 0.6391$ [11].

2.2. DIPTD-Modeled Processes

Consider the Double Integrating Plus Time Delay (DIPTD) model

$$G(s)=K\frac{e^{-Ls}}{s^2} \quad (K>0) \quad (10)$$

where K expresses the gain and L represents the time delay. Similar to the previous subsection, suppose that controller is considered as (2) and μ in the range $(0,2)$. In this case by using the results of [10], the positive stability region is specified by the following boundaries

$$K_p=0, \quad (11)$$

and

$$\begin{cases} K_p=\dfrac{-1}{KL^2}\dfrac{(\varphi)^2\sin(\varphi-\frac{\mu\pi}{2})}{\sin(\frac{\mu\pi}{2})} \\ K_d=\dfrac{1}{KL^{2-\mu}}\dfrac{(\varphi)^{2-\mu}\sin(\varphi)}{\sin(\frac{\mu\pi}{2})} \end{cases}, \quad (12)$$

where $\varphi=\omega L$ and ω changes from 0 to ∞. It can be easily verified that the effective interval for φ which specifies the positive stability region is as follows.

$$\varphi \in [0, \mu\pi/2]. \tag{13}$$

This effective part starts from the point $(K_d, K_p) = (0,0)$ in the $K_d - K_p$ plane and ends at the point $(K_d, K_p) = \left(\frac{1}{K}(\frac{\mu\pi}{2L})^{2-\mu}, 0\right)$. The positive stability regions have been depicted in Figure 2 for different values of μ. Two different situations can be possible for this part. It can have a monotonic behavior in K_d when φ is increasing (For example, the parts corresponded to orders $\mu = 0.9378, 1.0378, 1.1378, 1.2378$ in Figure 2) or can have a non-monotonic behavior in K_d when φ is increasing (For example, the parts corresponded to orders $\mu = 1.3378, 1.4378, 1.5378$ in Figure 2). Lemma 2 specifies the maximum range for order μ which guarantees the monotonic behavior in K_d when φ is the range (13).

Lemma 2. Consider the positive stability region in control of DIPDT process (10) by the fractional-order PD controller (2) in the $K_d - K_p$ plane. The boundary of the positive stability region which is achieved by (12) has a monotonic behavior in K_d when φ is increasing if and only if $0 < \mu \le \bar{\mu}$, where $\bar{\mu}$ is the smallest positive root of equation $\tan(\frac{\mu\pi}{2}) + \frac{\mu\pi}{2(2-\mu)} = 0$ ($\bar{\mu} \approx 1.2378$).

Proof. From (12),

$$\frac{dK_d}{d\varphi} = \frac{1}{KL^{2-\mu}} \frac{((2-\mu)\varphi^{1-\mu}\sin(\varphi) + \varphi^{2-\mu}\cos(\varphi))}{\sin(\frac{\mu\pi}{2})}. \tag{14}$$

To achieve a monotonic behavior in K_d when φ is increasing, condition (8) should be satisfied. (8) and (14) result in

$$(2-\mu)\sin(\varphi) + \varphi\cos(\varphi) \ge 0. \tag{15}$$

It can be easily verified that inequality (15) is valid for all $\varphi \in [0, \mu\pi/2]$ if and only if $0 < \mu \le \bar{\mu}$ where $\bar{\mu}$ is the smallest positive root of the equation $\tan(\frac{\mu\pi}{2}) + \frac{\mu\pi}{2(2-\mu)} = 0$. ■

In the next sections for simplicity in defining the integral limits, it is assumed that the differentiator order μ in the fractional-order PD controller belongs to the range $(0,1)$ for the IPTD-modeled process, and belongs to the range $(0,1.2378)$ for the DIPTD-modeled process.

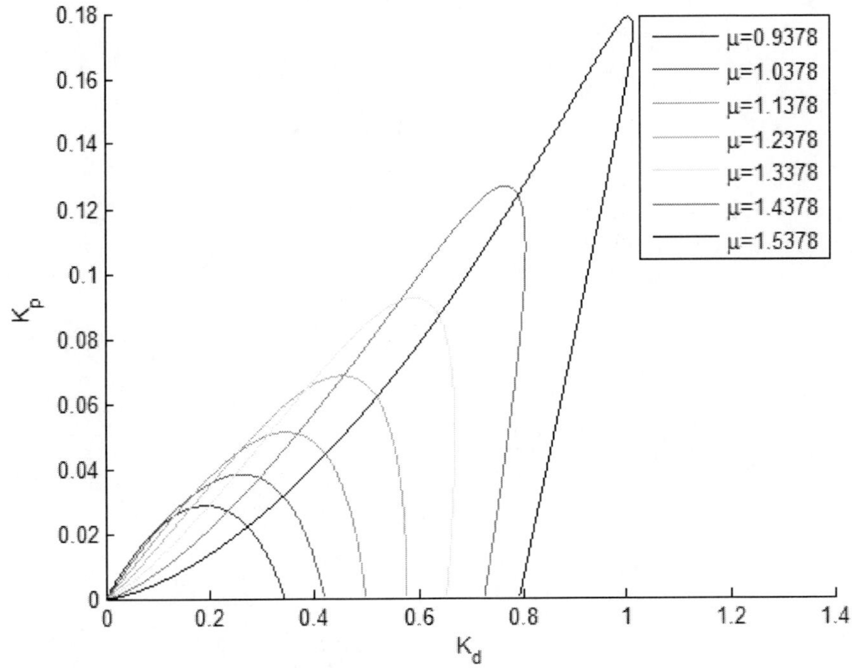

Figure 2. Boundaries specifying the positive stability region for different value of μ where $K=1$ and $L=4$ in model (10).

Considering this assumption, the curved part in the boundary of the positive stability region will have a monotonic behavior in K_d when φ is increasing.

3. Non-Fragile Tuning of a PD^{μ} Controller with a Fixed μ

3.1. IPTD-Modeled Processes and $\mu \in (0,1)$

As mentioned before, utilizing the centroids of the admissible regions in controller parameter space has been considered as an idea to reach non-fragile rules for tuning two-parameter controllers [7, 8]. In this section, such an idea is applied to achieve a non-fragile tuning rule for tuning PD^{μ} controllers in controlling IPTD-modeled processes. To achieve this aim, suppose that the fractional-order controller (2), whose differentiator order μ is fixed and is between 0 and 1, is used to control a process described by transfer function (1). Also, we assume that the admissible region is considered as the positive stability region [7, 8]. If $(K_{p-centroid}, K_{d-centroid})$ denotes the centroid of the positive stability region in this case, then

$$K_{p-centroid} = \frac{\iint K_p dK_p dK_d}{\iint dK_p dK_d} \tag{16}$$

and

$$K_{d-centroid} = \frac{\iint K_d dK_p dK_d}{\iint dK_p dK_d} \quad (17)$$

where the double integrals in (16) and (17) are calculated over the positive stability region [12]. Let us define new variables

$$\overline{K}_p = KK_p L \quad (18)$$

and

$$\overline{K}_d = KK_d L^{1-\mu}. \quad (19)$$

By substituting (18) and (19) in (16) and (17), we have

$$K_{p-centroid} = \frac{1}{KL}\frac{A(\mu)}{B(\mu)} \quad (20)$$

and

$$K_{d-centroid} = \frac{1}{KL^{1-\mu}}\frac{C(\mu)}{B(\mu)}, \quad (21)$$

where

$$A(\mu) = \iint \overline{K}_p d\overline{K}_p d\overline{K}_d, \quad (22)$$

$$B(\mu) = \iint d\overline{K}_p d\overline{K}_d, \quad (23)$$

and

$$C(\mu) = \iint \overline{K}_d d\overline{K}_p d\overline{K}_d. \quad (24)$$

Note that the double integrals in (22)-(24) are calculated over the positive stability region. According to (5), (6), and Lemma 1, $A(\mu)$, $B(\mu)$, and $C(\mu)$ are calculated as follows.

$$A(\mu) = \int_{\frac{\pi}{2}}^{\frac{\pi}{2}(1+\mu)} \frac{-1}{2\sin^3(\frac{\mu\pi}{2})}\varphi^{2-\mu}\cos^2(\varphi - \frac{\mu\pi}{2}) \times ((1-\mu)\cos\varphi - \varphi\sin\varphi)d\varphi \quad (25)$$

$$B(\mu)=\int_{\frac{\pi}{2}}^{\frac{\pi}{2}(1+\mu)}\frac{-1}{\sin^2(\frac{\mu\pi}{2})}\varphi^{1-\mu}\cos(\varphi-\frac{\mu\pi}{2})\times((1-\mu)\cos\varphi-\varphi\sin\varphi)d\varphi \tag{26}$$

$$C(\mu)=\int_{\frac{\pi}{2}}^{\frac{\pi}{2}(1+\mu)}\frac{1}{\sin^3(\frac{\mu\pi}{2})}\varphi^{2-2\mu}\cos(\varphi-\frac{\mu\pi}{2})\times((1-\mu)\cos^2(\varphi)-\varphi\sin(\varphi)\cos(\varphi))d\varphi \tag{27}$$

$A(\mu)$, $B(\mu)$, and $C(\mu)$ have been numerically computed and the corresponding results have been depicted in Figures 3-5. According to these results, the following approximations

$$A(\mu)\approx a_1e^{b_1\mu}-c_1\mathrm{e}^{d_1\mu}, \tag{28}$$

$$B(\mu)\approx a_2\mathrm{e}^{b_2\mu}-c_2\mathrm{e}^{d_2\mu}, \tag{29}$$

and

$$C(\mu)\approx -a_3\mathrm{e}^{b_3\mu}+c_3\mathrm{e}^{d_3\mu}, \tag{30}$$

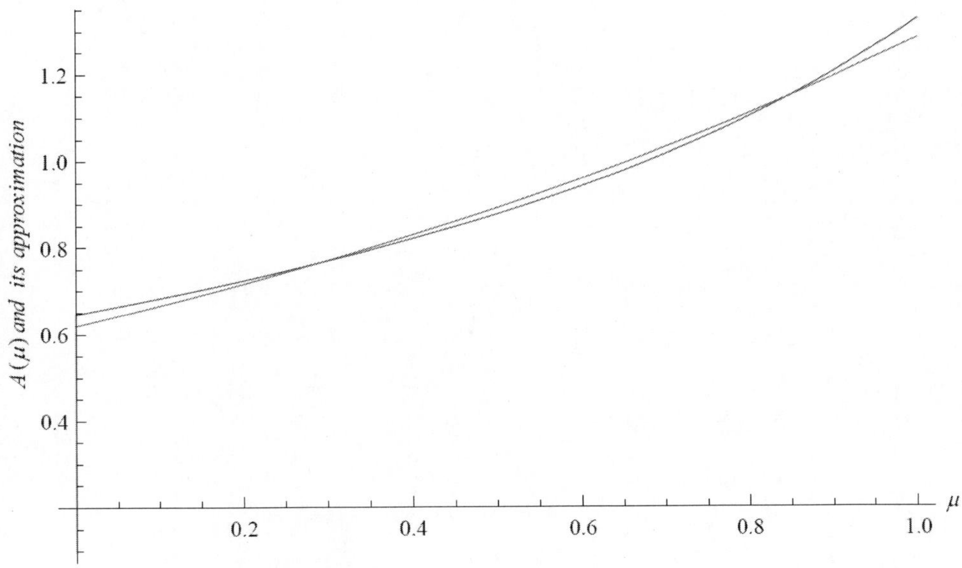

Figure 3. $A(\mu)$ and its approximation (Subsection 3.1).

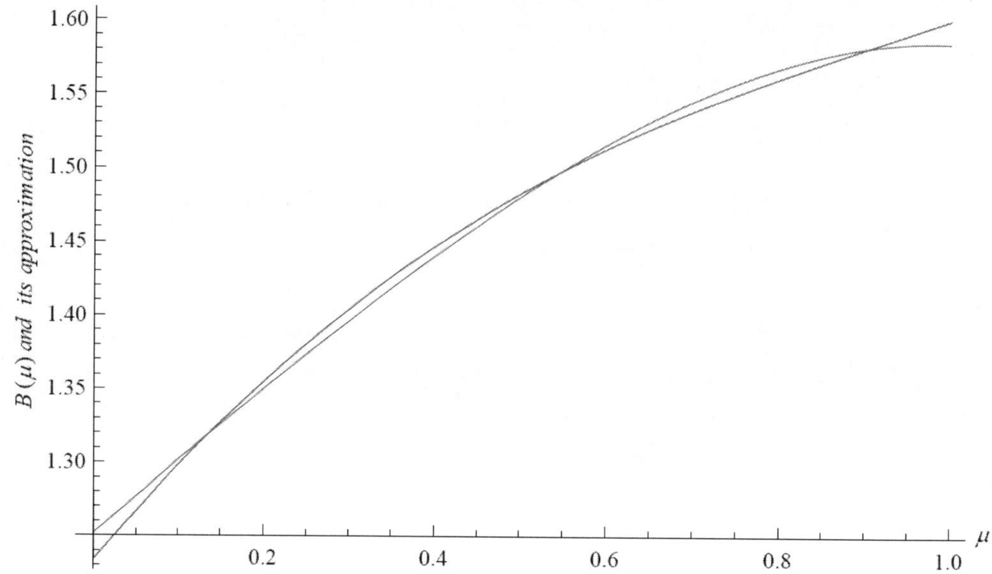

Figure 4. $B(\mu)$ and its approximation (Subsection 3.1).

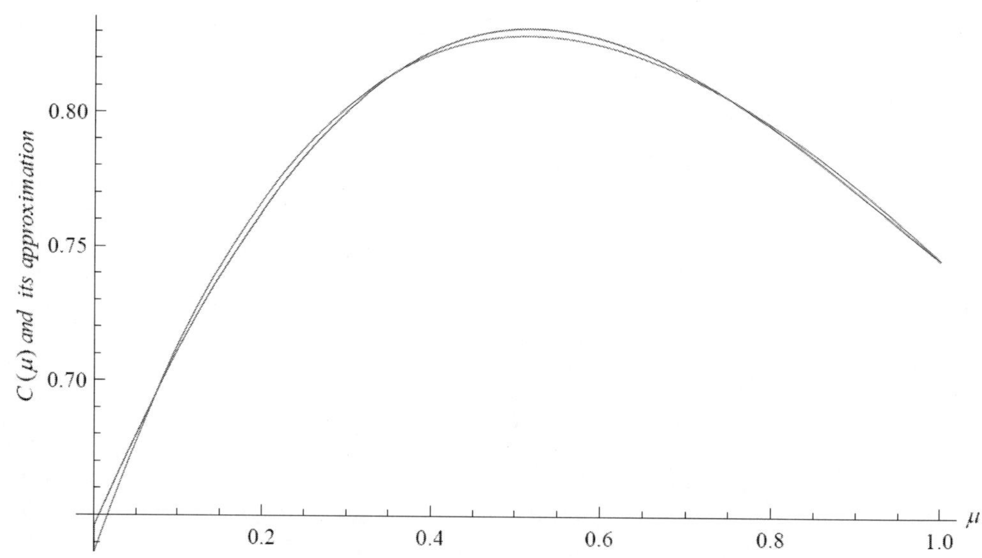

Figure 5. $C(\mu)$ and its approximation (Subsection 3.1).

where constants a_i, b_i, c_i, and d_i for $i = 1,2,3$ are given in Table 1, can be proposed for computing $A(\mu)$, $B(\mu)$, and $C(\mu)$ (See Figures 3-5). Consequently, the following equalities

$$K_p = \frac{1}{KL} \frac{a_1 e^{b_1 \mu} - c_1 e^{d_1 \mu}}{a_2 e^{b_2 \mu} - c_2 e^{d_2 \mu}} \tag{31}$$

and

$$K_d = \frac{1}{KL^{1-\mu}} \frac{-a_3 e^{b_3\mu} + c_3 e^{d_3\mu}}{a_2 e^{b_2\mu} - c_2 e^{d_2\mu}} \tag{32}$$

can be proposed as a non-fragile rule for tuning a PD^{μ} controller with a fixed $\mu \in (0,1)$ to be used in control of a process described by (1).

Constants a_i, b_i, c_i, and d_i for $i = 1,2,3$ in tuning rule (31)-(32) are specified in Table 1. As an illustration, assume that $K = 1$, $L = 1$, and $\mu = 0.6$.

Table 1. The values of the constants in approximations (28)-(30)

Constant	Value
a_1	1.4459
b_1	0.7290
c_1	0.8269
d_1	0.7290
a_2	6.0598
b_2	0.8194
c_2	4.8080
d_2	0.9284
a_3	3.5242
b_3	1.2797
c_3	4.2104
d_3	1.1565

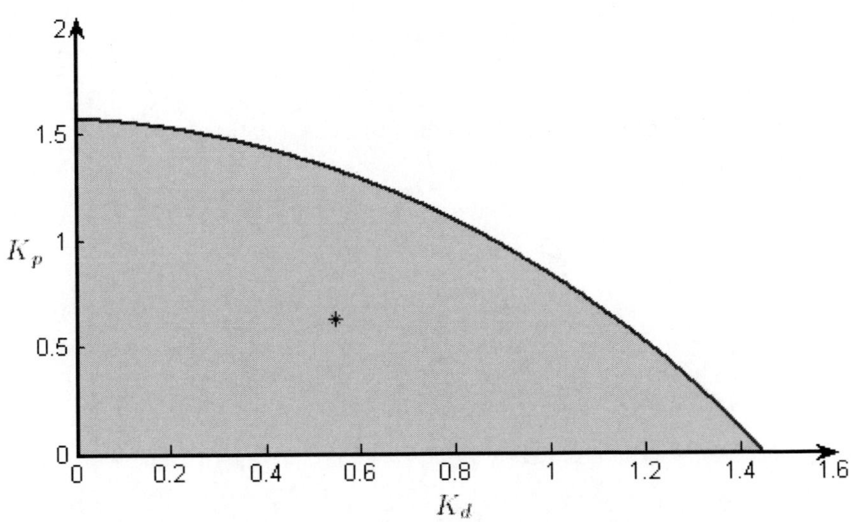

Figure 6. Positive stability region and its centroid where $\mu = 0.6$, $K = 1$, and $L = 1$ (Subsection 3.1).

For such a case, the tuning rule (31)-(32) proposes the following controller parameters $K_p = 0.62$ and $K_d = 0.55$ (See Figure 6).

3.2. DIPTD-Modeled Processes and $\mu \in (0,1.2378)$

In this subsection, it is assumed that μ is fixed and is between 0 and 1.2378. By the same approach used in the previous subsection, a non-fragile tuning rule for fractional-order PD controller (2) in order to be used for controlling a DIPTD-modeled process is proposed. If $(K_{p-centroid}, K_{d-centroid})$ denotes the centroid of the positive stability region in this case, then

$$K_{p-centroid} = \frac{\iint K_p dK_p dK_d}{\iint dK_p dK_d} \tag{33}$$

and

$$K_{d-centroid} = \frac{\iint K_d dK_p dK_d}{\iint dK_p dK_d}, \tag{34}$$

where the double integrals in (33) and (34) are calculated over the positive stability region [12]. Let us define new variables

$$\overline{K}_p = KK_p L^2 \tag{35}$$

and

$$\overline{K}_d = KK_d L^{2-\mu} \tag{36}$$

By substituting (35) and (36) in (33) and (34), the following relations

$$K_{p-centroid} = \frac{1}{KL^2} \frac{A(\mu)}{B(\mu)}, \tag{37}$$

and

$$K_{d-centroid} = \frac{1}{KL^{2-\mu}} \frac{C(\mu)}{B(\mu)}, \tag{38}$$

where

$$A(\mu) = \iint \overline{K}_p d\overline{K}_p d\overline{K}_d, \tag{39}$$

$$B(\mu) = \iint d\overline{K}_p d\overline{K}_d, \tag{40}$$

and

$$C(\mu) = \iint \overline{K}_d d\overline{K}_p d\overline{K}_d, \tag{41}$$

are achieved. The double integrals in (39)-(41) are computed over the positive stability region. According to Lemma 2, (12), and (13), $A(\mu)$, $B(\mu)$, and $C(\mu)$ are computed as follows.

$$A(\mu) = \int_0^{\frac{\mu\pi}{2}} \frac{1}{2\sin^3(\frac{\mu\pi}{2})} \varphi^{5-\mu} \sin^2(\varphi - \frac{\mu\pi}{2}) \times ((2-\mu)\sin\varphi + \varphi\cos\varphi) d\varphi \tag{42}$$

$$B(\mu) = \int_0^{\frac{\mu\pi}{2}} \frac{-1}{\sin^2(\frac{\mu\pi}{2})} \varphi^{3-\mu} \sin(\varphi - \frac{\mu\pi}{2}) \times ((2-\mu)\sin\varphi + \varphi\cos\varphi) d\varphi \tag{43}$$

$$C(\mu) = \int_0^{\frac{\mu\pi}{2}} \frac{-1}{\sin^3(\frac{\mu\pi}{2})} \varphi^{5-2\mu} \sin(\varphi - \frac{\mu\pi}{2}) \times ((2-\mu)\sin^2\varphi + \varphi\sin\varphi\cos\varphi) d\varphi \tag{44}$$

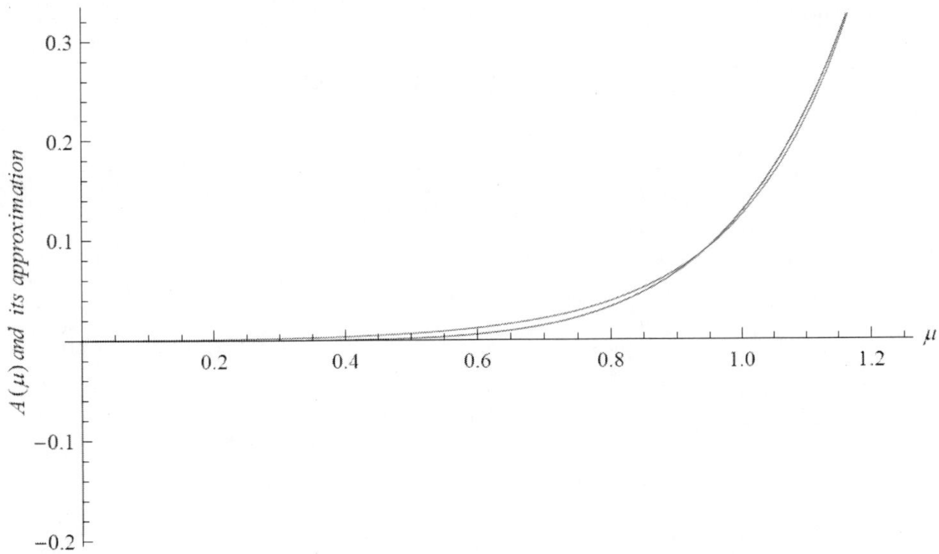

Figure 7. $A(\mu)$ and its approximation (Subsection 3.2).

$A(\mu)$, $B(\mu)$, and $C(\mu)$ have been numerically computed and the corresponding results have been shown in Figures 7-9. According to these results, the following approximations

$$A(\mu) \approx a_1 e^{b_1 \mu}, \tag{45}$$

$$B(\mu) \approx a_2 \mathrm{e}^{b_2 \mu}, \tag{46}$$

and

$$C(\mu) \approx a_3 \mathrm{e}^{b_3 \mu}, \tag{47}$$

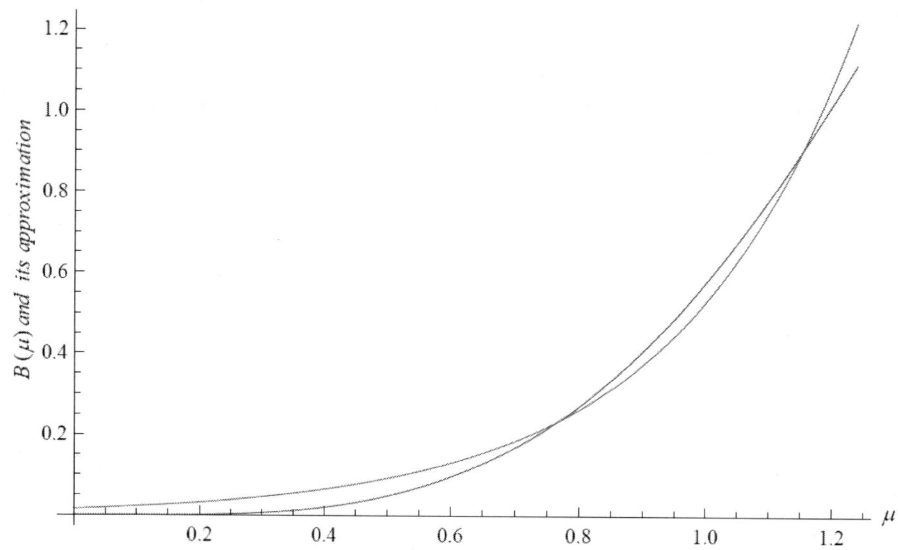

Figure 8. $B(\mu)$ and its approximation (Subsection 3.2).

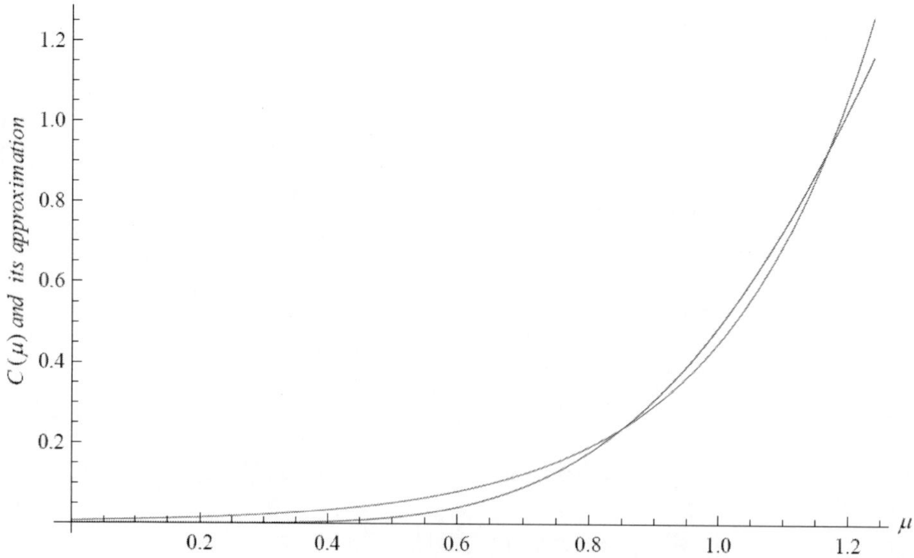

Figure 9. $C(\mu)$ and its approximation (Subsection 3.2).

where constants a_i, b_i, c_i, and d_i for $i=1,2,3$ are given in Table 2, can be suggested for computing $A(\mu)$, $B(\mu)$, and $C(\mu)$ (See Figures 7-9). Consequently, the following rule

$$K_p = \frac{1}{KL^2}\frac{a_1 e^{b_1\mu}}{a_2 e^{b_2\mu}} = \frac{1}{KL^2}\frac{a_1}{a_2} e^{(b_1-b_2)\mu} \tag{48}$$

$$K_d = \frac{1}{KL^{2-\mu}}\frac{a_3 e^{b_3\mu}}{a_2 e^{b_2\mu}} = \frac{1}{KL^{2-\mu}}\frac{a_3}{a_2} e^{(b_3-b_2)\mu}, \tag{49}$$

can be suggested as a non-fragile rule for tuning a PD^{μ} controller with a fixed $\mu \in (0,1.2378)$ to be used in control of a process described by (10). Constants a_i, b_i, c_i, and d_i for $i=1,2,3$ in tuning rule (48)-(49) are specified in Table 2. As an illustration, assume that $K=1$, $L=4$, and $\mu=1$. For such a case, the tuning rule (48)-(49) suggests the following controller parameters $K_p=0.015$ and $K_d=0.21$.

4. Non-Fragile Tuning of a PD^{μ} Controller with a Tunable Order

In this section, the aim is to present an extension for the tuning rules suggested in the previous section such that they can be used where the differentiator order μ is tunable. In this case, three controller parameters μ, K_p, and K_d should be determined such that the controller will be non-fragile.

Table 2. The values of the constants in approximations (45)-(47)

Constant	Value
a_1	0.0004
b_1	5.8214
a_2	0.0158
b_2	3.5115
a_3	0.0060
b_3	4.3168

For such a case, the idea of using the centroid could be spread to the idea of using the center of mass of the three-dimensional admissible region where this space is assumed as a mass [7, 8]. Note that increasing of the differentiator order leads to the noise amplification [13]. Thus, in practical applications, the differentiator order should not be chosen greater than an allowed value. Assume that we are allowed to select the order μ in the range $(0,\mu^*)$. Thus, the admissible region for selecting the controller parameters μ, K_p, and K_d in the

$\mu - K_p - K_d$ space is the space in which $0 < \mu < \mu^*$, and $K_p > 0$ and $K_d > 0$ are such that the stability of the closed-loop system is guaranteed.

4.1. IPTD-Modeled Processes

In this subsection, it is assumed that the process in modeled in the form (1) and the maximum allowed value for the controller parameter μ is not greater than 1 (i.e., $\mu^* \le 1$). By considering the admissible region as a mass, its center of mass denoted by $(\mu_{cg}, K_{p-cg}, K_{d-cg})$ is specified as follows.

$$\mu_{cg} = \frac{\int_0^{\mu^*} \frac{1}{KL} \cdot \frac{1}{KL^{1-\mu}} \mu B(\mu) d\mu}{\int_0^{\mu^*} \frac{1}{KL} \cdot \frac{1}{KL^{1-\mu}} B(\mu) d\mu} = \frac{\int_0^{\mu^*} L^{\mu} \mu B(\mu) d\mu}{\int_0^{\mu^*} L^{\mu} B(\mu) d\mu} \tag{50}$$

$$K_{p-cg} = \frac{\int_0^{\mu^*} \frac{1}{KL} \cdot \frac{1}{KL} \cdot \frac{1}{KL^{1-\mu}} A(\mu) d\mu}{\int_0^{\mu^*} \frac{1}{KL} \cdot \frac{1}{KL^{1-\mu}} B(\mu) d\mu} = \frac{1}{KL} \frac{\int_0^{\mu^*} L^{\mu} A(\mu) d\mu}{\int_0^{\mu^*} L^{\mu} B(\mu) d\mu} \tag{51}$$

$$K_{d-cg} = \frac{\int_0^{\mu^*} \frac{1}{KL^{1-\mu}} \cdot \frac{1}{KL^{1-\mu}} \cdot \frac{1}{KL} C(\mu) d\mu}{\int_0^{\mu^*} \frac{1}{KL} \cdot \frac{1}{KL^{1-\mu}} B(\mu) d\mu} = \frac{1}{KL} \frac{\int_0^{\mu^*} L^{2\mu} C(\mu) d\mu}{\int_0^{\mu^*} L^{\mu} B(\mu) d\mu} \tag{52}$$

By substituting the approximations of $A(\mu)$, $B(\mu)$, and $C(\mu)$ from (28)-(30) in (50)-(52), and calculating the integrals, the following non-fragile tuning rule is achieved.

$$\mu = \frac{\frac{a_2}{(b_2 + Ln(L))^2} - \frac{c_2}{(d_2 + Ln(L))^2} + L^{\mu^*}\left(\frac{a_2 e^{b_2 \mu^*}(\mu^*(b_2 + Ln(L)) - 1)}{(b_2 + Ln(L))^2} - \frac{c_2 e^{d_2 \mu^*}(\mu^*(d_2 + Ln(L)) - 1)}{(d_2 + Ln(L))^2} \right)}{-\frac{a_2}{b_2 + Ln(L)} + \frac{c_2}{d_2 + Ln(L)} + L^{\mu^*}\left(\frac{a_2 e^{b_2 \mu^*}}{b_2 + Ln(L)} - \frac{c_2 e^{d_2 \mu^*}}{d_2 + Ln(L)} \right)} \tag{53}$$

$$K_p = \frac{1}{KL} \frac{-\frac{a_1}{b_1 + Ln(L)} + \frac{c_1}{d_1 + Ln(L)} + L^{\mu^*}\left(\frac{a_1 e^{b_1 \mu^*}}{b_1 + Ln(L)} - \frac{c_1 e^{d_1 \mu^*}}{d_1 + Ln(L)} \right)}{-\frac{a_2}{b_2 + Ln(L)} + \frac{c_2}{d_2 + Ln(L)} + L^{\mu^*}\left(\frac{a_2 e^{b_2 \mu^*}}{b_2 + Ln(L)} - \frac{c_2 e^{d_2 \mu^*}}{d_2 + Ln(L)} \right)} \tag{54}$$

$$K_d = \frac{1}{KL} \frac{\frac{a_3}{b_3 + 2Ln(L)} - \frac{c_3}{d_3 + 2Ln(L)} + L^{2\mu^*}\left(\frac{-a_3 e^{b_3 \mu^*}}{b_3 + 2Ln(L)} + \frac{c_3 e^{d_3 \mu^*}}{d_3 + 2Ln(L)} \right)}{-\frac{a_2}{b_2 + Ln(L)} + \frac{c_2}{d_2 + Ln(L)} + L^{\mu^*}\left(\frac{a_2 e^{b_2 \mu^*}}{b_2 + Ln(L)} - \frac{c_2 e^{d_2 \mu^*}}{d_2 + Ln(L)} \right)} \tag{55}$$

As an example, suppose that $K = 1$, $L = 1$, and the permissible range for the choosing differentiator order μ is (0,1) (i.e., $\mu^* = 1$). For such a case, the conservative tuning rule (53)-(55) proposes the controller parameters $\mu = 0.52$, $K_p = 0.62$, and $K_d = 0.54$.

4.2. DIPTD-Modeled Processes

In this subsection, similar to the approach which was taken for the IPTD models in the previous subsection, a non-fragile tuning rule will be obtained for DIPTD-modeled processes. To achieve such a rule, assume that $\mu \in (0, \mu^*)$ is tunable, and $\mu^* \leq 1.2378$. In this case, the center of mass for the stability region is obtained as follows.

$$\mu_{cg} = \frac{\int_0^{\mu^*} \frac{1}{KL} \cdot \frac{1}{KL^{1-\mu}} \mu B(\mu) d\mu}{\int_0^{\mu^*} \frac{1}{KL} \cdot \frac{1}{KL^{1-\mu}} B(\mu) d\mu} = \frac{\int_0^{\mu^*} L^{\mu} \mu B(\mu) d\mu}{\int_0^{\mu^*} L^{\mu} B(\mu) d\mu} \tag{56}$$

$$K_{p-cg} = \frac{\int_0^{\mu^*} \frac{1}{KL} \cdot \frac{1}{KL} \cdot \frac{1}{KL^{1-\mu}} A(\mu) d\mu}{\int_0^{\mu^*} \frac{1}{KL} \cdot \frac{1}{KL^{1-\mu}} B(\mu) d\mu} = \frac{1}{KL} \frac{\int_0^{\mu^*} L^{\mu} A(\mu) d\mu}{\int_0^{\mu^*} L^{\mu} B(\mu) d\mu} \tag{57}$$

$$K_{d-cg} = \frac{\int_0^{\mu^*} \frac{1}{KL^{1-\mu}} \cdot \frac{1}{KL^{1-\mu}} \cdot \frac{1}{KL} C(\mu) d\mu}{\int_0^{\mu^*} \frac{1}{KL} \cdot \frac{1}{KL^{1-\mu}} B(\mu) d\mu} = \frac{1}{KL} \frac{\int_0^{\mu^*} L^{2\mu} C(\mu) d\mu}{\int_0^{\mu^*} L^{\mu} B(\mu) d\mu} \tag{58}$$

By substituting the approximations of $A(\mu)$, $B(\mu)$, and $C(\mu)$ from (45)-(47) to (56)-(58), and evaluating the integrals, the following non-fragile tuning rule is obtained.

$$\mu = \frac{1 + e^{b_2 \mu^*} L^{\mu^*} (-1 + \mu^* (b_2 + Ln(L)))}{(-1 + e^{b_2 \mu^*} L^{\mu^*})(b_2 + Ln(L))} \tag{59}$$

$$K_p = \frac{1}{KL^2}\frac{a_1(-1+e^{b_1\mu^*}L^{\mu^*})(b_2+Ln(L))}{a_2(-1+e^{b_2\mu^*}L^{\mu^*})(b_1+Ln(L))} \quad (60)$$

$$K_d = \frac{1}{KL^2}\frac{a_3(-1+e^{b_3\mu^*}L^{2\mu^*})(b_2+Ln(L))}{a_2(-1+e^{b_2\mu^*}L^{\mu^*})(b_3+2Ln(L))} \quad (61)$$

As an illustration, assume that $K=1$, $L=4$, and the allowable range for the choosing the differentiator order μ is (0,1.2378) (i.e., $\mu^*=1.2378$). For such a case, the tuning rule (59)-(61) suggests the controller parameters $\mu=1.04$, $K_p=0.02$, and $K_d=0.25$.

Conclusion

In this chapter, some algebraic rules were suggested for non-fragile tuning of fractional-order PD controllers to be used in controlling IPTD-modeled and DIPTD-modeled processes. The first group of the tuning rules which was presented in Subsections 3.1 and 3.2, could be used when the fractional-order PD controller has a non-tunable order. Also, the second group of the tuning rules, which was presented in Subsections 4.1 and 4.2, is applicable when the order of the fractional-order PD controller is tunable.

References

[1] C. Xiang, Q. G. Wang, X. Lu, L. A. Nguyen, and T. H. Lee, "Stabilization of second-order unstable delay processes by simple controllers," *Journal of Process Control*, 2007, Vol. 17, 675-682.

[2] A. O'Dwyer, *Handbook of PI and PID Controller Tuning Rules*, 3rd Edition, Imperial College Press, 2009.

[3] A. Ali and S. Majhi, "PID controller tuning for integrating processes," *ISA Transactions*, 2009, Vol. 49, 70-78.

[4] M. Huba, "Setpoint versus disturbance response of the IPDT plant," *IFAC Conference on Advances in PID Control, Brescia* (Italy), 2012, March 28-30.

[5] Y. Luo and Y. Q. Chen "Fractional-order [Proportional Derivative] Controller for Robust Motion Control: Tuning Procedure and Validation," *American Control Conference*, Hyatt Regency Riverfront, St. Louis, MO, USA, 2009, June 10-12, 1412-1417.

[6] D. J. Wang and X. L. Gao, "H_∞ design with fractional-order PD^μ controllers," *Automatica*, 2012, Vol. 48, 974-977.

[7] M. A. Rahimian and M. S. Tavazoei, "Application of Stability Region Centroids in Robust PI Stabilization of a Class of Second-Order Systems," *Transactions of the Institute of Measurement and Control*, 2012, Vol. 34, 487-498.

[8] M. A. Rahimian and M. S. Tavazoei, "Constrained Tuning of Two-Parameter Controllers: A Centroid Approach," *Proceedings of the Institution of Mechanical Engineers, Part I: Journal of Systems and Control Engineering*, 2012, Vol. 226, No. 5, 685-698.

[9] M. Bahavarnia and M. S. Tavazoei, "A new view to Ziegler-Nichols step response tuning method: Analytic non-fragility justification," *Journal of Process Control*, 2013, Vol. 23, 23-33.

[10] S. E. Hamamci and M. Koksal, "Calculation of all stabilizing fractional-order PD controllers for integrating time delay systems," *Computers and Mathematics with Applications*, 2009, Vol. 59, 1621-1629.

[11] M. S. Bahavarnia, M. S. Tavazoei and A. Mesbahi, "Non-Fragile Tuning of Fractional-Order PD Controllers for IPD-Modelled Processes," *6th IFAC Workshop on Fractional Differentiation and Its Applications*, Grenoble, France, February 4-6, 2013.

[12] G. B. Thomas and R. L. Finney, *Calculus and analytic geometry*, 8th ed., Reading, MA: Addison-Wesley, 1999.

[13] G. Lin, B. F. Feeny and T. Das, "Fractional derivative reconstruction of forced oscillators," *Nonlinear Dynamics*, 2009, Vol. 55, 239-250.

In: Fractional Calculus: Theory
Editors: R. Abi Zeid Daou and X. Moreau

ISBN: 978-1-63463-002-3

Chapter 11

ON DISCRETE, FINITE-DIMENSIONAL APPROXIMATION OF LINEAR, INFINITE DIMENSIONAL SYSTEMS

***Milan R. Rapaić*[1,*]*, Tomislav B. Šekara*[2] *and Mihailo P. Lazarević*[3]**

[1]Faculty of Technical Sciences, University of Novi Sad, Novi Sad, Serbia
[2]Faculty of Electrical Engineering, University of Belgrade, Belgrade, Serbia
[3]Faculty of Mechanical Engineering, University of Belgrade, Belgrade, Serbia

Abstract

Many phenomena are naturally described in terms of dynamical systems of infinite order. Such phenomena cannot be adequately described by an interconnection of a finite number of accumulating elements, i.e., by means of differential or difference equations of finite order. Among the well-known examples are distributed parameter systems, which are usually described by partial differential equations, and fractional order systems, which are described by fractional differential equations. In order for an infinite-dimensional system to be simulated or implemented using a digital computer, it must be approximated by a finite-dimensional model. Numerous methods for finite-dimensional approximations of infinite dimensional systems have been considered in literature. If spatial distribution of variables is of interest, distributed parameter systems are often simulated by means of the finite elements method (FEM). If the spatial distribution of variables is not of interest, as it is the case with fractional order models, an approximating ordinary differential equation of sufficiently high order is used for approximation. A novel, flexible and numerically efficient method for rational, finite-dimensional approximation of linear, infinite-dimensional systems is presented in the current chapter. The proposed method uses the least-squares (LS) procedure to interpolate frequency domain response of a fractional order system using a finite number of incident frequencies. An adequate comparative analysis has also been carried out through corresponding examples by applying several other known approximation methods.

Keywords: discretization; fractional order systems; rational approximations

[*] E-mail address: rapaja@uns.ac.rs; Address: Faculty of Technical Sciences, University of Novi Sad, Novi Sad, Serbia (Corresponding author)

Introduction

Design of dynamical control laws involving integral and differential actions [1, 2] often requires formulation of a discrete model of the process and/or the controller. Various discretization methods have been proposed in literature [3-20]. Some of these methods are based on matching of pulse, step or ramp responses of the continuous time system and its desired approximation. These methods are usually denoted as *response invariant methods* (including impulse-invariant, step-invariant and ramp-invariant methods). Other methods involve direct substitution of the differential operator by a suitable finite-diference approximation. Well-known *substitution methods* are forward and backward differentiation method (the so called First and Second Euler approximation), the bilinear method (the Tustin approximation), and others. There are, however, many other types of discretization methods in use. All these methods differ amongst each other in terms of complexity, precission for various sample values, , stability guarantees, and numerous other features.

Many phenomena are naturally described in terms of dynamical systems of infinite order. Such phenomena cannot be adequately described by an interconnection of a finite number of accumulating elements, i.e., by means of differential or difference equations of finite order. Among the well-known examples are distributed parameter systems, which are usually described by partial differential equations, and fractional order systems, which are described by fractional differential equations. An important fractional order system is the Fractional PID regulator, proposed by Podlubny [1, 2], which utilizes fractional integral and derivative action. In order for an infinite-dimensional system to be simulated or implemented using a digital computer, it must be approximated by a finite-dimensional model. Numerous methods for finite-dimensional approximations of infinite dimensional systems have been considered in literature. If spatial distribution of variables is of interest, distributed parameter systems are often simulated by means of the finite elements method (FEM). If the spatial distribution is not of interest, as is the case with fractional order models, an approximating ordinary differential equation of sufficiently high order is used for approximation. Whatever the case may be, an exact description of an infinite-dimensional system requires a transfer function of a non-rational form.

Since a stationary process can, in general, be represented by a transfer function $G_p(s)$ which is not a rational function [21, 22], the problem of rational approximation, and discretization in general, becomes complex. Special care must be taken in order for the fundamental system properties (such are steady-state gain and settling time, as well as basic properties in the frequency domain) to be preserved. In the process of discretization of linear, fractional order systems (LFOS) where, in general, fractional order integral and differential actions belong, one can make use of the well-known mapping of s-domain into z-domain in the complex plane.

$$z = e^{sT} \tag{1}$$

where T is the sampling time. Transform (1) maps left half-plane of the s-plane to interior of the unit circle in the z-plane. This means that stability of the continuous-time system is preserved, and that all poles of the discrete approximation are located within the unit circle, provided that the original continuous-time system has been stable. The basic goal of discretization is to obtain a model suitable for practical realization of the corresponding control law or filterwhich is almost fully equivalent to the original continuous-time system over a sufficiently wide frequency range.

A novel, flexible and numerically efficient method for rational, finite-dimensional approximation of linear, infinite-dimensional systems is presented in the current chapter. The proposed method relies on the interpolation of the frequency characteristic of the system on a predefined set of target frequencies. The approach has been originally proposed in [25]. The present manuscript extends the development of [25, 26] by suggesting the possibility of using the least-squares approximation on a wider frequency range.

The chapter is organized as follows. In the following section entitled "Motivating Examples" some well-known (and not so well-known) infinite dimensional systems are discussed shortly. The proposed method for rational approximation of FOS has been outlined in the subsequent section entitled "Rational Approximations of Transfer Functions of LFOS". The final section is devoted to some numerical examples and conclusions.

Motivating Examples

A number of motivating examples will be shortly considered in the present section. These examples illustrate the utility of infinite dimensional systems in describing natural processes, but also their applicability in modern control and estimation procedures.

A Diffusion Process

Consider a process described by the classical diffusion equation (also referred to as the heat equation), which is ubiquitous in science and engineering since it simultaneously describes a number of transfer phenomena, including heat-transfer and various other diffusion-like processes, including diffusion of mass (mechanical diffusion), diffusion of momentum (viscosity), diffusion of electrical potential (in long lines, when inductivity is negligible), and many others. One-dimensional diffusion equation is a partial differential equation of the form

$$\tau \frac{\partial^2 \rho}{\partial z^2} = \frac{\partial \rho}{\partial t}, \quad \tau > 0 \tag{2}$$

describing the process of transport (diffusion) of a quantity ρ along the z axis in time t. For simplicity, let us address only the diffusion within a semi-infinite medium, where both space and time variable take arbitrary positive values. Let us assume also that the process can be controlled by acting on the cross-section $z = 0$, and that the process output is taken (measured) at the cross-section $z = L$. The dynamics of the process is influenced by the diffusion time constant $\tau = \tau(z,t)$, which is, in general, a function of both space and time. However, in a variety of practically interesting cases this coefficient can be approximated by a constant.

Without loss of generality, assume that (2) describes a heat conduction process schematically shown in Figure 1. Let us obtain its transfer function. In this particular case,

$\rho = \rho(z,t)$, is the temperature of the cross section defined by space coordinate z evaluated at time instant t. Let $\tilde{\rho} = \tilde{\rho}(z,s)$ denote the Laplace transform of ρ, where the Laplace transform is taken with respect to the time variable t and the space variable z is considered as a parameter,

$$\tilde{\rho}(z,s) = \int_0^\infty \rho(z,t)e^{-st}dt \,. \qquad (3)$$

By applying the Laplace transform to equation (2), one obtains general solution

$$\tilde{\rho}(s,z) = C_1(s)e^{-z\sqrt{s/\tau}} + C_2(s)e^{z\sqrt{s/\tau}} \,. \qquad (4)$$

Since any heat conduction process is stable, the Laplace transform of the temperature in any cross-section must be bounded, i.e.

$$\lim_{z\to\infty} \tilde{\rho}(s,z) = const. \qquad (5)$$

has to be satisfied, thus C_2=0 and equation (4) takes the form

$$\tilde{\rho}(s,z) = C_1(s)e^{-z\sqrt{s/\tau}} \,. \qquad (6)$$

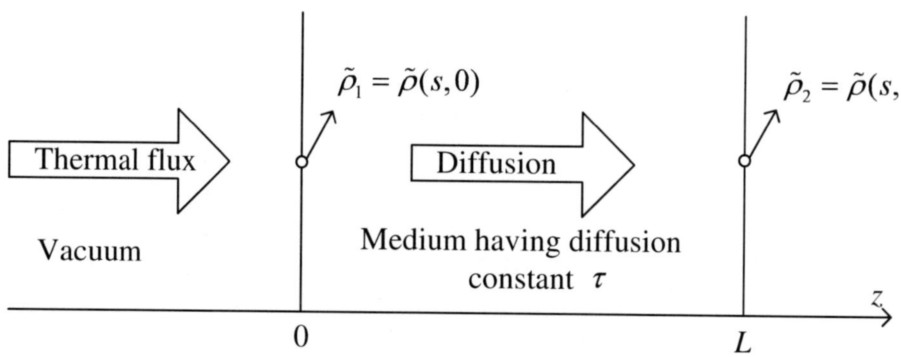

Figure 1. A sketch of the process of heat conduction by diffusion.

Integration "constant" C_1 as well as the conduction function are determined from the known (or given) boundary conditions. In view of this, the most frequent cases in practice are:

Case 1. Heat conduction without any convective exchange of heat with the environment and fixed temperature at the "left" boundary. In this particular case, the temperature of the cross-section $z = 0$ could be controlled directly, and considered as the input of the process, while the dependent temperature of the cross-section $z = L$ could be considered as the output. The left boundary condition for this case is $\tilde{\rho}(s,0) = C_1(s)$, and the transfer function takes the form

$$G_a(s) = \frac{\tilde{\rho}_2(s,L)}{\tilde{\rho}_1(s,0)} = e^{-L\sqrt{s/\tau}} = e^{-\sqrt{Ts}}\,,\ T = L^2/\tau\,. \tag{7}$$

Case 2. Heat conduction without any convective exchange of heat with the environment and fixed thermal flux at the "left" boundary. The process is influenced by gradient of quantity ρ at z = 0 (this is the boundary surface of the medium of Figure 1), the input quantity of the process being thermal flux through the boundary surface (again without any convective exchange with the environment)

$$\psi = -\lambda \frac{d\tilde{\rho}(s,z)}{dz}\bigg|_{z=0} \tag{8}$$

and the process (output) quantity is $\rho_2 = \rho(s, \mathrm{L})$. The transfer function is in this case

$$G_b(s) = \frac{\tilde{\rho}_2}{\psi} = \frac{K}{\sqrt{s}} e^{-\sqrt{Ts}}\,,\ T = L^2/\tau\,,\ K = \sqrt{\tau}/\lambda \tag{9}$$

Case 3. Heat conduction without any convective exchange of heat with the environment. The last characteristic case is when the convection is no longer neglected. Now, the process is influenced by a linear combination of the thermal flux and temperature at the "left" boundary

$$u = -\lambda \frac{d\tilde{\rho}(s,z)}{dz}\bigg|_{z=0} + \eta\,\tilde{\rho}(s,z)\,, \tag{10}$$

with output $\rho_2 = \rho(s,L)$, and the transfer function is

$$G_c(s) = \frac{\tilde{\rho}_2}{u} = \frac{K}{1+\sqrt{T_1 s}} e^{-\sqrt{Ts}}\,, K = \frac{1}{\eta}\,, T_1 = \frac{\lambda^2}{\eta^2 \tau}\,, T = \frac{L^2}{\tau}\,. \tag{11}$$

Fractional PID Regulator

Having established the fact that many physical systems are well-described by fractional order models, a logical question arises whether it is possible to formulate fractional order control laws and what would be their contribution to process control. Among many modern control strategies utilizing fractional order calculus, Podlubny's Fractional order PID [1, 2] regulator is emphasized here. Classical PID is arguably the most utilized control strategy in use today. By replacing classical integral and differential actions by their respective fractional order analogues, the flexibility and applicability of the PID regulator can be greatly increased. Transfer function of the fractional order PID is of the form

$$\mathrm{PI}^{\lambda}\mathrm{D}^{\mu}(s) = k + k_{\mathrm{i}} s^{-\lambda} + k_{\mathrm{d}} s^{\mu}\,,\ \lambda, \mu \in [0,1]\,. \tag{12}$$

The reader should notice that the implementation of Fractional order PID requires direct implementation of fractional order integrator and differentiator. Similar is also true for other types of fractional order regulators, as it can be seen from [25-27]. Such regulators are typically implemented as high order FIR or IIR filters [28]. Realization of fractional order control laws involving an adequate discretization is possible thanks to the fast modern computers. It is known that in the regulator design two approaches are possible. In the so called "direct approach", a discrete-time model of the process is developed first, and then a discrete-time regulator is designed and implemented directly with respect to this model.In the so called "indirect approach", the entire design procedure is conducted in the continuous-time domain, and the obtained controller is finally discretized for implementation purposes. Obviously, discretization is required by both approaches. However, the discretization procedure is not straightforward when fractional order systems are in question, a problem which has been causing a considerable research interest over the past years.

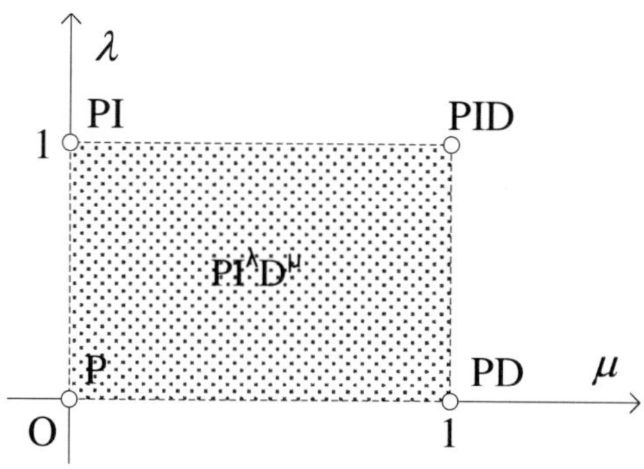

Figure 2. A sketch of the process of heat conduction by diffusion.

Some Additional Examples

In the examples above, the fractional derivative and integral operators have appeared in a number of contexts. It should be mentioned that other forms of fractional order transfer functions do emerge during investigations of different transfer phenomena. In fact, a stationary linear system of infinite dimension can be described by more-or-less arbitrary transfer function. In the analysis of axial diffusion, i.e., diffusion from the axis of the cylinder towards its lateral surface or vice-versa, one meets transfer functions originating from the Laplace transforms of Bessel functions, which have the form

$$G(s) = \frac{K}{\sqrt{1 + sT}} \tag{13}$$

From this example, transfer functions given by equations (7), (9), (11), and (12) belong to the fractional order systems having transfer functions which belong to the class of irrational functions [23, 24].

Rational Approximations of Transfer Functions of LFOS

Let us consider a rational transfer function

$$\frac{B(s)}{A(s)}=\frac{b_{n-1}s^{n-1}+\cdots+b_1s+b_0}{a_ns^n+a_{n-1}s^{n-1}+\cdots+a_1s+a_0} \tag{14}$$

which is used to approximate transfer function $G(s)$ of some linear fractional order system. For ($G(0)\neq0$, $b_0=1$) or ($G(0)=0$, $a_0=1$) there are $2n$ real coefficients which should be determined from $2n$ equations obtained from the condition of matching the frequency characteristics of the original $G(s)$ and its approximation (14) in the selected discrete frequency points $\omega\in[\omega_0,\omega_2,\cdots,\omega_{n-1}]$, i.e.,

$$G(\mathrm{i}\omega_k)-B(\mathrm{i}\omega_k)/A(\mathrm{i}\omega_k)=0,\ k=0,n-1, \tag{15}$$

or for $G(0)\neq0$, $b_0=1$ one obtains

$$\mathrm{Re}\big(G(\mathrm{i}\omega_k)A(\mathrm{i}\omega_k)-B(\mathrm{i}\omega_k)\big)=0,\ k=0,n-1, \tag{16}$$

$$\mathrm{Im}\big(G(\mathrm{i}\omega_k)A(\mathrm{i}\omega_k)-B(\mathrm{i}\omega_k)\big)=0,\ k=\overline{0,n-1}. \tag{17}$$

Note that $G(\mathrm{i}\omega_k)$ is a constant complex number for any fixed ω_k. For fixed ω_k both numerator and denominator polynomials are linear combinations of the unknown process parameters. Thus, the set of equations (16), (17) represents a linear system of equations having $2n$ unknown coefficients. By solving this system of $2n$ linear equations, one obtains $2n$ coefficients of rational approximation (14).

It is convenient to represent this system of equations in the matrix form. This form is particularly suitable for the numerical evaluation of the unknown coefficients. Consider equations (16) and (17). Assuming the case $G(0)=0$, $a_0=1$ (other cases can be dealt in the similar fashion), for any $k=\overline{0,n-1}$ one obtains

$$G(\mathrm{i}\omega_k)\big(a_n(i\omega_k)^n+...+a_1i\omega_k+1\big)-b_{n-1}(i\omega_k)^{n-1}+...+b_0=0,$$

or, introducing $R_{n,k}=\mathrm{Re}\{(i\omega_k)^n\}$,

$I_{n,k}=\mathrm{Im}\{(i\omega_k)^n\}$, $GR_{n,k}=\mathrm{Re}\{G(i\omega_k)(i\omega_k)^n\}$, $GI_{n,k}=\mathrm{Im}\{G(i\omega_k)(i\omega_k)^n\}$, one obtains

$$a_nGR_{n,k}+...+a_1GR_{1,k}-b_{n-1}R_{n-1,k}-...-b_0R_{0,k}=-GR_{0,k} \tag{18}$$

$$a_nGI_{n,k}+...+a_1GI_{1,k}-b_{n-1}I_{n-1,k}-...-b_0I_{0,k}=-GI_{0,k} \tag{19}$$

The obtained equations are conveniently rewritten in the following matrix form, which is easily solved using some of the modern computer algebra packages, in particular, introducing

$$\mathbf{M} = \begin{bmatrix} GR_{n,0} & \cdots & GR_{1,0} & -R_{n-1,0} & \cdots & -R_{0,0} \\ \vdots & \ddots & \vdots & \vdots & \ddots & \vdots \\ GR_{n,n-1} & \cdots & GR_{1,n-1} & -R_{n-1,n-1} & \cdots & -R_{0,n-1} \\ GI_{n,0} & \cdots & GI_{1,0} & -I_{n-1,0} & \cdots & -I_{0,0} \\ \vdots & \ddots & \vdots & \vdots & \ddots & \vdots \\ GI_{n,n-1} & \cdots & GI_{1,n-1} & -I_{n-1,n-1} & \cdots & -I_{0,0} \end{bmatrix}, \quad \mathbf{b} = \begin{bmatrix} GR_{0,0} \\ \vdots \\ GR_{0,n-1} \\ GI_{0,0} \\ \vdots \\ GI_{0,n-1} \end{bmatrix}$$

one easily obtains the desired system of linear equations in matrix form

$$\mathbf{Mx} = \mathbf{b}, \tag{20}$$

where $\mathbf{x}$ is the vector of unknown parameters, $\mathbf{x} = \begin{bmatrix} a_{n-1} & \cdots & a_1 & b_{n-1} & \cdots & b_0 \end{bmatrix}^{\mathrm{T}}$.

It is important to mention that the selected set of incidence points $\omega \in [\omega_0, \omega_2, \cdots, \omega_{n-1}]$ can produce a singular matrix of the set of equations (16), (17). In such a case, another, more appropriate set of points should be used. It is also significant to note that it is also possible to use more than n incident points in the selected set. The exact solution cannot be found in such a case. However, the best approximation, in the least-squares sense, can be found by means of pseudo-inversion.

$$\mathbf{x} = (M^T M)^{-1} M^T \mathbf{b}, \quad k \geq n-1 \tag{21}$$

Numerical Examples

Let us select several LFOS transfer functions and compare their Bode characteristics and responses to Heaviside excitation with those of the corresponding rational approximations determined on the basis of the set of linear equations (16) and (17). The exact characteristics are shown in blue line, while the approximations are plotted in red. For each of the examples considered below, a set of target (or incident) frequency values used in (16) and (17) has been specified also.

Example 1. Consider fractional order system described by transfer function $G_1(s) = 1/(s^{3/2} + 1)$, and consider its rational approximation obtained by interpolating frequency response in target points $\omega \in [0.01, 0.1, 0.5, 1, 5, 10, 100]$. The resulting approximation is illustrated by Figure 3.

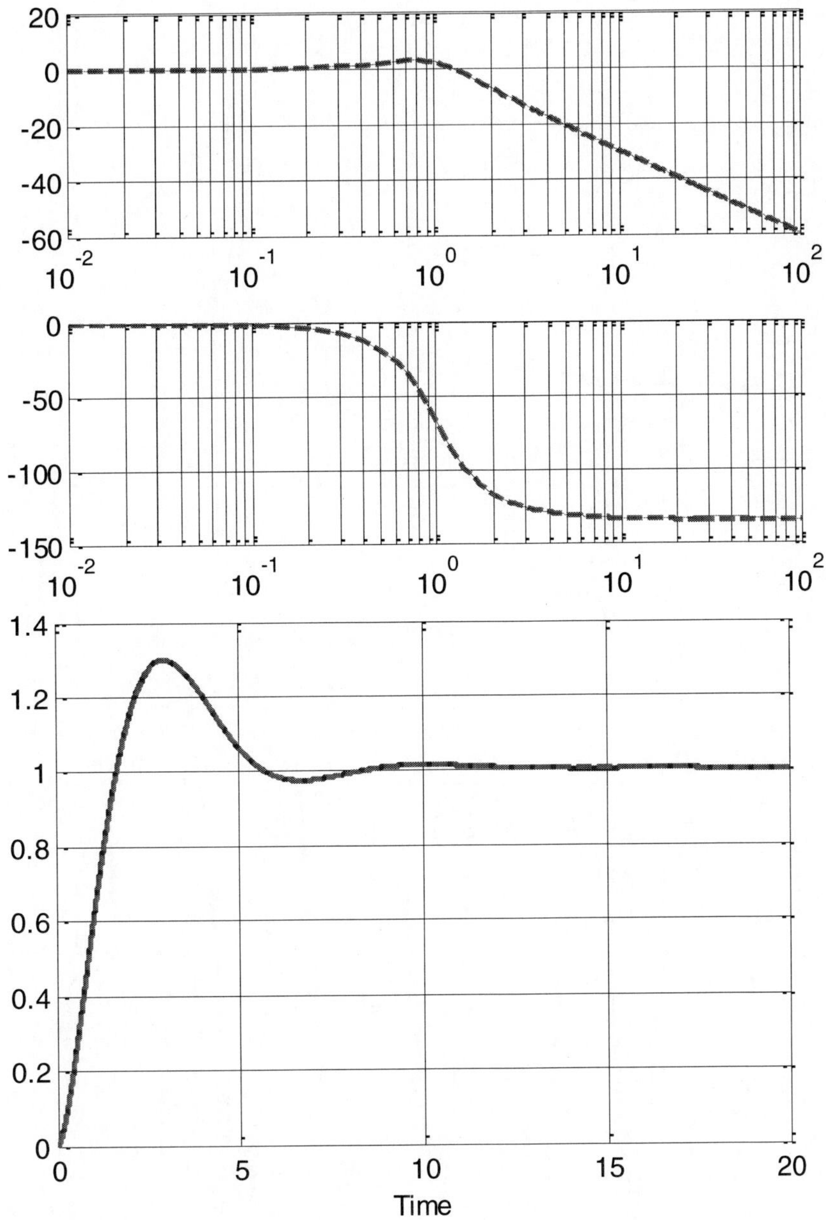

Figure 3. Frequency and time responses $G_1(s)$ (red) and $B_1(s)/A_1(s)$ (blue).

Example 2. Consider the fractional oscillator characterized by a strong resonant peak at unit angular frequency, $G_2(s) = 1/(s - \sqrt{2s} + 1)$, with $\omega \in [0.01, 0.1, 0.5, 1, 5, 10, 100]$.

The resulting approximation is illustrated by Figure 4.

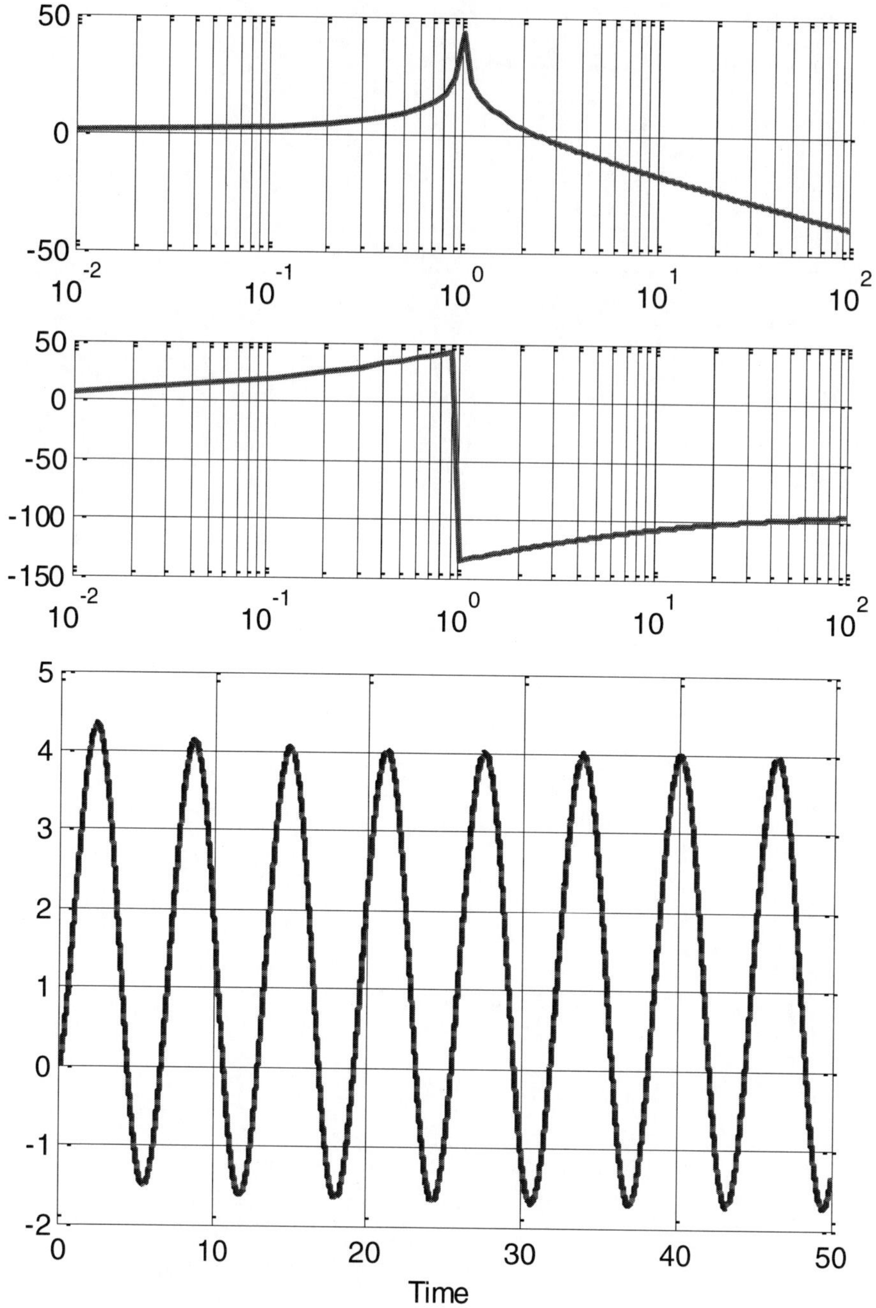

Figure 4. Frequency and time responses G2(s) (red) and B2(s)/A2(s) (blue).

Example 3. Consider a process with $G_3(s) = \exp(-\sqrt{s})$, with $\omega \in [0.01, 0.1, 1, 2, 10, 50, 100]$. The resulting approximation is illustrated by Figure 5.

Processes with such a transfer function are common in analysis of distributed parameter systems, particularly those involving heat and mass transfer; see for example [22, 23]. Similar phenomena are, in a generalized form, also studied in [21].

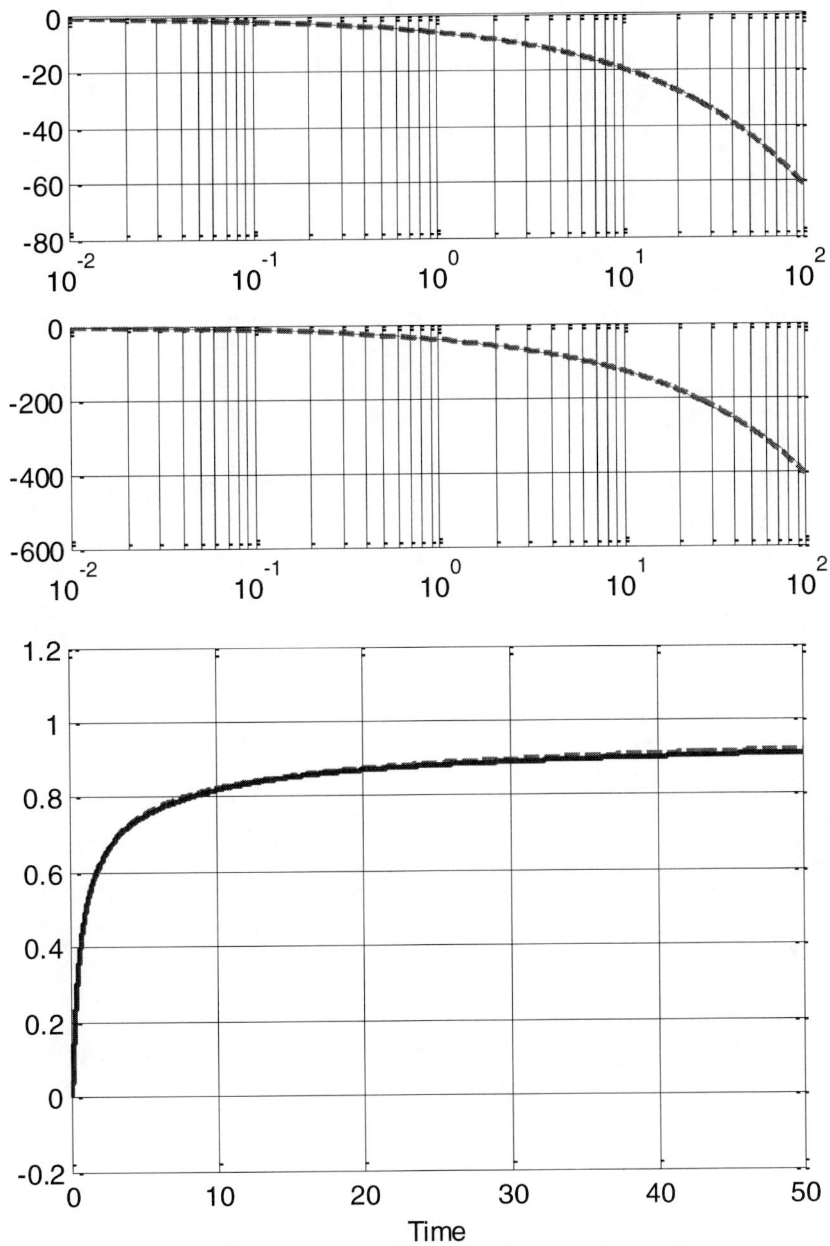

Figure 5. Frequency and time responses $G_3(s)$ (red) and $B_3(s)/A_3(s)$ (blue).

Example 4. Fractional logarithmic filter with transfer function $G_4(s)=\ln(s)/s$ appears in the study of adaptive fractional systems, as can be seen from [24]. The incident frequencies in this particular case where $\omega\in[0.001,0.01,0.1,0.5,1,5,50]$. The resulting approximation is presented in Figure 6.

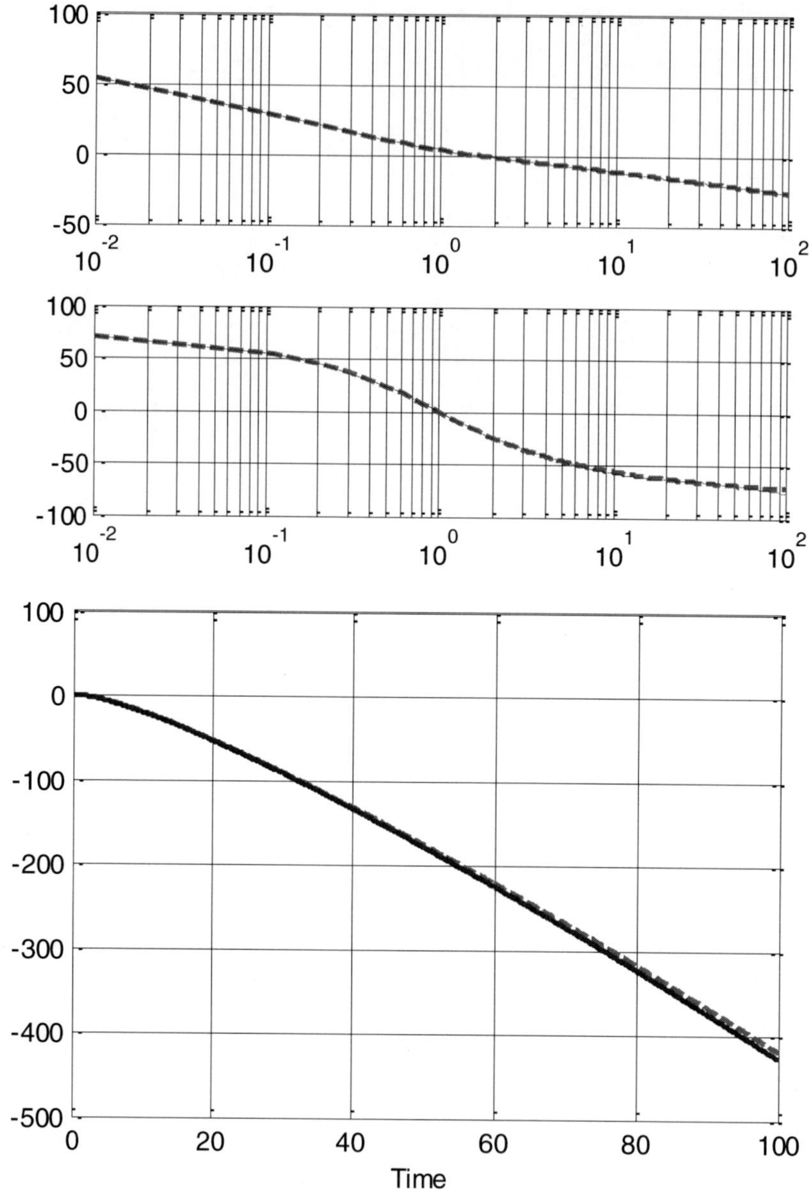

Figure 6. Frequency and time responses $G_3(s)$ (red) and $B_3(s)/A_3(s)$ (blue).

Example 5. The following example demonstrates a case when the process output is equal to the fractional semi-integral of a IIR filtered input. The actual IIR filter is, in fact, a differential compensator.

Consider fractional order system described by transfer function $G_5(s) = ((s+1)/(0.1s+1))^{0.5}$, with incidence frequencies $\omega \in [0.01, 0.1, 1, 3, 5, 10, 100]$. The resulting approximation is illustrated in Figure 7.

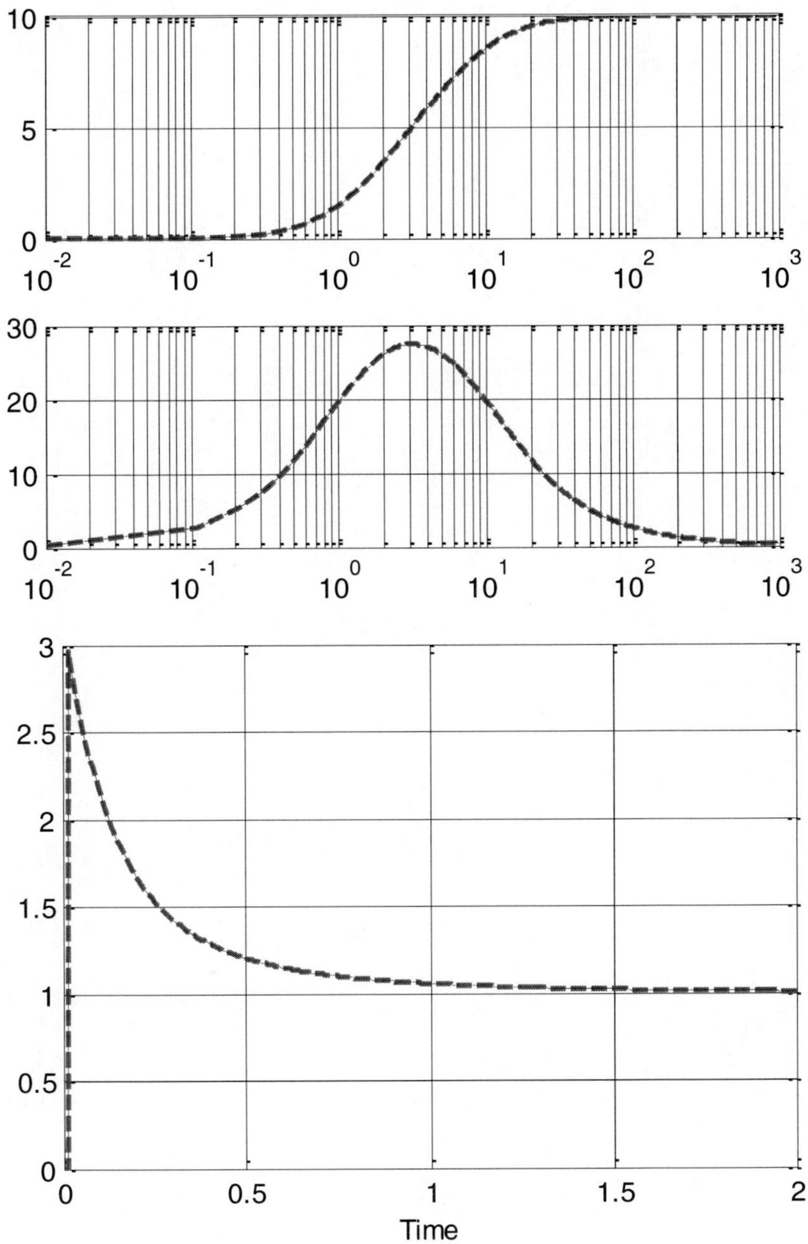

Figure 7. Frequency and time responses $G_5(s)$ (red) and $B_5(s)/A_5(s)$ (blue).

Example 6. Consider also the fractional order system $G_6(s) = (1 + 1/s + s^{1.2})/(0.1s + 1)^{1.2}$, with incidence frequencies $\omega \in [0.5, 0.8, 1, 2, 5, 30, 100]$. The performance of the resulting approximation is given by Figure 8.

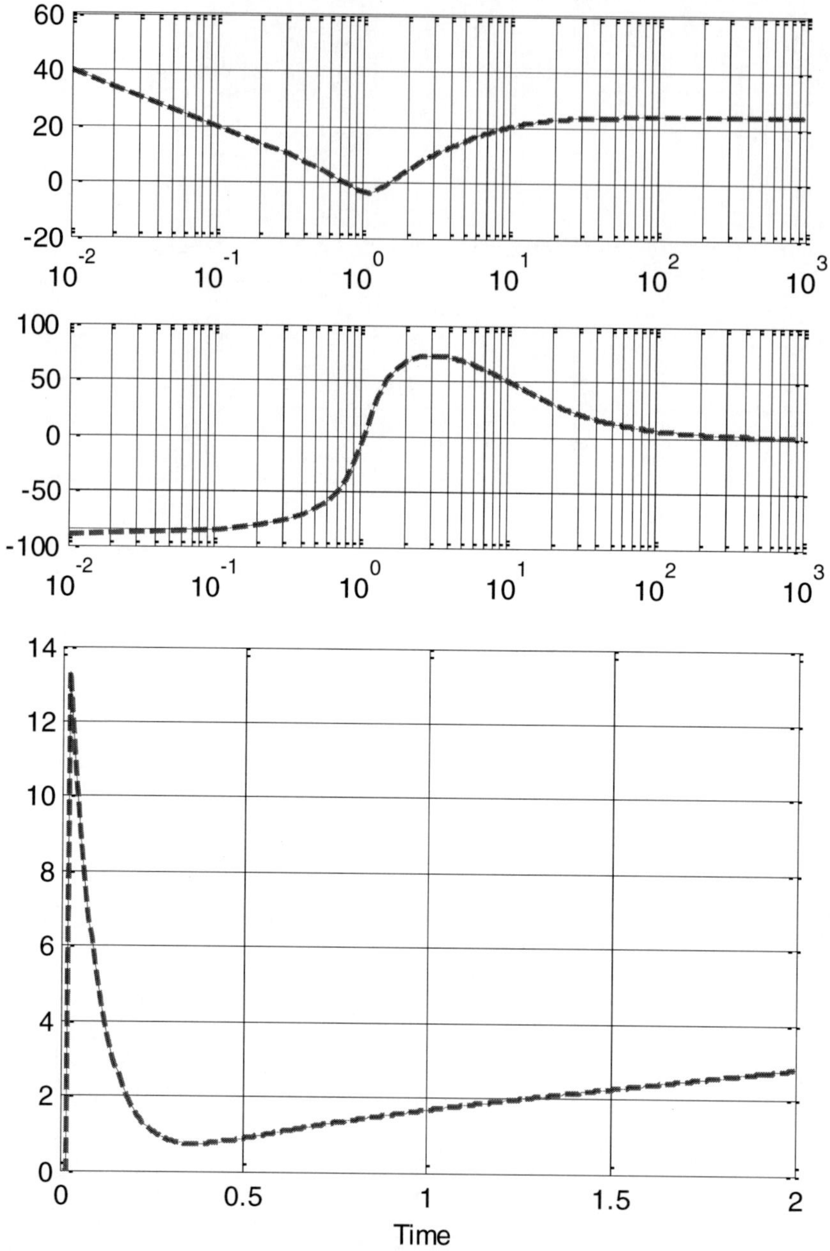

Figure 8. Frequency and time responses $G_6(s)$ (red) and $B_6(s)/A_6(s)$ (blue).

Example 7. Finally, consider the process described by transfer function $G_7(s)=\frac{(s+2)e^{-s}+s+3}{6s+1+(0.5s+1)e^{-5s}}$ which was also considered in [27]. The incident frequencies in this particular case where $\omega\in[0.01,0.1,1,4,5,10,100]$

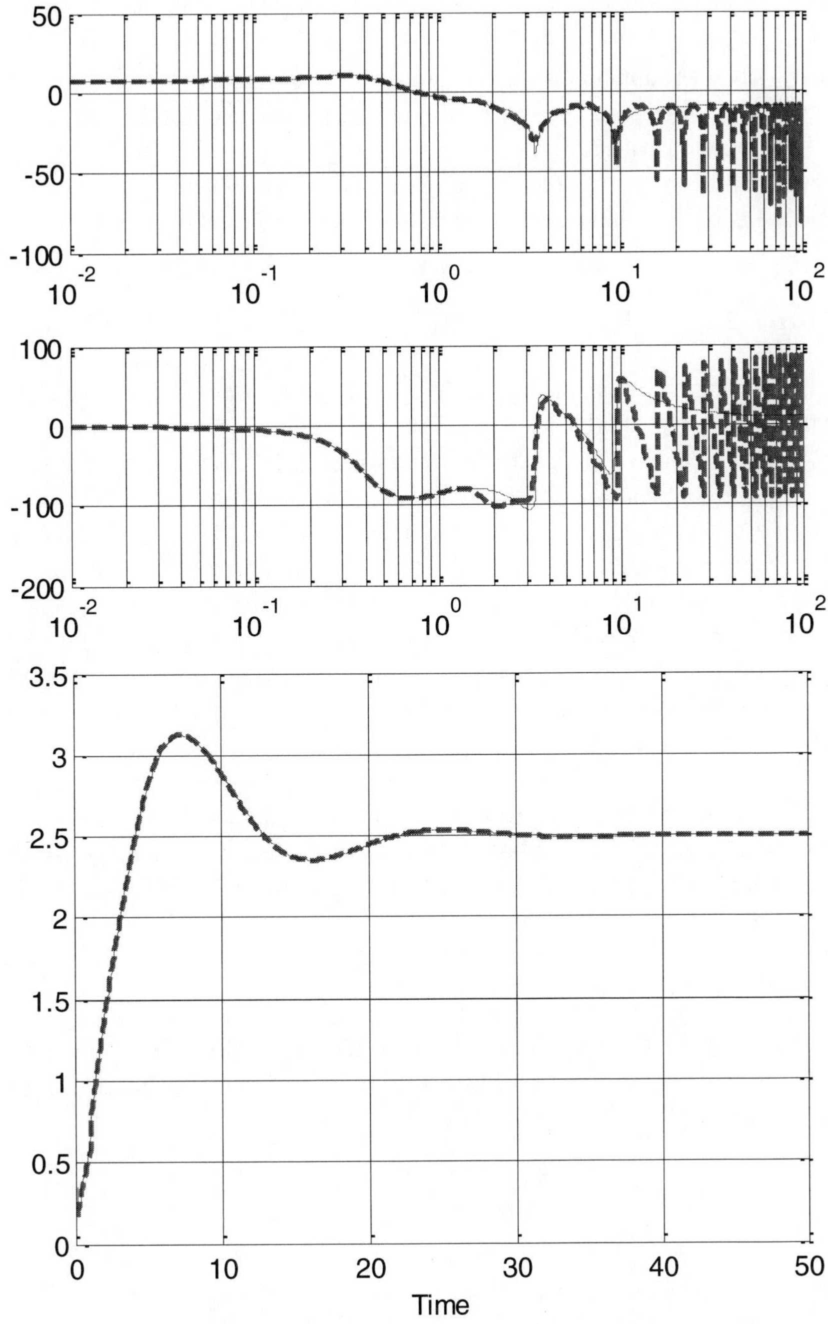

Figure 9. Frequency and time responses $G_7(s)$ (red) and $B_7(s)/A_7(s)$ (blue).

As the previous examples show, the rational approximations give adequate matching for a wide range of LFOS. It should especially be mentioned that the corresponding frequency points are selected on the basis of the knowledge of Bode characteristics of LFOS transfer functions. In all previous examples the selected order $n = 7$ can be lower, but under condition that the frequency characteristics in the selected frequency range are not violated. Also, for the selected set of frequencies in all these examples, the matrices of the sets of equations (16), (17) have been regular.

Obviously, since a rational transfer function is in question, applications of all techniques of discretization are possible; consequently adequate discrete models of LSF transfer functions within certain frequency range are available.

All of the time-domain responses presented above are obtained by means of direct integration in the complex domain. The interested reader is referred to [14].

Conclusion

In the present chapter, the rational approximation method applied previously for discretization of fractional order systems has been extended to the class of arbitrary, linear, stationary, infinite dimensional models. It should be emphasized that, since the proposed technique results in a rational, continuous-time, the subsequent discretization can be performed by any among the well-established discretization techniques. .

The most important feature of the proposed method is its computational efficiency. The method is, in fact, very simple both conceptually and computationally. The obtained results are, as it can be seen from the previous examples, quite satisfactory. The main drawback of the proposed method is that it is not possible to guarantee stability *a priori*, in other words no constraints on the coefficients are enforced. Indeed, the form of these constraints would be so complex, that their introduction would impair the established efficiency of the presented method.

A possible solution to this problem is to use the obtained coefficients as the initial guess for the more elaborate, non-convex optimization procedures. Such an approach would lose the desired computational efficiency, but would be able to give stronger stability guarantees.

Acknowledgment

The authors would like to gratefully acknowledge the financial support of Serbian Ministry of Science and Education, under grants TR32018 and TR33013 (M.R.R), III41006 and TR33020 (T.B.Š), and TR 35006 (M.P.L.).

References

[1] Podlubny, *Fractional Differential Equations*, Academic Press, London, 1999.

[2] Podlubny, Fractional-Order Systems and $PI^{\lambda}D^{\mu}$ Controllers, *IEEE Transaction on Automatic Control,* vol. 44, 1999, pp. 208–214.

[3] R. Boxer, A Simplified Method of Solving Linear and Nonlinear Systems, *Proceedings of the IRE,* vol. 44, 1956, pp. 89-101.

[4] R. Boxer, A Note on Numerical Transform Calculus, *Proceedings of the IRE,* vol. 43, 1957, pp. 228-229.

[5] K. J. Åström, P. Hagander, J. Sterbnby, Zeros of sampled systems, *IEEE Automatica* vol. 20, 1984, pp. 31-38.

[6] K. J. Åström, B. Wittenmark, *Computer Controlled Systems: Theory and Design*, 3. ed., Prentice-Hall, 1997.

[7] M. R. Rapaić, T. B. Šekara, Novel direct optimal and indirect method for discretization of linear fractional systems, *Electrical Engineering*, vol 93, 2011, pp. 91-102.

[8] J. M. Smith, *Mathematical Modeling and Digital Simulation for Engineers and Scientists,* 2. ed., Wiley, New York, 1987.

[9] T. Šekara, New Transformation Polynomials for Discretization of Analogue Systems, *Electrical Engineering,* vol. 89, 2005, pp. 137-147.

[10] T. B. Šekara, M. R. Stojić, Application of the α-approximation for discretization of analogue systems, Facta Universitatis, *Ser: Elec. Energ* vol. 3, pp. 571-586, 2005. http://factaee.elfak.ni.ac.yu/fu2k53/Sekara.pdf

[11] T. B. Šekara, L. S. Draganović, M. S. Stanković, Novel Approximations for Discretization of Continuous Systems, In: *Proceedings of the XLVI ETRAN Conference,* vol. 1, Bosnia and Herzegovina, 2002, pp 220-223 (in Serbian).

[12] T. B. Šekara, M. S. Stanković, Novel Transformation Polynomials for Discretization of Continuous Systems. In: *Proc. of XLVII ETRAN Conference,* vol. 1, 2003, pp. 247-250, Montenegro (in Serbian).

[13] T. B. Šekara, Indirect application of transformation polynomials for discretization of fractional integrators (differentiators).In: *Proc. of XLIL ETRAN Conference*, vol.1, 2005, pp.234-237. Montenegro, (in. Serb.)

[14] T. B. Šekara, Fractional Transformations with Applications to Control Systems and Electrical Circuits. PhD Thesis, Faculty of Electrical Engineering, University of Belgrade (in Serbian), 2006.

[15] T. Dostál, J. Pospíšil, *Switched circuits* I, Rersearch Report No. FE-58, Department of Radioelectronics, TU Brno, 1985.

[16] G. Zhang, X. Chen, T. Chen, Digital Redesign via Generalized Bilinear Transform, *International Journal of Control*, vol. 82, 2009, pp. 741-754.

[17] K. Zaplatilek, M. Lares Efficient Algorithms of Direct and Inverse First-Order s-z Transformations. *Radioengineering*, vol. 10, 2001, pp. 6-10.

[18] Y. Choo, Computing Transformation Matrix for Bilinear s-z Transformation. *IEICE Trans. Fundam.*, vol. E90-A, 2007, pp. 872-874.

[19] E. W. Bai, Z. Ding, Zeros of sampled data systems represented by FIR models, *Automatica*, vol. 36, 2000, pp. 121-123.

[20] J. Le Bihan, Novel Class of Digital Integrators and Differentiators, *Electronic Letters,* vol. 29, 1993, pp. 971–973.

[21] M. R. Rapaić, Optimalno i suboptimalno upravljanje klasom sistema sa raspodeljenim parametrima, doktorska disertacija, Fakultet tehničkih nauka, Novi Sad, 2011.

[22] J. L. Schif, *The Laplace transform – Theory and applications*, Springer 1999.

[23] T. B. Šekara, *Frakcioni sistemi upravljanja*, Akademska misao, 2011.

[24] M. R. Rapaić, A. Pisano, Variable-order fractional operators for adaptive order and parameter estimation, *IEEE Trans. On Automatic Control*, 2014 (in press), doi: 10.1109/TAC.2013.2278136.

[25] T. B. Šekara, M. R. Rapaić, M. P. Lazarević, *Indirect application of a method for discretization of linear fractional order systems*, INDEL 2012.

[26] T. B. Šekara, M. R. Rapaić, M. P. Lazarević, An efficient method for approximation of non-rational transfer functions, *Electronics*, vol. 17, 2013, pp. 40–44.

[27] L.B. Xie, L.S. Shieh, J.S. H. Tsai, Y. Zhang , Approximated modeling and minimal realization of transfer function matrices with multiple time delays, *Journal of Process Control*, vol. 23, 2013, pp. 3–11.

INDEX

E

F

G

H

I

J

K

L

M

N

O

P

Q

R

S

T

U

V

W

Y